Systems Engineering for Ethical Autonomous Systems

Other related titles you might find of interest

SysML for Systems Engineering: A Model-Based Approach, 3rd Edition by John Holt and Simon Perry, 2018. ISBN 978-1-78561-554-2
Swarm Intelligence: Volume 1: Principles, current algorithms and methods by Ying Tan (Ed), 2018. ISBN 978-1-78561-627-3
Swarm Intelligence: Volume 2: Innovation, new algorithms and methods by Ying Tan (Ed), 2018. ISBN 978-1-78561-629-7
Swarm Intelligence: Volume 3: Applications by Ying Tan (Ed.), 2018. ISBN 978-1-78561-631-0
Enhanced Living Environments: From Models to Technologies by Rossitza Ivanova Goleva, Ivan Ganchev, Ciprian Dobre, Nuno Garcia and Carlos Valderrama (Eds.), 2017. ISBN 978-1-78561-211-4
Practical Robotics and Mechatronics: Marine, Space and Medical Applications by Ikuo Yamamoto, 2016. ISBN 978-1-84919-968-1
Optimal Adaptive Control and Differential Games by Reinforcement Learning Principles by Draguna Vrabie, Kyriakos G. Vamvoudakis and Frank. L Lewis, 2012. ISBN 978-1-84919-489-1
Control Theory: A Guided Tour, 3rd Edition by James Ron Leigh, 2012. ISBN 978-1-84919-227-9

For more information about our books please visit www.theiet.org/books

Systems Engineering for Ethical Autonomous Systems

Tony Gillespie

The Institution of Engineering and Technology

Published by SciTech Publishing, an imprint of The Institution of Engineering and Technology, London, United Kingdom

The Institution of Engineering and Technology is registered as a Charity in England & Wales (no. 211014) and Scotland (no. SC038698).

© The Institution of Engineering and Technology 2019

First published 2019

The Institution of Engineering and Technology
Michael Faraday House
Six Hills Way, Stevenage
Herts, SG1 2AY, United Kingdom

www.theiet.org

British Library Cataloguing in Publication Data
A catalogue record for this product is available from the British Library

ISBN 978-1-78561-372-2 (hardback)
ISBN 978-1-78561-373-9 (PDF)

Typeset in India by MPS Limited
Printed in the UK by CPI Group (UK) Ltd, Croydon

Contents

List of figures

List of tables

Foreword

Autonomous systems are becoming an accepted part of life, but questions are often raised about the ethics of their use. The debates can be at the philosophical level, with arguments about which ethical theory or moral code should be applied, or even if technologies for specific applications should be banned. These debates are usually very general even when discussing systems which are the subject of this book: those which interact with humans and have both some level of autonomy and a physical structure. More practical arguments arise about the legal framework that should be used when the system is in use, usually demonstrating that this new technology requires new laws or new interpretations of the existing ones. These debates and arguments have an extensive literature and are often given wide publicity, especially when artificial intelligence or autonomous road vehicles are included.

Whilst these arguments continue, the engineers and research workers developing the technologies and systems have to make practical decisions about the form and behaviour of their product. In general, the technologists are under pressures to deliver. The pressure may be from their companies' management or investors, or academic pressures. The result is that the products usually meet applicable regulations, but rarely with consideration of the wider ethical background and potential changes to the law or changes in its interpretation.

This book was originally intended mainly for systems engineers and those working on autonomous systems. However, there are many other professions currently involved with autonomous systems and ensuring that they are acceptable to society. Consequently, the chapters on systems engineering are supplemented by a fairly detailed description of product design, work packages and other topics which may be well known to engineers but not to other professions.

The book shows that it is possible to invert the process of an unacceptable product being developed, then it being heavily regulated or banned with resultant waste of effort and finance. Instead, the process becomes one of identifying criteria for acceptability and using these to set requirements before the product is designed. It allows non-technical constraints such as ethics to be introduced at the initial concept phase. If the acceptability criteria cannot be met, then the product should not be made. This is not only morally laudable, it also makes economic sense as there will be no investment beyond the concept assessment stage.

Chapter 1's title is *The art of the acceptable, not the art of the possible*, because engineers can deliver almost anything that does not break scientific laws and has sufficient demand – the art of the possible. It is a smaller set of products which are acceptable to society and can be made – the art of the acceptable. The phrase arose during meetings about unmanned air vehicles (UAVs) cooperating with manned aircraft in targeting operations. The engineering problems became clear to me when analysing how a country could specify, and use, an unmanned military aircraft to attack targets in difficult circumstances. It was clear that ethics, as articulated in International Humanitarian Law (IHL), placed more constraints on concepts and architectures than technological ones. The technology issues could all be solved, the problem being their relative importance, with the answers dictated by IHL, the accepted interpretation of ethical principles in warfare.

The first solution to the engineering problem was published in 2010, applying systems engineering techniques to ensure unmanned combat aircraft can comply with the Geneva Conventions [1]. Following retirement from the UK Ministry of Defence, I developed the technique into a more general methodology as a visiting professor at the University College London. The methodology has been applied to civilian applications, leading to the concept of authorised power [2]. This gives a method to control the behaviour of an autonomous system in a clearly defined way that should satisfy the needs of both engineers and lawyers.

When starting in the field, I found a gap in both knowledge and literature about the ethical and legal issues surrounding autonomous systems. There was little that was useful for practical application to the engineering problems associated with systems that were questioned on legal or ethical grounds. To a large extent, this gap is still there and compounds the common problem of different professions not understanding each other's' viewpoints. Similarly, there is also a gap in the literature for other professions describing the practical issues which dictate realistic autonomous systems concepts and their performance when produced. This book is intended to fill both gaps.

There is no common vocabulary for key issues in autonomous systems. The word 'autonomous' itself is only one example: consider the difficulty of translating the legal term 'reasonable precautions' (in attack) into engineering probabilities of correct object identification in a cluttered scene. It is necessary for engineers to have this cross-disciplinary translation skill as well as translating customer requirements into design specifications.

This book examines autonomous systems in the context of their wider ethical and legal environment with equal weight given to both engineering and legal principles. It is complementary to those by Sarah Spiekermann [3], Bill Boothby [4] and Reg Austin [5]. They examine separately ethical, legal and detailed design aspects of specialised aspects of autonomous systems making them too narrow for general use. This book takes a pragmatic but wide view of the whole area, with chapters giving self-contained summaries of the relevant engineering and legal processes involved in bringing an autonomous system from concept to reality.

This volume should be useful to several types of reader: engineers designing highly automated systems; analysts and marketers specifying them; lawyers who need to understand the technology; and students and non-specialists who need a more detailed understanding of autonomous systems without going into detailed design issues. Engineering, regulatory and legal aspects are covered in sufficient detail so that all these audiences can understand the others' problems and approaches. The book has a bias to aerospace applications but shows how system engineering principles are applied to highly automated systems in other fields such as autonomous vehicles and robots.

It is a pleasure to thank many people who have contributed directly or indirectly to this book. I would particularly like to thank Robin West formerly of DSTL and Steve Hailes of UCL who stimulated my ideas through many technical debates and arguments, and Lambert Dopping-Hepenstal for many discussions and the description of ASTRAEA in Panel 6.1. There are several colleagues and friends who have given detailed and helpful comments on draft chapters, improving their quality and content. There are also indirect contributions from numerous others at or around conferences, meetings and workshops.

Despite comments and information from others, all errors are my responsibility, as are any, hopefully few, failures to make correct references and acknowledgements to those in the field. Any opinions expressed are entirely those of the author and do not represent those of any organisation or UK government policy.

References

[1] Gillespie T., West R. 'Requirements for autonomous unmanned air systems set by legal issues'. *International C2 Journal*. 2010, vol. 4(2): pp. 1–32.
[2] Gillespie A.R., Hailes S. In preparation.
[3] Spiekermann S. *Ethical IT innovation: A value-based system design approach*. Boca Raton: CRC Press; 2016.
[4] Boothby W.H. *The law of targeting*. Oxford: Oxford University Press; 2012.
[5] Austin R. *Unmanned Aircraft Systems: UAVS Design, Development and Deployment*. Aerospace Series, Chichester, Wiley; 2010.

List of acronyms

3Ds	Dull Dirty or Dangerous
5Cs	Congested, Cluttered, Contested, Connected and Constrained
ACAS	Airborne Collision Avoidance System
ACL	Autonomous Control Level
AFRL	(US) Air Force Research Laboratories
AI	Artificial Intelligence
AIS	Automatic Identification System
ALARP	As Low As Reasonably Practicable
ALFUS	Autonomy Levels For Unmanned Systems
API	1977 Protocol additional to the Geneva Conventions of 12 August 1949, and relating to the protection of victims of international armed conflict (Protocol 1)
APII	1977 Protocol additional to the Geneva Conventions of 12 August 1949, and relating to the protection of victims of non-international armed conflict (Protocol 2)
APIII	2005 Protocol additional to the Geneva Conventions of 12 August 1949, and relating to the Adoption of an Additional Distinctive Emblem (Protocol 3)
ASTRAEA	Autonomous Systems Technology Related Airborne Evaluation and Assessment
ATC	Air Traffic Control
ATI	Assisted Target Indication
ATI	Assisted target Identification
ATI	Automatic Target Indication
ATI	Automatic Target Identification
ATRA	Autonomy and Technology Readiness Assessment
AWS	Autonomous Weapon System, or Autonomous Weapon Systems
BCE	Before Common Era
C2	Command and Control
C4I	Command, Control, Communications, Computers and Intelligence

CAC	Contextual Autonomous Capability
CADMID	Concept, Assessment, Development, Manufacture, In-service & Disposal
CapDEM	Collaborative Capability Definition, Engineering and Management
CBP	Capability Based Planning
CEP	Circular Error Probable
CCD	Camouflage, Concealment and Deception
CCW Protocol	1980 Convention on Prohibitions or Restrictions on the Use of Certain Conventional Weapons Which May Be Deemed to Be Excessively Injurious or to Have Indiscriminate Effects
CIWS	Close In Weapon System
CMMI	Capability Maturity Model Integration
Comms	Communications
COTS	Commercial Off The Shelf
COTS	Customized Off The Shelf
CPD	Continuous Professional Development
CPU	Central Processing Unit
CWS	Complex Weapon System
D3A	Decide, Detect, Deliver and Assess
DAS	Defensive Aids Suite
DEF STANS	Defence Standards
DLODs	Defence Lines Of Development
DNAW	Day Night All Weather
DOD	Department Of Defense
DODAF	Department Of Defense Architecture Framework
DOODA	Dynamic Observe, Orient, Decide and Act
DTC	Defence Technology Centre
DTED	Digital Terrain Elevation Database
EA	Enterprise Architecture
EFB	Electronic Flight Bag
EU	European Union
EW	Electronic Warfare
F2T2EA	Find, Fix, Track, Target, Engage and Assessment
F3EAD	Find, Fix, Finish, Exploit, Analyse and Disseminate
F3EA	Find, Fix, Finish, Exploit and Analyse
FAR	False Alarm Rate

FMS	Flight Management System
FIAC	Fast Inshore Attack Craft
GCS	Ground Control Station
GFE	Government Furnished Equipment
GFI	Government Furnished Information
GGE	Group of Government Experts
GNC	Guidance, Navigation and Control
GPS	Global Positioning System
HALE	High Altitude Long Endurance
H&S	Health and Safety
HI	Human Independence
HRI	Human Robot Interface
HRI	Human Robot Interaction
HRL	Human Rights Law
HUMINT	Human Intelligence
IADS	Integrated Air Defence System
ICAO	International Civil Aviation Organization
ICRC	International Committee of the Red Cross and Red Crescent
IED	Improvised Explosive Device
IHL	International Humanitarian Law
IHRL	International Human Rights Law
IMO	International Maritime Organization
INCOSE	Chapter of the International Council On Systems Engineering
IR	Infra-Red
IPR	Intellectual Property Rights
IRL	Integration Readiness Level
ISO	International Organisation for Standards
IT	Information Technology
ITU	International Telecommunications Union
JSP	Joint Service Procedure
LAWS	Lethal Autonomous Weapon System
LOAC	Law Of Armed Conflict
LOI	Levels of Interoperability
M3	MODAF Meta Model
MAA	Military Aviation Authority
MAD	Mutually Assured Destruction

MAWS	Missile Attack Warning System
Mil Specs	(US) Military Specifications
MITL	Man In The Loop
MOTL	Man On The Loop
MMIC	Monolithic Microwave Integrated Circuits
MOD	Ministry Of Defence
MODAF	Ministry Of Defence Architecture Framework
MOTS	Modified Off The Shelf
MRL	Manufacturing Readiness Level
MTBF	Mean Time Between Failures
MTI	Moving Target Indication
NAF	NATO Architecture Framework
NASA	National Aeronautics and Space Agency
NATO	North Atlantic Treaty Organization
NCIA	NATO Communications and Information Agency
NCTI	Non-Cooperative Target Identification
NCW	Network Centric Warfare
NGO	Non-Government Organisation
NIIRS	National Imagery Interpretability Rating Scale
OODA	Observe, Orient, Decide and Act
OPLAN	Operation Plan
OSI	Open Systems Interconnect
PACT	Pilot Authorisation and Control of Tasks
QA	Quality Assurance
QC	Queen's Counsel
R&D	Research & Development
RAC	Root Autonomous Capability
RAG	Red, Amber, Green (In charts used in trade studies and project management. They may use additional colours)
RCS	Radar Cross Section
RCS	Real-time Control System
RF	Radio Frequency
RFQ	Request For Quotation
RPDM	Recognition-Primed Decision-Making
RPV	Remotely Piloted Vehicle
SAM	Surface to Air Missile
SE	Systems Engineering

SEAS	Systems Engineering and Autonomous Systems
SEI	Software Engineering Institute
SF	Special Forces
SHOR	Stimulus, Hypothesis, Option, Response
SIGINT	Signals Intelligence
SIPRI	Stockholm International Peace Research Institute
SME	Small or Medium-sized Enterprise
SOLAS	International Convention for the Safety of Life at Sea 1974
SRL	System Readiness Level
STANAG	STANdardization AGreement (NATO standard)
SUMO	Suggested Upper Merged Ontology
SysML	Systems Modelling Language
Ts&Cs	Terms and Conditions
TDMA	Time Domain Multiple Access
TCAS	Traffic alert and Collision Avoidance System
TDP	Technology Demonstrator Programme
TLCM	Through Life Capability Management
TRL	Technology Readiness Level
TST	Time Sensitive Target
TTP	Tactics, Techniques and Procedures
UAV	Unmanned Air Vehicle
UAS	Unmanned Air System
UGS	UAV Ground Station
UML	Universal Modelling Language
UMS	UnManned System
UMV	Unmanned Maritime Vessel
UN	United Nations
UNCLOS	United Nations Convention on the Law of the Sea of 10 December 1982
UNGA	United Nations General Assembly
UUV	Unmanned Underwater Vehicle
UON	Urgent Operational Need
UOR	Urgent Operational Requirement
V&V	Validation and Verification
VSE	Very Small Entities
WGS-84	World Geodetic System 1984
WWII	World War II

Chapter 1

The art of the acceptable, not the art of the possible

1.1 Introduction

This book covers a wider range of topics than is usual in an engineering text. Systems engineers normally anticipate looking at how a system interacts with those around it, including humans, but do not expect ethical and legal issues to play a central role. The ability of a system to make decisions and act on them does raise ethical issues about their design, not just their use. Early public debate centred on smart munitions and remote bombing using unmanned air vehicles (UAVs). Now ethical questions are raised about the acceptability of many other types of autonomous system whether they use artificial intelligence (AI) or not.

Cross-disciplinary debates about autonomous systems involve users, regulators, lawyers, insurance companies and politicians as well as engineers. Engineers must understand the interaction between their work and that of these other communities. Conversely, these other communities need to understand how engineers produce sophisticated systems and the need for quantification and precision in setting requirements. Without mutual understanding and respect, progress will be difficult.

This chapter gives an overview of the engineering processes and techniques, as well as the legal and human rights background described and developed in more detail elsewhere in the book. This should then provide the time-pressed and possibly non-specialist reader with some impressions of the whole, whilst enabling them to concentrate on their specific problem using the relevant chapter.

Chapter 14 gives some final topics for consideration drawing on results from other chapters.

1.2 Technologies and their acceptance

Technological advances have always brought changes in the society. The development of writing brought precision to the transmission of information across distances and time. The invention of the printing press in the fifteenth century made books more common and accessible. Public access to scripture and knowledge, with consequent questioning of religious authority, made a large contribution to The Reformation with its wars and changes in the distribution of power in European society.

New concepts which may change society can originate in science, engineering or any other field of human activity. Technology can turn these concepts into a product which is useful to one or more sections of society. The product will only succeed if it, and its use, is accepted by the relevant sections of the society. Autonomous systems have the potential to change society, but need to be widely accepted.

Success may bring completely unforeseen changes, for example, the small computers developed by IBM in the 1970s were only expected to have a market in the tens of thousands but their successors are now produced by the millions if not billions. They have evolved into sophisticated machines connected to a vast world-wide infrastructure and become an essential tool for many aspects of modern life.

Late twentieth- and twenty-first-century science and technology offers a vast range of potential products and services from genetic engineering to AI. A reaction against a new technology is to be expected as it always threatens some part of the *status quo*, but it usually becomes accepted when the benefits are seen to outweigh the risks and disadvantages. A technology which brings radical changes nearly always leads to local rules, national legislation or international treaties. The issue for engineers, ethicists, lawyers and market-makers is deciding which products are acceptable, not what is possible. The problem which they must solve together is one of determining what is acceptable in a market and then see if it is possible from technical, legal and ethical perspectives.

New technologies lead to new industries. Domestic use of PCs was an unexpected development but, arguably, led to the new market for computer games, which itself created a new industry for computer-games software. The Internet and smart phones have given almost everyone access to huge volumes of data, social networks and other facilities, also giving new industries. Arguments such as those about the social responsibilities of service providers do not destroy public acceptance of the underlying technology of fast processors, probably due to the perceived benefits outweighing the risks.

1.3   Machines that think

Machines that think for themselves have been the subject of science fiction stories for decades. Their behaviour may be portrayed as either beneficial or harmful to humanity depending on the aim of the author. Now that sophisticated PCs and more specialist machines are pervasive in all aspects of modern life, there is widespread discussion about robots, AI and the 'technological singularity'. The singularity is when AI develops to a stage where the machines take over the world and humanity dwindles away. Reference [1] gives a readily accessible discussion of some of the topics related to the replication of human intelligence through technological advances.

The autonomous systems which are discussed in this book are those which are highly capable but also mobile, with a clear physical instantiation. They are often referred to as robots, but this term is too restrictive for our purposes. Autonomous systems offer tremendous opportunities to society, forming a significant part of the fourth industrial revolution, with benefits ranging from autonomous vehicles to improved healthcare. However, there is a risk that one or both of the product and the designers will be out of control in some way and the autonomous systems will act in an unacceptable way.

Control of all aspects of an autonomous system is essential for their continued acceptance. Definitions of control and autonomy are introduced in Chapter 3. Legal and regulatory means and new standards to achieve ethical control of their design process are discussed in Chapters 6 and 8. There is considerable activity in this area, generating new laws, regulations and standards which will form the background for future technical developments. The methods developed in Chapters 10 and 13 show how an autonomous system can be designed in a way that ensures that it will only behave in defined, controlled ways.

1.4   What is autonomy?

Engineers, in common with other professions, have specialist words which are already in general use, but have a very specific meaning in an engineering context. They may even have different meanings in different disciplines within engineering. Definitions of critical words and terms are given in standards or, if necessary, in a contract for the product. The equivalent for the legal profession is legal dictionaries so that courts have a consistent set of definitions for their terms.

Ontological arguments are an essential part of setting standards for systems engineering. The reader can refer to many of the standards and documents referenced in the relevant chapters for the definitions of specific terms and use them for their own work in their sector. However, it is necessary and important to understand the concepts of autonomy and automatic. The word 'Autonomy' is in widespread use by the public and media with no consistency in its use. The engineering profession also has difficulties defining it, so this section defines the terminology used in this book, with a wider discussion and explanations in Chapter 3, after the discussion of decision-making processes in Chapter 2.

Autonomy, autonomous and automatic are important concepts when describing a system or product used to support or help humans in some way. When asked the question: what do these words mean, it is easy to turn to a dictionary and use the definition given there. Although this works for ordinary life, more precision is needed when the words are used in the context of a specific field of activity.

The *Oxford Online Dictionary* gives the following definitions [2]:

Autonomy	Freedom from external control or influence
Autonomous	1.1 Having freedom to act independently
	1.2 Denoting or performed by a device capable of operating without direct human control
Automatic	(Of a device or process) working by itself with little or no direct human control.

These definitions seem clear when applied to a person or nation and lawyers use dictionary definitions of words for this reason. The implicit controls of law and influences of upbringing and history for individuals and nations can be assumed to apply equally within nations so the definitions also apply equally.

The above definitions, except 1.2, are not adequate when applied to an engineered system which has powers to make decisions and act on them. Even using definition 1.2 above, no society would allow an autonomous robot to operate without any external control or way of influencing it. There must be a fundamental precept that every machine must be under some form of human control or at least only allowed to operate on its own under controlled conditions and when authorised by a human.

The term 'highly automated' is probably the most accurate generic description of both the systems and products which are commonly described as autonomous and those which are automatic but with complex control systems. Although accurate, it does not necessarily imply any ability to make decisions and act on them so we must use more suitable terminology.

The most relevant, but limited definitions given in standards answering our question are:

Autonomy:	The ability to perform intended tasks based on current state and sensing, without human intervention, From the British Standard BS 8611:2016;
Autonomous system:	System which has the ability to perform intended tasks based on current state, knowledge and sensing, without human intervention. From the International Organisation for Standards ISO 8373;
Automatic operation:	State in which the robot executes its task program as intended. From the International Organisation for Standards ISO 8373.

The two standards were intended mainly for the restricted field of industrial robots operating in a carefully controlled environment, but the writers were aiming at the wider field of autonomous systems. These definitions all have the word *intended*. This is because every system is designed to fulfil one or more purposes which have been determined at some stage before its use. An autonomous system must be designed so that it can be given a goal which it achieves by choosing and completing tasks. The first two definitions of the three given above are sufficiently general that they can be used here, whilst the third describes automatic operation if the reference to robots is removed.

A more extensive discussion of definitions and their implications for autonomous systems is in Chapter 3 with the main ones used in this book, shown in Panel 3.1, which is reproduced here as Panel 1.1. For completeness, the panels include definitions of authorised power from Section 10.8.1, and an autonomous weapon system (AWS) from the International Committee of the Red Cross (ICRC) discussed in Section 5.4.1.

There have been many attempts to describe autonomous systems with increasingly complex behaviours using autonomy levels. These are discussed in

Chapter 3, but it will be shown that their use is very domain specific, and even contradictory, so their use is problematic. Instead, the concept of authorised power is developed in Chapters 10 and 13.

It will be shown in later chapters that engineers and lawyers can agree that autonomy should only be defined for the functions performed by the system, not the system as a whole. These functions will be implemented in the design by one or more identifiable subsystems within the autonomous system.

Panel 1.1:   Definitions used in this book

Automatic operation

The state in which a system operates according to commands from an operator and where the resultant actions are predictable.

Autonomy (From ISO 8373)

The ability to perform intended tasks based on current state, knowledge and sensing, without human intervention.

Autonomous system (From BS 8611)

A system which has the ability to perform intended tasks based on current state, knowledge and sensing, without human intervention.

Autonomous operation (Based on ISO 8373)

The state in which an autonomous system executes its task program as intended.

Authorised power

The range of actions that a node is allowed to implement without referring to a superior node; no other actions being allowed. (A node is an entity within a control system which performs a specific, defined task.)

Authorised entity (From DOD document: *Unmanned systems safety guide for DOD acquisition*, 27 June 2007)

An individual operator or control element authorized to direct or control system functions or mission.

Autonomous weapon system (AWS) (From ICRC)

An AWS is a weapon system that can select (i.e. search for or detect, identify, track) and attack (i.e. use force against, neutralize, damage or destroy) targets without human intervention.

1.5   Maintaining control

Control of a complex process or operation has evolved from direct human control to the human operator exerting control through steadily more complex systems. Many control systems deliver capabilities which cannot be performed effectively by a human, such as an automatic braking system in a car or the part of the flight control system controlling the position of the ailerons of an aircraft. Automation is essential because a human cannot respond in the timescale needed to maintain stable operation of the car or aircraft. These systems can be described as autonomous according to the definitions in Panel 1.1; the time for which they are autonomous is restricted to the time between different commands coming from the brake pedal or control column.

Control theory is a well-developed branch of engineering. Control systems use data from measurements at different parts of the system and use them either in feedback or feed-forward processes, or a combinations of both. These are discussed in Chapter 3.

Control of human activity is achieved in the military through the Command and Control (C2) infrastructure and personnel who are trained in tactics, techniques and procedures (TTP). Authority to act autonomously is generally delegated to the lowest practical level, giving freedom to deviate from standard TTPs. These have been established from long experience to control human behaviour in warfare. Chapter 10 will show how the C2 infrastructure can be extended to automated and autonomous systems.

There is no comparable infrastructure for civilian autonomous system applications, with the state's legal structures providing the equivalent framework. Chapter 6 covers this and explains the roles of regulations and standards. Chapter 13 shows how an autonomous system can be designed meeting applicable laws and that ensures human control of its behaviours, either through design or fail-safe procedures.

The operator of an autonomous system should maintain control of it until a time when he or she instructs it to carry out a task without further intervention. Further observation of the system with the possibility of intervention by the operator may be a requirement in some cases. The operator must also ensure that the autonomous system returns to a safe, human-controlled mode after the task is completed. This means that any autonomous system's use can be split into the three phases, requiring different levels of human control, which are summarised below.

Phase 1   Under automatic control

> The autonomous system can be either inoperative or operating autonomously until given an instruction to come under the control of an operator. This phase starts when that instruction is obeyed.
>
> The operator must have a control link to the system which may be by cables, radio or by any other means. This link can be defined using standard engineering specifications including reliability, security and range. There will be a signal, probably with accompanying data to start autonomous operation.

An autonomous system that is in motion, or already in use under the direct control of an operator in this phase, may be classified as automatic, but the system is likely to have some level of autonomy. One example is where a UAV will be given a set of waypoints for its flight but it calculates its own route between them and flies it whilst still under the direct control of a ground-based pilot or operator[1] for all other functions. In this mode, the UAV is more properly called a remotely piloted vehicle (RPV).

The phase ends when the operator sends a command to start autonomous operation and the system starts to operate in that mode. Safety rules will require that the operator must be told by a reliable method that the second phase has started, and that the plan the system will execute will complete the operation that it was instructed to carry out. There must be a procedure for the event of failures in any part of this process.

Phase 2 Autonomous operation

This phase starts when the system accepts the command for operation, and begins to execute it. The assigned task must be within its capabilities with the procedural authorisation to complete it.

The system will respond to defined inputs which could be from its own sensors and inputs over authorised data links. The level of human intervention during its use will be defined; the operator may need to supervise it and give final authorisation for an action which is beyond its authority to carry out. There may be situations when another human may need to intervene, but this will require authorisation which can be validated by the system. Without this, hijacking by hostile or mischievous operators will be possible.

There must be a clearly defined procedure for the system when it has completed its task or is unable to do so. Without a 'fail-safe' mode, the autonomous system may be out of control, giving possibly unpredictable and serious consequences. For example, an autonomous car which continues to try to reach its destination using its original route even though there are unexpected diversions and contra-flows becomes a hazard to others; similarly an out-of-control weapon system can be described as a terror weapon and illegal under International Humanitarian Law (IHL).

The phase ends when control passes back to the original or other authorised operator. This may happen because the operator takes control for one of the following reasons: at a pre-planned point; or for direct intervention due to circumstances; or the autonomous system requests transfer of control.

---

[1] The person flying a UAV, whether a drone or a large UAV, will need some knowledge of the rules of flying as well as the skill to operate it. It is usual to expect the person flying a large UAV to be a qualified pilot so that his or her reactions are based on the flying experience as well as the aviation regulations.

Phase 3 Post task control

The details of this phase will depend on whether there will be a further phase of autonomous operation or whether the system has to be returned to its base or other safe storage. In all cases, the autonomous system must be placed in a safe state and acceptable location if the operator has not resumed control.

The objectives and detail within the phases will be very different for different civilian autonomous system applications, but an AWS will have a common set of core processes that must be carried out. However, in all cases, performing an assessment of the required actions of the autonomous system during all phases of its task execution, gives valuable insight into the required behaviour and functions that must be part of the autonomous system's design. The walk-through for an AWS is discussed in Sections 8.4 and 8.5, whilst the walk-through for a generic civilian autonomous system is in Section 13.6.

The operator(s) must be given the information they need to control and supervise the system's operation. Their interface is called the human–machine interface. The term 'human–machine interface' includes all the hard, firm and software necessary to present the operator with the required information for the mission. It can be as small as the hand-held controller for a miniature drone, or a large console enabling one or more operators to access many resources in real time. Whatever form it takes, its design is crucial as it is an important aspect of the mutual trust needed between a human and an intelligent machine.

1.6   Responsibilities

According to our definition, an autonomous system performs intended tasks without human intervention. An important question is that of the responsibility for any consequences after the start of its autonomous phase. In civil applications, this can be addressed as a question of legal liability for prosecution or compensation. This will probably also be the case when military forces use autonomous systems in peacetime applications away from operational areas. Chapter 6 discusses the overlap between engineering, standards, regulation and the law for autonomous systems. In war, IHL applies with its engineering implications examined in Chapter 7.

When humans make a decision and act on it, they are responsible for the consequences. Despite proposals in some countries about giving robots some form of citizenship, there is general acceptance that an autonomous system is not capable of taking legal responsibility for its actions as any sanctions against it would be meaningless. The problem has become one of deciding if the manufacturer, supplier or operator has responsibility. This problem has not yet been resolved although the role of trials, testing, Validation and Verification (V&V) and the level of tests is an important issue in proving that an autonomous system's behaviour is understood and limited. Different aspects of these are discussed in Sections 4.13, 5.4, 11.5, 13.4 and 13.9.

It is important to make several distinctions which help clarify where responsibility may lie. There is a difference between a decision and an action. This is clear when there is a delay between the two. When a time delay can be inserted, it may be utilised for authorising the action if this is necessary. If the authorisation has to be by a human, then the autonomous system is not fully autonomous. It may not be classed as automatic either if it uses learning algorithms as the decisions may not be fully predictable. Regardless of whether the autonomous system has a predictable output or not, it is classified as a decision-aid because it is aiding a decision by a human decision-maker. When an autonomous system takes action based on its decision, without human intervention, it must be designed and operated so that it has authority for that action. A car's satnav system is a decision-aid as the driver makes the decisions about following the route. Cars that drive themselves based solely on their sensors and navigation systems are still not generally available.

An autonomous system will have complex responses to the changes in its environment, so it will be difficult or even impossible to predict its exact behaviour. Performing tests under realistic conditions may give a good indication of its behaviour, but they cannot be exhaustive. Restrictions on system behaviour can be designed in, if specified during the concept development phase. Computer modelling and simulations will give information about performance under conditions which may not be available during tests with the real system, but will be limited in other ways. The whole question of validation of the design concept and verification of its actual performance is one of the key issues to be solved in the design of autonomous systems of all types. The issues are addressed at several points in this book for the procurement and design processes as well as at entry into service and subsequent upgrades (Sections 4.13, 5.4.3, 5.4.4, 13.4.5 and 13.8.9).

Responsibility for the reliability of an autonomous system must lie with the companies or organisations delivering it, and is an important issue for the supply and design chain. Reliability of mechanical, electrical and electronic components is a well-known subject with many standards and well-proven test techniques. Reliability criteria for a system are set by the consequences of a system failure which set the level of criticality of the system.

Good engineering practice should lead to designs that have low failure rates, but high-reliability and safety-critical[2] systems cost more to design, build and test than normal systems. It is usual for safety-critical systems to have three or more parallel processes with the response being determined by a voting choice at the end. Three or more are necessary because if there are only two parallel systems, and they give different outputs it is not clear which is correct. Referring to a human may resolve the dilemma but this is not usually an option. Systems with parallel-processes are called multiply-redundant systems.

---

[2]Safety-critical software is mandatory in applications where there is a direct, high risk to human life from a failure of the system. The necessary checks in its production and support make it one of the most expensive items in any system.

When the software is relatively simple and with only thousands of lines of code, it is still possible to have multiply-redundant systems. When the code is measurable in millions of lines of code this approach becomes less practical. Alternatives approaches are still the subject of research programmes, but a common approach is to check the output from the software to ensure that it is within reasonable limits and have a fail-safe response if not.

The challenges in taking the approach of testing the algorithm outputs can be seen from the work of Arkin and co-workers [3]. Their project was to develop an ethical governor that would test the decisions of an AWS in order to allow or inhibit actions that it assessed on the basis of compliance to the laws of war.

1.7   Autonomous Weapon Systems (AWS)

Several chapters in this book cover the legal framework, design and the use of AWS. Strictly speaking those weapon systems are automated rather than autonomous but AWS is generally used for them and so is also used in this book.

Autonomy in weapon release is controlled by IHL, as discussed in Chapters 7 and 8. The Red Cross definition in Panel 1.1 shows the detail needed to define an AWS, although this definition is not universally accepted. There is another term Lethal Autonomous Weapon System (LAWS) which also does not have an agreed definition but is generally taken to mean an AWS that specifically targets humans or leads directly to loss of life. Despite their automation and lethality, there is a universal agreement that naval and land mines are not AWS; they have specific treaties under IHL.

Interpretation of IHL has shaped the development of autonomy in weapon systems over at least the last three decades. This is arguably a longer time than for civilian autonomous systems with the result that many of the solutions to military problems may be applicable to civilian autonomous systems. This is the basis of Chapter 13.

Armed forces of all nations seek advantage over their opponents whether this is in combat or not. The perception that an opponent has superior firepower over your own forces can act as a deterrent to aggression and preserve peace. Increasing automation in weapon systems is one way to improve the effectiveness of your armed forces or counter specific threats. Phalanx is a naval gun system which detects incoming missiles and responds with very rapid machine-gun fire without human intervention. It can be classified as an AWS according to the definitions in Panel 1.1 and meets IHL as its operational use is restricted to a clearly defined area with prohibition of unauthorised access to it. The engineering changes necessitated by its use on land in compliance with IHL are explained in Section 7.14.

IHL places responsibility for the effect of a weapon on the person who fires or releases it. This is straightforward when the weapon operator or commander has direct sight of the target and its surroundings or a video link from a local surveillance sensor. This responsibility still applies to a more automated system such as a weaponised UAV with an on-board camera and a video link to the operator.

Acceptance of this principle is the reason that armed UAVs are not now seen to be in breach of IHL.

Placing all responsibility on the operator is only acceptable if he or she knows how the weapon will operate, the likely failures and the probability of that failure occurring. Soldiers know the accuracy of their guns and can ensure a good chance of hitting the target and not a nearby non-combatant. A more automated armed system with well-defined properties such as the Phalanx close-in defence system or a sentry gun can be deployed with confidence if they are placed in locations where there should be no non-combatants in their line of fire even if the weapon has a failure.

The widely accepted fear about an AWS is that they will be out of control after release and may fire when they detect what they, and not a trained person, decide is a legitimate target. These fears are addressed in this book. Legal constraints on the use of weapons are combined with the procurement processes, from identified need to in-service and upgraded systems. The methods given here should ensure that AWSs are used legally with human authorisation of their actions in an ethically acceptable way.

Reviews of the legality of new 'means or methods of warfare', in this case, AWS, are mandatory for all nations and are known as Article 36 Reviews. They are the subject of Chapter 11. How they should be conducted, and the evidence required for them is a subject of debate in the legal profession. This book explains how the engineering part of the procurement process can deliver suitable evidence.

Operational use of a weapon system is governed by rules-of-engagement. These are written before and during a campaign and are specific for a time, location and campaign phase. The process is not the same as the Article 36 reviews for the system but will need similar engineering evidence. Test and trials plans should reflect this dual use otherwise setting rules-of-engagement may require new trials to be performed in very short timescales.

This book gives an introduction to acceptable, as against possible but unacceptable AWS used for military purposes. New engineering techniques have been developed in conjunction with lawyers and ethicists to demonstrate how acceptable systems can continue to be developed instead of the unacceptable ones that autonomous technologies offer. These techniques can be applied to autonomous systems in other fields as is shown in Chapter 13.

1.8 Concept to product

Developing an idea into a product is not necessarily a simple process. However, there are well-developed design processes which facilitate it whilst identifying and controlling risks. They can be summarised as refining the concept, stating the requirements, deriving product specifications, development, test, and product release. Disposal may also be considered as part of the design 'cycle'. This sequence applies to all products, but clearly simple products will have fewer steps and detail. Section 13.4 shows the differences between civil and military design processes for complex autonomous systems that must meet ethical and legal constraints. Both civil and military complex systems will have iterations around the specifications and design due to emerging understandings of both the design and its interaction with its environment. These changes are often the cause of cost and time overruns.

An autonomous system may be considered to be self-contained if it does not rely on specialised interfaces with people or other systems. Products of this type may be readily upgraded with new releases of software. Hardware upgrades will be more difficult. However, it is likely that many autonomous systems will only have a life of a few years and be replaced with a superior model after a few years. This means that upgrades and long-term support may not an issue for the designers of short lifetime products.

Autonomous systems that are part of large platforms such as aircraft, and infrastructure which have lifetimes of decades face extra complexities. There are usually few suppliers and the choice of product and supplier implies a relationship that lasts for the lifetime of the product. The designers must assess their customers' likely needs for equipment and services for decades ahead. Predictions of this nature are fraught with errors but long-term trends can be identified. The result is that refining the concept and stating requirements usually take many months or years to complete. Even then there are usually further refinements and changes during design and build.

All procurements for long-lifetime products follow very similar processes. In this book, the military process is described as one example. It is described in some detail in order to illustrate the type of problem that arises for all procurements of this nature. It should then be possible for the reader to see the engineering implications of increasing the autonomy of an existing capability delivered by a system of systems which may already have autonomy in some functions. Most modern systems of systems have autonomy in decision aids if not in taking actions.

Military organisations use the techniques of operational analysis[3] to identify their needs. These are initially expressed as capabilities rather than as specific platforms or hardware so that there is flexibility in finding solutions to the needs. This is similar to a civilian organisation's approach to developing a product for a known market. Modelling enables combinations of possible new and existing equipment to be tested in simulations of realistic scenarios. The new products can be assessed in use against extant laws and regulations, and possible new ones. Autonomous systems are particularly problematic as there will be trade-offs between their autonomous capabilities, their control and certainty of performance. This process is examined in Chapter 4. A military example is developed in Chapters 11 and 12 to illustrate the problems. A hypothetical autonomous system is used in Chapter 13 to illustrate the problems for civilian autonomous systems and how military techniques and the use of authorised power can solve many of them.

Analysis is followed by deriving detailed requirements so that the procurement agency can obtain tenders from industry and run a competition for delivery. In practice, this is an iterative process with detailed options from different suppliers

---

[3]Operational analysis is virtually the same as operational research (OR), the main difference being that operational analysis is used for military analysis.

being assessed by the analysts, leading to a down-selection to two or three suppliers for final assessment. Finally, a contract is placed on a supplier or combination of suppliers. The design process in the supply chain is described in Chapter 5.

Turning IHL requirements into specifications that ensure that a weapon system can be used legally is not a huge problem when the time between the human's lethal decision and its end effect is short. An AWS may have minutes or hours between the lethal decision and its effect so there is a significant problem. The target area may have changed considerably in this time, or more data may show that the target is no longer a legitimate one. This is exemplified by the scenario of an attack on a bridge which was empty at weapon release, but has a non-military train on it when the weapon hits it. There may also be scenarios where an AWS spends long periods patrolling a large area with the potential to release a weapon when it detects a legitimate target, but none is detected even though there was at least one when the AWS was despatched.

Chapters 9 to 12 show how IHL requirements can be made an integral part of procurement of autonomous systems. This is achieved by deriving IHL requirements as capabilities which can be interpreted as engineering specifications. A weapon system which meets these specifications will then be capable of legal use and pass an Article 36 Review.

1.9 Future regulations

Autonomous systems in general will be subject to new regulations as they evolve. The ISO, the IEEE and British Standards Institute (BSI) have all produced new standards for robotic systems which are applicable to autonomous systems and new ones will evolve with the technology. These may apply to autonomous weapon systems whilst in use for trials and training along with other national and international laws.

Autonomous weapons, their use and morality are the subject of debate in international fora with implications for their place in IHL. They may become a technology included under a future protocol to the *1980 United Nations convention on prohibitions or restrictions on the use of Certain Conventional Weapons which may be deemed to be excessively injurious or to have indiscriminate effects* (CCW protocol). Procurement and design teams should take suitable legal advice about this protocol as IHL reviews should take account of likely changes to the law.

Changes in the law and regulations are part of the influences on autonomous system technologies discussed in Section 6.5 for autonomous systems and Section 9.10 for AWS.

1.10 Principles applicable to non-military systems

Much of this book is about military applications of autonomous systems, specifically weapon control systems. This may not seem immediately relevant to civilian uses as the military environment is very controlled with command structures and formal authorisation at all levels for both human and machine nodes in the C2 chain. Conversely, use of autonomous systems by the military has moved from scepticism and hostility to general acceptance in about ten years. Arguably this is due to the military establishment demonstrating its adherence to IHL and that the technologies it uses are understood and the systems are under close human control. It is lessons from this approach of adherence to law and regulations with strict control of the system's behaviour that can be applied to civil applications. This is the subject of Chapter 13.

The same three-part model of human decision-making applies to both military and civilian autonomous systems. The use of the 4D/RCS (NSTIR6910) standard with its three-part model is introduced in Chapter 10 and can also be applied to both. It enables the critical elements in the decision-making and acting processes to be identified for both human and machines. Combining this with an architectural framework, as described in Chapter 4 gives a solid base for a walk-through of an autonomous system's actions whilst carrying out an instruction.

The legal frameworks for military and civilian autonomous systems are different, but it is possible to derive some generic ethical and legal principles which are likely to form the basis of future regulations. Once these are identified, the civilian equivalent of the military concept to product process is derived and shown in Figure 13.1.

High-level requirements for each part of the decision-making process are derived as well as describing the role of authorised power, defined in Panel 1.1, in limiting the autonomous system's behaviour to taking only acceptable actions.

References

[1]   Shanahan M. *The technological singularity*. The MIT Press Essential Knowledge Series, Cambridge, Massachusetts: MIT Press; 2015.
[2]   en.oxforddictionaries.com [accessed 18th September 2018].
[3]   Arkin R.C. *Governing lethal behaviour: Embedding ethics in a hybrid deliberative/reactive robot architecture*. Georgia Institute of Technology technical report GIT-GVU-07-11, 2007.

Chapter 2

Decision-making

2.1 Freedom of action

It is taken for granted that as humans we can all take decisions and act on them. Automation replaces human actions with automated or autonomous systems which make decisions and carry out the consequent action. The decision may be based on the output of the comparator in a simple feedback loop, or it could be the output of a high-level Artificial Intelligence (AI) system with many inputs. Once a decision is made, action should only follow if the human or machine making that decision is authorised to act or order execution of that action.

This book examines the introduction of increased automation and autonomy into systems of humans and machines, so it is important that we understand how humans make decisions. Human cognition and decision-making are usually modelled as a three-part process of awareness, understanding and deliberation. The same three-part model is used to analyse requirements and design for autonomous systems. It is shown in Figure 2.1 and is used at several points in this chapter. An example of the interaction of multiple three-part models with humans is given in Section 2.6. Later sections show how freedom of action for an autonomous system may be restricted by the limitations of autonomous decision aids in different parts of the model.

Individuals are allowed to take almost any action they choose, unless it is forbidden by law or regulation. Every individual will have a moral or ethical code which guides the actions they take. In general, ethics places more restrictions on actions than the law but there are few fundamental conflicts between them. (Occasionally, someone's personal ethical code will clash with the law of the land; pacifists' conscientious objections to military service is a well-known example of this conflict.) The design of an autonomous system that replaces a human decision-maker and takes actions based on its decisions must include constraints based on the relevant laws and regulations.

Military commanders' authority is clearly defined with responsibilities set by national and international law. Firing or releasing a weapon against a military target with people in the area must be authorised with decisions based on reliable information. The requirements for authorisation and reliability of information also apply for actions by any autonomous systems that will interact with humans. The military targeting process is described later in this chapter as this is based on human decision-making processes and illustrates the type of information and decisions made by any autonomous system that interacts with humans.

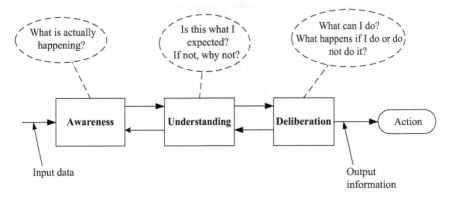

Figure 2.1   The three-part cognition model

Panel 2.1:   The trolley problem

There are several variants of the trolley problem, but the basic ethical dilemma is shown in Figure 2.2.

There is a trolley, which cannot be stopped, travelling along a railway track which splits into two at a set of points. The points are currently set so that the trolley will run into and kill a group of five people. The alternative track has only one person who would be killed if the points are changed.

The dilemma is that of the person who has control of the points. If they do nothing, five lives will be lost. If they change the points, five lives will be saved but one person will die who would otherwise have lived. Does the person at the points do nothing and let fate takes its course or do they choose to kill one

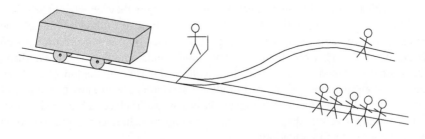

Figure 2.2   The trolley problem

person to save five others? Since they have the ability to save the five lives, have they chosen to kill the five and save one by doing nothing?

There are many variants on the problem with positive or negative 'values' placed on the lives of the different people on the two tracks such as criminals, peace-brokers or Nobel Prize winners.

The development of autonomous systems which can assess their surroundings, make decisions, and act on them has led to much public and professional debate about the legitimacy of these actions. The 'trolley problem' shown in Panel 2.1 is an often-quoted example of the dilemmas which a human decision-maker can face. These can become intractable when used to question how an autonomous system should act, leading to potentially impossible guidelines for the designers.

One well-known trolley problem is that of an autonomous car when faced with a decision to either allow its driver to be hurt in an accident, or swerve and kill a pedestrian. Analogous military dilemmas can be devised if an autonomous weapon system were allowed to choose between alternative targets based on their military importance and the collateral loss of live. Military decisions which potentially involve deliberate harm to a human are called lethal decisions, can only be taken by a human, and are subject to International Humanitarian Law (IHL). There is an important legal difference between civilians and military decision-makers. As stated above, civilians basically have freedom of action within the law, but clearly not to kill another person. This is part of Human Rights Law (HRL). Military personnel, in wartime, act under IHL, but within the constraints of their delegated authority.

2.2   Skills, behaviours and automation

Rasmussen's 1983 paper [1] on human skills and interactions with machines has had a large influence on models of human decision-making. The paper looks at how humans react to process control systems when presented with their sensor outputs. A human recognises a feature from sensor and other inputs, and that a response is needed to achieve a stated goal. The relevant results for our purposes are the modelling of human performance to achieve the goal. The model uses three levels of behaviour, each level depending on the person's familiarity with the situation.

Figure 2.3 shows Rasmussen's original three levels of behaviour and the complexity of the decisions made at that level for achieving the goal. The three behaviour levels are not the same as the parts of the three-part models of decision-making used elsewhere in this book. They can be considered as orthogonal to them, and are included in Figure 2.3 as awareness, understanding and deliberation which lead to actions. These are directly equivalent to the phases of Observe, Orient, Decide and Act from the OODA loop discussed in Section 2.7 and are also shown in Figure 2.3. The three skill levels are discussed in the following sections.

Skill-based behaviour

This is the most automatic form of human behaviour and applies to a person who has considerable training for the tasks they are performing. A task is needed in order to achieve the desired goal, but it requires very little conscious thought by the human. The task is probably a feed-forward process and could be a general type of task which is needed to achieve several goals. The human response to an input or cue has little or no conscious effort and is

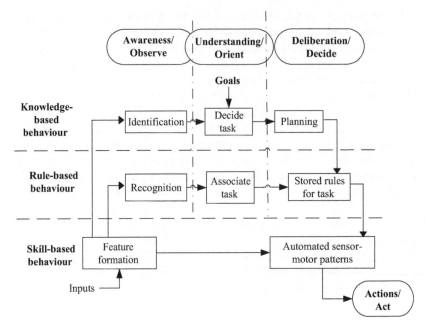

Figure 2.3   Human cognition and performance. Based on Rasmussen's (1983) skills levels with three-part cognitive processes and OODA loop added

a feed-forward response. The person has a fairly detailed mental model of the process and acts in accordance with their predictions of the system's future state. *Features* are anticipated and recognised when seen, enabling very rapid, anticipatory behaviour.

A difference between anticipated and actual features will cause feedback with the person making a corrective response. The feedback loop can be considered to be part of the *automated sensor-motor pattern* function in Figure 2.3. If the response to the feedback is still within the person's skill level, it will require only a small amount of conscious effort. If not, the behaviour becomes rule-based.

The important points are that the environment is constrained and fits the mental model with only anticipated types of deviation which can be corrected in a straightforward manner. Skilled behaviour is based on taking action by selecting suitable responses from a wide range of automatic ones.

Rule-based behaviour

A human will encounter situations where the skilled behaviours will not give an adequate response. This is the *Recognition* part of Figure 2.3 when the scope of the task has to be considered and suitable rules identified. It is highly likely that known, stored rules will give an adequate response when followed. These rules will be more general than the automatic responses and will require some interpretation in their application. The rules may be the ones learnt during training or ones that come from positive or negative experience with similar situations. There is considerable overlap between skill and rule-based behaviour. The boundary will be different for different individuals.

Knowledge-based behaviour

When a human is in an unfamiliar situation, it is likely that the existing rules are inadequate or they will not have any specific rules to achieve the goal. Their response is to consider the goal of the operation and move mental processing to the conceptual level. The goal is formulated in terms that relate to the processes and the facilities that are available to perform a suitable task. Solutions are sought by considering options and their chances of achieving the goal. The person will have a mental model of the environment and the changes needed to achieve his or her aim. They also have a mental model of their options for responses and how they will affect the environment. One or more implementation plans are considered, based on previous experience. One is chosen and acted on, possibly with more detailed planning preceding action.

Each function for knowledge-based behaviour is more complex and sophisticated than the equivalent ones for rule-based behaviour. This level is essentially where the human has to use judgement to reach the desired goal, i.e. the human has to have autonomy in one or more functions. Deciding the functions which can be automated and meet legal requirements is a task for the design team and their legal colleagues.

The original paper does not consider machine behaviours, but the three human levels could form a hierarchy of automation levels or technologies. Level 1 is very similar to the response of a classic control system with a feedback loop. The system design will probably be based on standard techniques using Laplace and other transforms to model its responses to a defined range of inputs. The analogy for Level 2 is where the automated system can identify objects and features, and relate them to its environment in a way that allows calculation of a set of options for action. It can then choose an action according to simple pre-set criteria. Level 3 is where the environment's dynamics are taken into account and complex responses are needed. The authorisation to act on the proposal, or not, is defined within the system design at all levels.

Figure 13.5 shows an architecture for a sophisticated autonomous system with complex interactions with humans. This has been derived using ethical and legal requirements for the autonomous system's behaviour and is based on a three-part model of decision-making. There are some close analogies with Figure 2.3 if the increasing levels are taken to be more autonomous functions in the system. The sensory-processing column is equivalent to the three left-hand (Awareness/Observe) blocks in Figure 2.3; world-modelling is equivalent to the centre (Understanding/ Orient) blocks; behaviour generation is equivalent to the right-hand (Deliberate/ Decide) blocks, leading to action by the actuators. Value judgement is based on the criteria for allowed action; it does not appear in Rasmussen's model as that assumes all actions are allowed.

2.3   Situational awareness

The term 'situational awareness' is widely used to describe a person's knowledge of their environment. Ideally, decisions are made with all relevant information and time to consider the options, i.e. full situational awareness. In practice, this is never the case. When discussing decision-making, it is normal to look at the person's situational awareness at the time. Their response to the situation is often discussed in terms of the available data, whether its importance was correctly assessed and the need for additional information. Workload and stress are factors which adversely affect situational awareness even if all necessary data is available to the person making the decision. System design for maximum operator situational awareness is important and there is an extensive literature describing it with at least one textbook [2].

Situational awareness can be defined as [3]:

> The perception of the elements in the environment within a volume of time and space, the comprehension of their meaning, and the projection of their status in the near future.

This definition is coupled with the diagram shown as Figure 2.4, which also comes from Reference [3]. It shows situational awareness and its relationship to decision-making, emphasising that the two are different but closely related.

Situational awareness is the white block inside the larger grey one. It is decomposed into three levels. These are not the same as the three parts of human cognition models, but are described as:

Level 1 Perception of elements in the current environment

> This is awareness of the critical aspects of the surroundings that the person must monitor closely and how to react to changes in them.

Level 2 Comprehension of the current environment

> This is the awareness that comes from the synthesis and integration of the available data including that from Level 1. The environmental understanding is related to the person's goals and how they can be achieved.

Level 3 Projection of future status

> This involves an understanding of the dynamics of the surroundings and how they will influence achieving the goal. Prediction of the future using mental models is an important part of this level.

Autonomous systems will be distinct from the operator. Although the operator may have a good interface with the system, the human and the machine will build up, and use, situational awareness in completely different ways. They will use different datasets; the autonomous system will have immediate access to its sensor data and so can detect changes more rapidly than the operator, but its available database and processing power will be more limited.

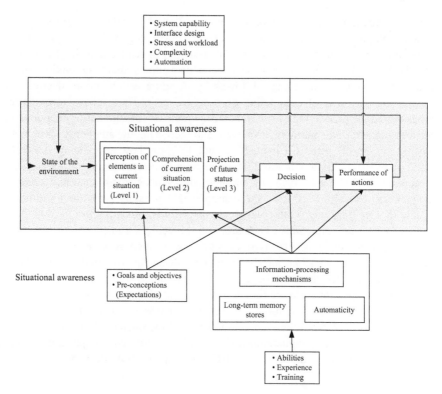

Figure 2.4 Model of situational awareness and decision-making

A single, independent operator will draw on his or her knowledge and experiences as well as the information from the human–machine interface and the autonomous system's response to commands. A number of autonomous systems and multiple, linked operators will have access to a wide range of data from external sources; these will be both hard digital data from IT systems and soft qualitative and interpreted data from humans. Bandwidth limitations will limit data flows between autonomous systems and operators, ensuring that situational awareness will be different in different locations.

Both the operator and the processors in the autonomous system can achieve high levels of situational awareness although the dynamic modelling of the two will be very different. The autonomous system's situational awareness will be formed from its sensors and on-board data by algorithms written according to firm specifications. The situational awareness will be expressed in digital form for on-board interpretation or transmission over the datalink to the operator. High-resolution images can be processed on board for scene changes or object recognition cues and/or sent to the operator for interpretation using more capable software and human judgement.

There are clear parallels between the three-part cognition model of Figure 2.1, the OODA loop of Figure 9.3, the processes at different skill levels in Figure 2.3, and the 'Situational awareness' part of Figure 2.4. The reader will probably not need a detailed understanding of the various psychological models of human decision-making processes. However, it is essential to realise that there is a split into the three phases involving successively: data inputs about the environment; understanding their significance; followed by deciding which action, if any, will provide the best way of completing an assigned task. These phases provide an initial step in the production of subsystem requirements for automating functions in the development of an autonomous system and increasing the autonomy level in an existing system-of-systems. Chapters 12 and 13 show how an existing standard, (NISTIR 6910, 4D/RCS Version 2.0) based on a three-part cognition model, can be applied to the development of autonomous systems in military and civilian applications. (NSTIR 6910 uses the example of a possibly-autonomous military vehicle in a military unit, with control starting with commands from headquarters leading to steering commands to actuators on the vehicle's wheels.)

The fundamentally different types of situational awareness on the vehicle and at the operator's human–machine interface station mean that the types of decision that can be made at the two locations will be different. There can be full flexibility in decision-making by the operator but more limited flexibility by the autonomous system's systems due to the more limited on-board information and algorithms. This difference is fundamental to the discussion of the compliance of the operators' and autonomous systems' behaviours with regulations and national laws for civilian applications, and targeting commanders' compliance with IHL for weapon-release decisions.

2.4   Human workload

There is an implicit assumption in many discussions, including the ones in this chapter, that the human making the decision is able to devote enough attention to it and so make rational decisions. When true, it is irrelevant which of the approaches described in Section 2.5 (Decision-making) is taken and the humans make their best choice at the time, with the situational awareness level they have. This is often expressed as the operator, or commander making their decision in good faith, based on the information available at the time. The decision should be defensible, even if it eventually turns out to have been the wrong one. (The benefit of 20/20 hindsight.) It is also not allowed in IHL to deny a commander information so that they inadvertently make illegal decisions.

Stress plays a role in decision-making. Any person making an important decision feels some stress, and when coping with a high workload the level of stress and chance of an error will increase. System designers may not be able to reduce the user's stress level directly, but can reduce the workload by introducing automation into mundane tasks. This can be considered as one of the main aims and justifications of automation. A second, complementary approach is to develop decision aids which extract information from data and present it in a user-friendly way. These are of growing importance in many fields, perhaps the best known example being satellite navigation in cars. These make recommendations that are usually clear and enable the driver to make more-informed decisions in unfamiliar locations.

The workload on the operator will depend on the design of the human–machine interface and specific autonomous system implementation that they are using. The designer and system integrator should have a good idea of the user's workload so that they can make suitable design decisions at a working level. Even with a user-friendly interface, their situational awareness at the time and the time available for the decision will be important. Good autonomous systems design takes human comprehension and reaction times into account at all stages of the system analysis.

There are several ways of measuring workload for many tasks and professions with some going into great detail. Some idea of the wide range of approaches and physiological measurements can be found in a literature review by Miller [4] in 2000 carried out for examining automobile driver workload.

It should be straightforward to specify, and provide, an acceptable operator workload for a new autonomous system which is not closely coupled to other systems and provides one task for an operator. The specification of an autonomous system that is upgrading functions in an existing system or process may be more problematic. There will need to be a workload assessment of the existing process, as well as including the operator's workload from other sources.

One method of measuring workload and its scale is presented here as it is both illustrative and straightforward to use. It is particularly useful when automation makes changes to the existing processes within an overall task for the person. This is the Bedford Pilot Workload Scale given in Table 2.1, which was developed in the 1980s and published in 1990 [5]. It was developed to measure the workload of pilots in existing aircraft under a range of conditions as part of programmes to prevent air accidents caused by pilot overload and/or fatigue. It is also relevant to new autonomous systems as it was used to assess the workload on pilots of new highly automated fast jets whilst in the design and prototype stage.

Table 2.1   The Bedford workload scale for pilots

Rating	Workload description
1	Workload insignificant
2	Workload low
3	Enough spare capacity for all desirable additional tasks
4	Insufficient spare capacity for easy attention to additional tasks
5	Reduced spare capacity Additional tasks cannot be given the desired amount of attention
6	Little spare capacity: level of effort allows little attention to additional tasks
7	Very little spare capacity, but maintenance of effort is the primary task set in question
8	Very high workload with almost no spare capacity Difficulty in maintaining level of effort
9	Extremely high workload. No spare capacity Serious doubt as to ability to maintain level of effort
10	Task abandoned Pilot unable to apply sufficient effort

The measurement principle is to put the human in an actual or closely representative process and present them with either the changed process and/or increasing work whilst operating their system. The tests were carried out on pilots in a large (tens of metres high) test rig whilst being subjected to high g-forces and vibration. The workload assessment was carried out by presenting the pilots with increasingly complex decisions whilst subject to representative physical and mental conditions. They were questioned about each decision immediately after the trial and their workload assigned a score using the flow diagram given in Figure 2.5. It is

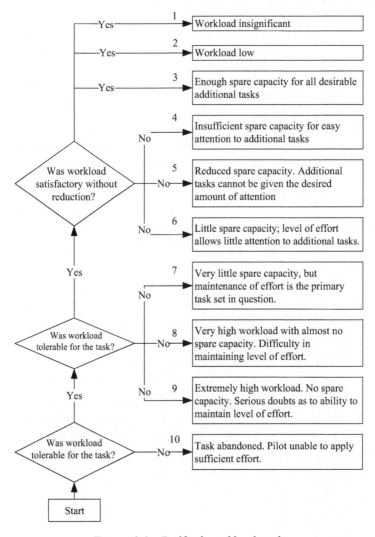

Figure 2.5   Bedford workload scale

a unidimensional subjective scale which is both easy to understand and gives a useful framework to discuss workload.

This type of scale and simple measurement methodology could give a method to evaluate human workloads in scenarios with mixed autonomous and human functions contributing to the operation, for example, measuring the workload on an operator controlling more than one UAV at the same time whilst evaluating multiple sources of information about their task goals and progress. The questions can be asked at the end of each of a set of exercises or simulations. The answers and scale point will give data which can be used in assessing the effectiveness of automating one or more critical functions in achieving acceptable levels of correct decisions under operational conditions.

2.5   Decision-making

The widespread application of processors in automation led to both an increase in the range of applications and flexibility in the control system. The human–machine interface is an important part of the design as the operator must have the correct information to make an optimum choice and act. It presents the operators with a range of choices depending on the specific circumstances and variables in the process. The operators also need skills appropriate to the complexity of the choices they must make so that they recognise the best one and have the authority to act on it.

Studies into human–machine interface design led psychologists and cognitive scientists to examine human decision-making and its basis. Both qualitative and quantitative models are used to understand how humans decide and act in situations of varying complexity.

A paper by Parasuraman *et al.* [6] discusses human understanding and its relationship to automation. It uses a four-part model of human decision-making which is the three-part model in Figure 2.1 with action selection added as the fourth. They consider systems with very low levels of autonomy by modern standards but they do address the problem of choosing the most appropriate functions to automate as a way of improving human capabilities. This is achieved by using ten levels of automation which are equivalent to the autonomy levels discussed in Chapter 3 and are included there.

This book develops a similar approach, considering the use of autonomous systems for specific functions and comparing them with the legal criteria for the process as a whole. Section 12.1, for weapons, and 13.3 for civilian autonomous systems summarise the approach. The results of taking this approach to system and system-of-systems design gives common ground between lawyers and engineers when considering the legality of its use.

It is often assumed that humans make decisions using a rational process when making a decision. This is: identifying options to perform the operation or solve the problem; assessing their suitability; choosing one option based on their relative performance against suitable criteria; and then acting using the chosen option. In practice, this is not the way that many decisions are made. Experience is used to

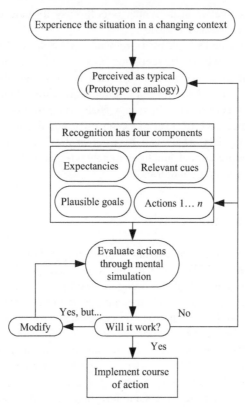

Figure 2.6   Klein's third variation – evaluate a course of action with changing circumstances (© 1998 Massachusetts Institute of Technology, by permission of The MIT Press)

recognise analogies between the current problem and an earlier one. Experience with that earlier problem is then used to find a solution to the current one. Reference [7] has a discussion of this intuitive approach for military commanders recommending its wider adoption, especially where there is only partial information.

The more intuitive decision process is known as Recognition-Primed Decision-Making (RPDM) and is due to Klein [8]. He considered three approaches to problems of increasing complexity. The first is where the problem has a simple known solution which can be implanted directly, the second has a problem which is not immediately recognisable and requires better diagnosis and probably more information. The third approach is shown in Figure 2.6 which is similar to the second, but has changing circumstances which complicate the problem and solution necessitating iterations of the solution until the problem is solved. This approach is the most appropriate to military and other complex scenarios. AI with learning systems is more akin to RPDM than rational decision-making.

Military operations always have an initial plan even if it is limited. Trials using two opposing teams [9] showed that effective decisions in dynamic conditions require the humans to have a model of the scenario which they can change over time. The plan can and does act as the starting model, but it must be adequate. With inadequate plans, i.e. with significant missing information, the teams failed to make correct dynamic decisions. Before contact with the enemy, operations followed the plan as a feed-forward system. After contact the system becomes one with feedback and only an adequate plan can produce an effective correction (Difference signal).

The rational and RPDM approaches to decision-making can be considered as analogous to different approaches to automation. The rational approach is analogous to taking a process as a set of linear steps, probably with feedback loops in and around them; each step in the process is then automated to replace a human or increase the currently automated performance. This is also known as weak AI. The RPDM process is analogous to having a machine-learning process replacing the human at any or all stages, also known as strong AI.

The two types of decision-making lead to a range of questions to be considered in designing an autonomous system. The two approaches will have very different problems for validation and verification of performance and these will have to be matched to the importance of the decision, i.e. what is the consequence of the decision being wrong?

Dynamic decision-making of the type shown in Figure 2.6 is necessary due to changing circumstances and exposes a key problem in automating decision-making. A human will probably use RPDM in these conditions, but rational decision-making is more amenable to an algorithmic solution and automation. It is difficult to validate or verify the results of strong AI analyses as they depend on the type of training and the data used for it. The results of an AI assessment of an image, for example, may be unexpected but accepted by users as correct. This may be acceptable or advantageous if the AI system is acting as a diagnostic aid supplementing human judgement. However, the false alarm rate[1] must be extremely low if the results are going to be implemented directly by another autonomous system.

A system-of-systems which bases its actions on the results from an AI-based system, and can have adverse effects on humans, must have close examination of the AI-based system before it can be accepted. The false alarm rate will need to be carefully specified and measured. There will also need to be careful verification of an acceptable level of correct identifications. There will need to be agreement between lawyers, engineers and others about quantifying these parameters in the medical and military fields among others.

2.6 Multiple humans and autonomous systems

It was stated in Section 2.1 that the same three-part model of decision-making can be used for both humans and autonomous systems. This is illustrated here by work of a team working on an air-defence problem as part of the UK MOD Systems

---

[1]False alarm rate is a radar term for the rate at which noise peaks or clutter is incorrectly declared as targets. In medical applications, it is the number of false positives from a diagnostic or screening process. It is only indirectly related to the fraction of real targets which are correctly identified as such.

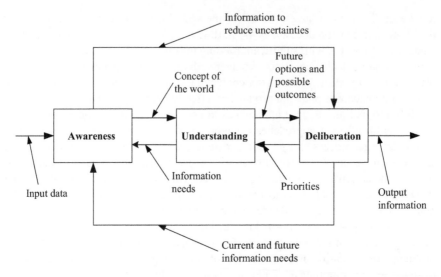

Figure 2.7   Information flows in three-part cognition model (From Thoms 2009)

Engineering for Autonomous Systems (SEAS) Defence Technology Centre (DTC) programme [10].

A three-part model with information flows is used for each human and then several are connected together to form a team. The proposition is then that functions within each of the three parts can be automated; understanding the information flows within each 'human' and across the team can then be used to identify the most suitable functions for automation.

The connections between the parts in a simple model for an autonomous system are shown in Figure 2.7. (A more complex model is shown in Section 10.2 and in a hierarchy in Figure 10.3.) Similar information flows could lend themselves to similar automation algorithms and human–machine interface specifications. Figure 2.8 shows an example of information flows around a team of three-part models.

The team in this case comprises the command centre, a package of defending fast-jet aircraft and a Surface-to-Air Missile (SAM) battery. Team members may be the pilots of multiple unmanned vehicles, but could also be commanders of the air, sea and land forces in a littoral operation. Extensive analysis and more details are given in Reference [10]. They are specific to other SEAS DTC applications but the principles given here are widely applicable.

In the case shown in Figure 2.8, there is an air defence operation underway. The command centre controls the packages of defending aircraft[2] and SAM batteries, each having radar and electro-optical sensors. Only one of each is shown in Figure 2.8 and only a sample of the intra-team information flows are shown. These

---

[2]A package of aircraft usually has four members. It comprises a package leader who is in communication with the air controller and who commands the actions of the three subordinate aircraft.

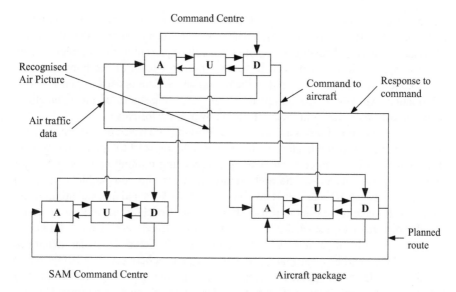

Figure 2.8   Example information flows between three-part cognition models, based on Thoms 2009 (A = Awareness, U = Understanding, and D = Deliberation)

are, from the left to right across the figure, for the response to an unknown aircraft entering the local airspace:

- The recognised air picture. This is the main reference for the current position and information about all aircraft in the operational area. It is distributed to all assets to enable their 'Understanding' functions;
- Air traffic data. The SAM command centre sends information about all aircraft seen by its sensors to the command centre. This is correlated there with other data to update the recognised air picture;
- Command to aircraft. This is the way that the defending aircraft receive instructions. They may be general, such as patrol area x or be more specific, such as investigate aircraft at position y, which are probably hostile. There are standing instructions to follow, rules-of-engagement[3], when unknown or hostile aircraft are detected;
- Response to command. This may include requests for more information as well as confirmation of understanding the instructions;
- Planned route. This is information about the defending aircraft package's interpretation of their command so that the SAM batteries will be aware of their presence and intentions.

In a real operation or an exercise, there will be large amounts of data and information flowing between assets. The Command and Control (C2) network and functions

---

[3]Rules-of-engagement are described in Section 7.12. They are the interpretation of IHL for the specific operation underway at the time.

will be described using architectures which are discussed in Chapter 10. Building an architecture can be complex but is essential for both an understanding of the information flows and the potential emergent properties that will be present in any complex system.

The system design will need to balance the skill level of the humans in the system, the information presented to them, and which data is used to generate that information. Models of the complete system must take note of the interactions between team members and their effects on overall performance.

Skill levels for military staff are well defined, but are based on current systems. If the new or revised architecture requires new skills for staff to operate the system, there will be large extra costs in creating and running training courses and major revisions of user manuals. The architecture must also be compatible with allied nations such as NATO member states. The use of NATO standards and architecture models will make this possible if used from the start of the procurement process.

An important aspect of decision-making is the timeliness of the information available to the local commander. The recognised air picture has been mentioned above and it should be clear that there will be time lags between data being sent to it and the receipt of an updated version by the front line assets. Similar issues arise for all data and information flows in modern network centric warfare (NCW). Modelling must include this aspect of the operation so that critical data and information has minimal latency and the skill and authority level of commanders can be evaluated. Decisions about the functions which should be automated and its level can be made on the basis of this information.

It can be seen that an equivalent analysis using the three-part model for both humans and autonomous systems can produce a detailed picture of the decision processes and the information needed to make them. The information flows can be classified as ones that need human interpretation for a decision or as ones that are reliable based only on machine analysis. This type of analysis is discussed in more detail in Chapters 12 and 13. The results of these analyses can then be used by the procurement and design teams to negotiate and agree which functions will be automated.

Complex automated systems and processes such as national power supplies and transport systems are, in principle, very similar to the military systems considered here. They have close interaction between trained human operators and automated processes with proposals for increasing the autonomy level of the automated processes. The same questions must be asked of them as are asked about military applications and similar techniques can be used to answer them as is shown in Chapter 13.

2.7   The Observe Orient, Decide and Act (OODA) loop

The OODA loop has been mentioned in Section 2.2 and the four parts illustrated in Figure 2.3. The OODA model does illuminate how humans, and especially military commanders, actually make decisions. OODA is a three-part model of decision-making, but is not a precise engineering concept. It is not strictly a loop, but is a feed-forward process and needs feedback extensions to become a loop. This section will describe it in more detail and discuss its shortcomings.

Colonel John Boyd, its originator, was a USAF pilot in the Vietnam War and later became a statistician. He examined air combat in that war in order to explain US superiority over their air opponents. His explanation was simply that US pilots observed their adversary, oriented (in the sense of pointing towards the enemy), made their decision and acted faster than their opponents. He formalised this into early versions of the OODA loop. Some lecture notes were published informally, but he evolved the OODA concept over time and never published it in a definitive form.

The concept is recognised by military officers and is taught in officer training courses. Engineers considering automating any part of the C2 chain and weapon-release decisions are well-advised to relate their proposals to it in discussions with military customers. Using it as a basis for capability analysis and system-of-systems design gives a good starting point to ensure that acceptable and legal targeting capabilities will be delivered by a new system.

The original application in air combat was generalised by Boyd to cover all forms of warfare by changing the orienting of an aircraft into a more general human-centred orientation process [11]. The now widely-used version of the loop is shown in Figure 2.9. The main advance was to expand the Orient part into the elements shown in the circle in Figure 2.9. Boyd also added some 'insights' to accompany the figure:

- Note how orientation shapes observation, shapes decisions, shapes action, and in turn is shaped by the feedback and other phenomena coming into our sensing or observing window.
- Also note how the entire 'loop' (not just orientation) is an on-going many-sided implicit cross-referencing process of projection, empathy, correlation, and rejection.

It should be noted that the original lectures including the OODA loop were entitled 'A discourse on winning and losing'. The main message being that you win if you can complete each part of OODA faster than your opponent. It is debateable

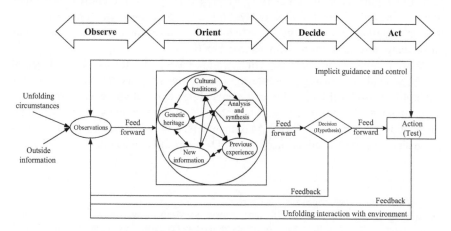

Figure 2.9   Boyd's OODA loop

whether Figure 2.9 is really a loop, or a feed-forward process to an action with feedback loops modifying the form of the action. Certainly in combat, after successful action, the combatant turns to a new adversary or returns to base. The process comes to an end at that point and a new process with a different goal is initiated.

Attempts to use the OODA loop for the design of C2 systems in the age of NCW have led to serious criticisms of it. Papers were published in 2005 by two independent groups looking at modern applications of the OODA loop.

1. Grant and Kooter [12] compare OODA with numerous other models to assess its applicability for modelling C2 systems and as the Operational View in the Department Of Defense Architecture Framework (DODAF)[4]. They develop the Observe, Decide and Act parts to the same level as Orient in Figure 2.9 and then make comparisons with several other C2 system models and Endsley's model of situational awareness discussed in Section 2.4. They conclude that the conventional OODA loop needs to be re-engineered into consistent notation and structure before it can be used for developing C2 systems.

    The main criticisms are that it is a centralised feed-forward process and consequently: mission planning is omitted; information flows are not specified; it is rule-based and only applicable to Levels 1 and 2 of the situational awareness model in Section 2.3 above.

2. Brehmer [13] investigates application of the OODA loop in dynamic situations in order to put research into C2 on a solid base. Cybernetic[5] approaches to C2 are discussed, one of which is Stimulus, Hypothesis, Option Response (SHOR) to illustrate that C2 is responsive with feedback and not just reactive. He identified similar shortcomings to those of Reference [12] and proposed a Dynamic OODA (DOODA) extension of OODA. It was shown to be suitable for developing C2 research as well as providing a more representative model of military decision-making.

The DOODA loop is a set of functions forming a dynamic decision loop. It is shown in Figure 2.10, which is from Reference [13] with the equivalent OODA functions added. There are time delays in the C2 processes which are known as friction in military terminology. Automation will change the time delays as well as introducing new separations within some functions.

Most campaigns in the twenty-first century have had extensive surveillance of the operational areas using longer-term observations from assets such as satellites and high- or low-level UAVs. Increased cross-force connectivity enables transfer of large amounts of data, its analysis by remote intelligence analysts and targeting information passed to the forward commanders. These developments have happened since the general adoption of the OODA loop and make targeting a wider and

---

[4]Architecture frameworks are described in Chapter 4.
[5]One of many definitions of cybernetics from http://www.asc-cybernetics.org/foundations/definitions.htm is: Cybernetics is defined as the science of communication and control. It maps the pathways of information by which systems may either be regulated from outside, or regulate themselves from within.

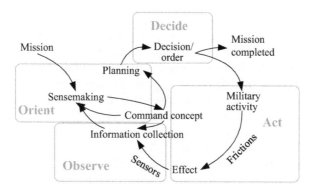

Figure 2.10   The dynamic OODA loop from Reference [13] with OODA functions added for illustrative purposes

more dynamic process. The term and terminology of OODA are still widely used, having a smaller but still not clearly-defined part in targeting. These modern versions are described and compared with each other and OODA in Section 8.4.

Despite these criticisms, in the author's experience, the OODA loop is still one of the most effective communication tools between engineers and military officers when discussing specific actions in a scenario, such as a strike on a bridge or air defence against intruders. It can clarify the required military functions and how they are implemented by items of equipment and the changes needed when the automation level is increased. This is because these activities are feed-forward operations in response to either an order in the former example or the detection of possibly hostile action in the latter. Time delays (friction) can be directly related to operational factors such as exposure to hostile action before weapon release.

2.8   Authority to act

2.8.1   Authority and responsibility

Section 2.1 discussed the freedom of an individual to take actions and the constraints on them. Although there are philosophical and ethical debates about a person's responsibilities which come from their rights, these are irrelevant for autonomous systems design. It is unlikely that society will allow the operation of autonomous systems which are allowed to operate with no human oversight even if it is remote. This means that every autonomous system will have an operator who is responsible for its safe operation whilst in use. The consequences of the autonomous systems' actions will be the responsibility of the operator. The division of these authorities and responsibilities between human and machine is one of the key issues for autonomous systems design and a central part of the techniques described in this book.

One starting point is to think about management responsibility and authority in an organisation. There is a hierarchy from the shareholders down to the lowest non-management level in the organisation. The authorities and responsibilities are formalised by job descriptions and company procedures. They can also be assigned to external bodies by the use of contracts on subcontractors. The whole structure must be coherent so that there is no ambiguity, and there must be no gaps (Colloquially, cracks between the floorboards), both of which bring uncertainty.

An organisation is liable to legal redress from the recipients of any unexpected or illegal actions by its products. With unambiguous processes, job descriptions and subcontracts, the responsibility for any mistake can be placed on an individual or group of individuals and their manager. This provides a robust defence for the organisation and individual, limiting their personal culpability, or can show that the product user was at fault. The aviation industry uses this identification of responsibility to check, and possibly change, the relevant international regulations. This then leads to enhanced safety levels in the industry. Many other industries rely on litigation for specific incidents, or the threat of it, to make the necessary changes.

The specification and design of an autonomous system is closely analogous to the design of a company. There must be clear understanding of the actions of the autonomous system in normal use, analogous to the operation of the company's product in normal use. This is called the autonomous system's behaviour to cover the more general case of an autonomous system compared with a product with no autonomy.

The first design step is to identify the functions which create the autonomous system's behaviour. Delivery of these functions is provided by the systems and subsystems created by decomposition and/or architectural analysis.[6] Functions are provided by sets of actions. Specifications must then be written which both allocate responsibilities for these actions to specific subsystems and ensure that the resultant functions can be tested. The difference between an automated system and an autonomous one is that, for the latter, the authority for a particular action can be assigned to either the machine or the human. If it is to the machine, then it must have sufficient information to allow it to make the decision and take action without reference to a human.

A function can be delivered by a combination of automated, autonomous and human generated actions. The design team and reviewers must ensure that increasing autonomy of one or more of the actions does not lead to a function taking actions autonomously when the decisions and authorisation should still be made by a human.

Capabilities and the architectural analysis that delivers them with autonomous systems are described in Chapter 4. Both civilian and military autonomous systems will have a functional architecture with nodes having responsibility for, and the authority to complete actions. The problem is to ensure that responsibility and authority for each action are both located in one node only, and that the necessary data is in a suitable form at that node. Absolute clarity of allocation of these

---

[6]Figure 5.1 shows the type of system and subsystem structure which is in all designs. Note that a function may be enacted by several subsystems which are not necessarily part of the same system in that figure.

parameters (Authority, responsibility and correct information) to each node has been fundamental to military organisations for centuries. There are variations between coalitions, nations and services, but the essential elements are common. These provide a useful reference framework for considering their allocation within an autonomous system that is able to carry out actions without direct human control. Their role in military decision-making is described in Section 2.8.2.

2.8.2 Military Command and Control (C2)

All military personnel operating in a campaign have a clearly defined position in a command chain with their commands flowing from decisions at higher levels in the chain. They obey these and may issue further commands to their subordinates. The details may vary between nations and even between individual services within a nation, but the principles are the same.

All military operations require decisive actions in order to succeed. An action is taken as a result of a decision, but a decision can only be followed by action if there is authority to take it. The procedures should also cover the case of deciding to take no action. The inputs to the decision may be orders from superiors, or information from sensors or subordinates. Anyone initiating an action is called a commander and the command chain should ensure that an action can only be taken by a commander who has the correct level of authority. This section will discuss the aspects of this authority relating to autonomous systems in the command chain.

Military activities in peacetime, and away from operational areas in wartime, are carried out following national laws. The use of force however is different. It is always subject to IHL. When a subordinate is ordered to carry out a task by a superior, authority to take the necessary actions is passed to the subordinate. This does not remove the requirement on the subordinate to follow IHL as defined by the rules-of-engagement.

Military situations are always dynamic and the flow of information to the commander comes from several sources with different latencies and reliabilities. The commander tries to remain in control, instructing his or her assets to take actions aimed at mission success, but must also respond to unfavourable decisions made by his or her opponent. Officer training is geared to making rapid decisions, often on the basis of incomplete information even though it may come from sophisticated Information Technology (IT) systems, supported by software analysis such as image analysis algorithms.

Object recognition in images is one example of a decision aid which are now an important, even essential, part of decisions and actions. The system-of-systems designer and legal adviser must consider whether these are autonomous and the level of human interpretation required before authorising release of a weapon on the basis of an algorithmic identification. Similar considerations apply when the local commander is supported by a remote analyst who may be using a simulation of local environment to take some of the workload off the local commander. Their position in the C2 chain must be clear and that there is correct authorisation of decisions made on the basis of information from these sources.

The terms 'command' and 'control' are sometimes used interchangeably by engineers especially as the C2 system comprises many items of equipment which now interface with many other communication and (engineering) control systems. In common with many other terms, there are agreed definitions in NATO Standard (STANAG) AAP-6 [14]. Some of these are given in Panel 2.2.

Panel 2.2:   Definitions from STANAG AAP-6

Chain of command:

It is the succession of commanding officers from a superior to a subordinate through which command is exercised.

Command:

The authority vested in an individual of the armed forces for the direction, coordination, and control of military forces.

Command:

An order given by a commander; that is, the will of the commander expressed for the purpose of bringing about a particular action.

Operational command (OPCOM):

The authority granted to a commander to assign missions or tasks to subordinate commanders, to deploy units, to reassign forces, and to retain or delegate operational and/or tactical control as the commander deems necessary.

Control:

The authority exercised by a commander over part of the activities of subordinate organizations, or other organizations not normally under his command, that encompasses the responsibility for implementing orders or directives.

Operational control (OPCON):

The authority delegated to a commander to direct forces assigned so that the commander may accomplish specific missions or tasks which are usually limited by function, time, or location; to deploy units concerned, and to retain or assign tactical control of those units. It does not include the authority to assign separate employment of components of the units concerned. Neither does it, of itself, include administrative or logistic control.

Command and control system:

An assembly of equipment, methods and procedures and, if necessary, personnel, that enables commanders and their staffs to exercise command and control.

The term 'commander' is used in IHL for the person who is responsible for a lethal decision[7], usually the release of a weapon. This is usually the person who releases the weapon but in a few restricted circumstances it may be someone else. An aircraft pilot is usually responsible for the release of its weapons, but where the pilot does not have 'eyes-on' the target and relies on instructions from a Forward Air Controller (FAC), then the FAC is responsible.

The C2 chain is structured so that everyone has a clearly defined level of authority and can act within it. In principle, when someone does not have the authority to act, they refer the decision up the chain for a decision and authority to act or take no action. If that authorisation is either not accessible or will take too long, then that person will take a decision either to take the action they consider best or not. There will be consequences of this choice which will involve further decisions and actions.

All lethal decisions can be considered to be part of targeting. In addition to the rules-of-engagement, military forces have targeting directives to place limits on their authority to act. These are usually classified, but some idea of their scope can be found in Section 2.6 of an Australian document [15].

The targeting directive, which may be standing, or issued for specific exercises/operations, specifies:

- approved, restricted and no-strike status of target categories;
- the collateral damage estimation methodology to be used;
- levels of risk authorised for use by designated commanders;
- targeting command, control and oversight arrangements and the responsibilities of supporting agencies; and
- national policies on legal issues affecting targeting.

The targeting directive is fundamental to the planning of the operation rather than the weapon release decision, but should be understood by the decision-maker for their interpretation of rules-of-engagement at weapon release.

An automated weapon system which responds in a clearly defined and well-understood manner has to be installed and operated in a way which meets the rules-of-engagement for that phase of the mission. Increased intelligence in the autonomous functions of a weapon control system will lead to the expression of some parts of the rules-of-engagement in a digital form that can be interpreted by the automated parts and place appropriate restrictions on the action it can take. This use of rules-of-engagement is discussed in more detail in Chapter 7 and is one of the considerations which led to the concept of 'authorised power' defined in Panel 1.1 and discussed as part of the 4D/RCS model in Chapter 10 and its applications in Chapters 12 and 13.

Software decision aids are now taken for granted in many applications, and users may become familiar with their shortcomings and biases. Linking them together in an automated process requires considerable analysis and understanding of these issues before specification and design can start. They are not normally considered to be part of the weapon release system, but they play a significant role

---

[7]A lethal decision is one to perform an action which will result in death or injury to an intended target.

in the decisions made by the last human in the C2 chain. It is essential to critically examine the time that the human has to assess the information from the decision-aids before making a decision. If the person's situational awareness is poor and workload high, they may simply accept the result and act on it. An example could be a target indication system which offers potential targets with some uncertainty for the commander's consideration. An uncritical weapon-release decision could result in unjustified civilian casualties and be illegal. This means that careful consideration should be given to whether decision-aids are within the scope of the 'Article 36' reviews described in Chapter 11.

One approach to ethical decision-making is that developed by Arkin *et al.* for weapon systems [16]. They use the concept of an ethical governor which assesses the robot's decision against ethical criteria to either allow or veto the action coming from the decision. Their approach is different to that taken by this book in that the ethical governor is checking a decision that is made by an autonomous robot whereas the approach here is to specify, design and test the robot according to ethical criteria so that the only checking needed is through the authorised controllers. The approach here also takes into account some uncertainties in the autonomous systems' situational awareness measurements and understanding of its environment rather than their assumption that it has near-perfect understanding of the scene.

When a military campaign is underway, personnel usually have a general authority to respond with lethal force in certain circumstances such as when there is imminent danger to their life. There is some debate about whether a UAV can legally defend itself against an attack from a manned platform as there is no human life at risk from the attack. The military counter-argument is that the UAV, which may have a capability which is vital to the operation would then be defenceless against any hostile manned asset. Resolving this argument is beyond the scope of this book and the design team. The legality of fitting self-defence systems to unmanned systems should be settled by the procurement authority and their legal advisors before any contract is placed. They should have followed the 'Article 36' processes described elsewhere in this book.

References

[1]   Rasmussen J. 'Skills, rules, and knowledge; signals, signs, and symbols, and other distinctions in human performance models'. *IEEE Transactions on Systems, Man, and Cybernetics*. 1983, vol. SMC-13(3): pp. 257–66.

[2]   Endsley M.R., and Jones D.G. *Designing for situational awareness*. 2nd edn. Boca Raton: CRC Press; 2016.

[3]   Endsley M.R. 'Theoretical underpinnings of situational awareness: A critical review' in Endsley M.R. and Garland D.J. (eds.). *Situational awareness analysis and measurement*. Mahwah N.J.: Lawrence Erlbaum Associates; 2000, pp. 1–24.

[4]   Miller S. *Literature review: Workload measures*. University of Iowa document ID N01-006, 2001.

[5] Roscoe A.H., and Ellis G.A. *A subjective rating scale for assessing pilot workload in flight: a decade of practical use.* Royal Aerospace Establishment, Technical Report TR90019, March 1990.

[6] Parasuraman R., Sheridan T.B., and Wickens C.D., 'A model for types and levels of human interaction with automation'. *IEEE Transactions on Systems, Man and Cybernetics-Part A: Systems and Humans.* 2000, vol. 30(3): pp. 286–297.

[7] Roth R.I. 'The rational analytical approach to decision-making: An adequate strategy for military commanders?' *The Quarterly Journal.* 2004, vol. 3(2): pp. 84–8.

[8] Klein G.A. *Sources of power: How people make decisions.* MIT Press; Figure 3.1, p. 25, 1999.

[9] Brehmer B., and Thunholm P. 'C2 after contact with the adversary: Execution of military operations as dynamic decision making'. *Proceedings of the 16th International Command and Control Research and Technology Symposium*; Quebec, Canada, June 2011.

[10] Thoms J. 'Understanding the impact of machine technologies on human team cognition', *Proceedings of the 4th SEAS DTC Technical Conference –* Edinburgh, Paper B7, 2009.

[11] Boyd J.R. 'The essence of winning and losing'. Unpublished lecture notes. 1996 Jan;12(23): pp. 123–5. Slides also available at http://pogoarchives.org/m/dni/john_boyd_compendium/essence_of_winning_losing.pdf Accessed 3 March 2018.

[12] Grant T., and Kooter B. 'Comparing OODA & other models as operational view C2 architecture'. *Proceedings of the 10th International Command and Control Research Technology Symposium.* Jun 2005.

[13] Brehmer B. 'The dynamic OODA loop: Amalgamating Boyd's OODA loop and the cybernetic approach to command and control'. *Proceedings of the 10th international command and control research technology symposium.* Jun 2005, pp. 365–368.

[14] NATO Standard. AAP-06 Edition 2015 *NATO Glossary of terms and definitions (English and French).* North Atlantic Treaty Organisation, NATO Standardisation Office (NSO) 2015.

[15] Australian Defence Doctrine Publication 3.14 2009, Publisher: Director, Defence Publishing Service, Department of Defence CANBERRA ACT 2600.

[16] Arkin, Ronald C. *Governing lethal behaviour in autonomous robots.* Florida: Chapman & Hall/CRC Press; 2009.

Chapter 3
Automated control and autonomy

3.1 Introduction

Autonomy, when applied to humans, is taken to mean freedom to make choices and take actions based on those choices without constraints from others. There is an implicit assumption that the autonomous person will be subject to the laws and other ethical constraints which apply to every other person. It is generally accepted that the current developments of robots and highly automated systems will continue. However, concerns are raised about whether there are risks to individuals and to society as a whole from these developments. These concerns can only be alleviated by showing that the new type of product can be trusted to behave in an ethical manner. This book presents engineering approaches to the problem.

The terms 'robot', 'intelligent system' and 'Artificial Intelligence (AI)' are frequently used, but with little consensus about their exact meaning. Clear definitions of terms are essential in both engineering and law. Section 3.3 and Panel 3.1 give definitions from several references and explains their use. Panel 3.1 gives the definitions used in this book. An autonomous system is defined there as 'a system which has the ability to perform intended tasks based on current state, knowledge and sensing, without human intervention'. Autonomy level was a specialist concept, but is now gaining more widespread use, so the range of definitions of the relevant terms, with a selection of industry-specific definitions, is explained in Appendix A3.1.

An important part of demonstrating acceptability of an autonomous system is proving that it is safe, reliable and always under human control. It is also essential to show law-makers and regulators that there are proven processes available to show safe operation of an autonomous system, and quantify the risks associated with its use. Later chapters discuss the relevant legal frameworks, but demonstrating that an autonomous system is under control and will not cause harm to people or property in normal use is the key issue. An indication of the methods used to control automated and autonomous systems is given in the latter part of the chapter.

One of the first widespread public debates about the ethics of an autonomous system was that about the use of UAVs or drones, as they became called at the time, for attacks on ground targets in Iraq and Afghanistan. The debates have subsided, possibly due to the demonstration by governments that their armed forces have the C2 structures described in Section 2.8.2 and processes which ensure that all lethal decisions are made by an authorised human commander[1]. These processes are discussed in Section 3.9. The risks which may come from technology developments are discussed in Chapter 9.

3.2   Automatic or autonomy – does the choice of word matter?

Automation is now a widely accepted part of everyday life. Examples include autopilots in ships and planes, automatic speed control in cars and financial assessments for 'instant' loans. These systems are mainly a result of the widespread implementation of increasingly complex algorithms on very capable computers and microprocessors. Hardware costs are relatively low, so the complexity, and hence the price of the automated system is set by market forces and the demand for automation in a specific application. A further consideration is the relative ease with which a new capability can be achieved by combining existing systems directly or with a new piece of software. Apps on mobile phones are a good example of the latter. Although we can all agree that these processes are automated, difficulties arise if we use the word autonomous.

The general use of the adjectives automatic, autonomous, or robotic to describe a particular system would appear to depend on the context or the aim of the speaker or writer. The latter two adjectives are widely used in marketing or by pressure groups to suggest a highly sophisticated system which acts without human intervention to achieve a higher-level goal. Marketing organisations use them as a positive attribute to impress potential customers, whilst some pressure groups use them with negative connotations, suggesting that the system is out of human control. The words are used across many fields from relatively limited manufacturing robots to high-level AI decision-making systems.

Sometimes the same word can be used by two or more professions with precise definitions, but with significant differences in meaning. An example is 'target' which in military terminology means [1] *The object of a particular action ....* *planned for capture, exploitation, neutralization or destruction by military forces.* In radar terminology, any object detected by a radar is called a target, with no implication of any military significance. This difference is an example of where two professions, in this case the engineering and military ones, must be absolutely clear which definition applies in the specific context where it is used.

Automatic and autonomous could be further examples of words with straightforward definitions as they imply different levels of control, but their widespread often-interchangeable use prevents this. We must both define them and explain their use in our context, recognising that there are already many definitions and meanings for them.

Whatever word is used to describe it, there are many ethical questions arising from the use of a machine which makes its own decisions and acts on them to achieve a general aim. A recent book by Wallach and Allen [2] suggests ways that

---

[1]Here, we assume that a weapon system is ethical if it can be demonstrated to comply with IHL.

it may be possible to ensure that machine-made decisions will be inherently moral. Reference [2] and other literature discussing these possibilities look further to the future than the next decade or so, which is the timescale considered here, but their ideas will become increasingly important.

Chapter 13 reviews current guidance for principles which can be applied when considering ethical and legal design of new autonomous systems. Military systems have much clearer guidance from their compliance with IHL. In both cases, the prevention or minimisation of harm is paramount regardless of whether the system or system of systems is described as automatic, autonomous or any other descriptor.

3.3 Definitions of autonomy and automatic

Despite their widespread use, there is no single definition of an automatic or autonomous system and the differences between them [3–5]. It is essential to understand the terms and identify the key differences so that the capabilities of a desired or existing system can be stated clearly for their design, use and regulation.

Section 1.3 in this book started with the *Oxford Online Dictionary* [6] definitions:

Autonomy –	Freedom from external control or influence
Autonomous –	1.1 Having the freedom to act independently
	1.2 Denoting or performed by a device capable of operating without direct human intervention
Automatic –	(Of a device or process) working by itself with little or no direct human control.

It was shown that direct application of the dictionary definitions to automated systems can cause confusion. Freedom to act independently can be viewed as a positive attribute, but does this mean that there is no external control at all? If so, the system will only stop if it thinks that it has achieved its aim even if this is impossible, so a human could only intervene in a disruptive way.

'Autonomous cars' is an example of the use of the word 'autonomous' where it is clear that there will always be limits on the cars' freedom due to their use on public roads. It is less clear when small drones (Weight less than, say, 10 kg) are discussed. Initially, hobby owners thought that they could fly almost anywhere, but now regulations are restricting that freedom of movement considerably, as well as privacy considerations giving the probability of more restrictions.

The above definitions are especially confusing when applied to weapon systems which must always be under the control of an authorised, responsible human with strict limits on their authority. The ethical and legal questions arising from the use of a 'Terminator' type of autonomous system would be myriad. However, careful consideration of the issues and the approach given in this book should make it possible to design highly automated systems that have an ethically and legally acceptable level of autonomy.

Appendix A3.1 to this chapter gives most of the commonly used definitions for the terms used in this book, mainly from BS 8611:2016 and ISO8373. These and other useful ones are given in Panel 1.1 which is repeated here for ease of reference as Panel 3.1. They are based on the level of human control, and will be used to describe systems or their subsystems as automatic or autonomous in the rest of this book.

A process or apparatus, without a human in direct control, which produces a defined and predictable response to a command is automatic. They will always respond in a completely predictable way with the same response to the same set of inputs whether these are from a human, another machine or from a computer algorithm, i.e. they are deterministic. The automatic braking system in a car is one example.

Panel 3.1:   Definitions used in this book

Automatic operation:

The state in which a system operates according to commands from an operator and where the resultant actions are predictable.

Autonomy (From ISO 8373)

The ability to perform intended tasks based on the current state, knowledge and sensing, without human intervention.

Autonomous system (From BS 8611)

A system which has the ability to perform intended tasks based on the current state, knowledge and sensing, without human intervention.

Autonomous operation (Based on ISO 8373)

The state in which an autonomous system executes its task program as intended.

Authorised power

The range of actions that a node is allowed to implement without referring to a superior node; no other actions being allowed. (A node is an entity within a control system which performs a specific, defined task.)

Authorised entity (From DOD document: *Unmanned systems safety guide for DOD acquisition*, 27 June 2007)

An individual operator or control element authorized to direct or control system functions or mission.

Autonomous weapon system (AWS) (From the ICRC)

An AWS is a weapon system that can select (i.e. search for or detect, identify, track) and attack (i.e. use force against, neutralize, damage or destroy) targets without human intervention.

3.4   Automated and Autonomous Weapon Systems (AWS)

A simple military example of an automatic process is firing a gun using its control system. The shell is launched towards the aim-point chosen by the artillery officer. The fire control computer will make corrections to the initial direction due to wind and atmospheric conditions, but these are based on analytic parameters and completely reproducible. In a similar way, when a cruise missile is launched, it flies to a pre-determined target location using a route which is planned according to a set of input parameters. Both are deterministic, so both are automatic systems but with a higher degree of automation. The difference between the two is the time from command to its direct consequence. This is not a difference of classification; they are both automatic but with different levels of sophistication in the projectile. Both have the same type of control discussed in Section 3.9 and shown in Figure 3.2, with the human at Location 3. The human at Location 2 has already given permission to release the weapon. (This may be the same person.)

The Phalanx close in weapon system, now in widespread service, is a useful exemplar when looking at the differences between automatic and autonomous systems from both social and engineering viewpoints. It is typical of a category of weapons called sentry guns. It was developed in the 1970s to defeat a hostile threat to a warship, such as a missile or shell, which suddenly appears and is travelling so fast that there is insufficient time for a human to respond and destroy it. The original concept is that the system detects the threat by radar, identifies it using a characteristic such as speed and/or direction and then makes an appropriate response. The response of the original system is to lock on to the threat and fire a low number of bullets with high accuracy and a rapid rate of fire.

Further developments have been made and the manufacturer's literature now discusses a development of the original design, Block 1B, as [7]:

> A self-contained package, the Phalanx weapon system automatically carries out functions usually performed by multiple systems: search, detection, threat evaluation, tracking, engagement and kill assessment. The Block 1B version of the system adds control stations that allow operators to visually track and identify targets before engagement.
>
> The 1B variant's configuration augments the Phalanx system's proven anti-air warfare capability by adding a forward looking infrared sensor. It allows the system to be used against helicopters and high-speed surface craft at sea while the land-based version helps identify and confirm incoming dangers.

Moral objections have been raised about close in weapon systems and sentry guns due to the lack of human control or oversight between detection and the kinetic response[2], but their use in restricted areas and circumstances negate most of

---

[2]A kinetic response is one which involves the use of a projectile with significant kinetic energy, whether in the form of explosive power or high momentum.

these. Section 7.14 shows how the Phalanx system was modified for use to defend land bases against similar targets, but near populated areas, in order to meet the requirements of IHL.

It is implicit in the terms *threat evaluation* and *kill assessment* that the system decides whether the object detected is a legitimate target or not, using pre-programmed rules. It is reasonable to assume that the system's use in operations has been subject to a legal review by the United States. It is also reasonable to argue that it is necessary to make a lethal response to a heavy object coming at high speed directly towards the defended platform. However, the conclusion is that this is an automatic system which can make decisions autonomously, but they are based on a deterministic balance of probabilities derived from human judgement. This is also an illustration of the concept that it is only sensible to discuss the autonomy of critical functions and not the autonomy of a whole system. Critical functions are those directly involved in the decision to fire the gun.

The development of Block 1B is interesting as threats such as helicopters and high-speed surface craft are detected at sufficient range for a human to respond and so must have a human operator who makes the lethal decision. Presumably the Phalanx gun becomes simply a convenient accurate and highly lethal weapon which can be used at very close range.

It has been argued [8] that Phalanx and similar systems are automatic rather than autonomous as they are defensive and only execute pre-programmed instructions more rapidly than a human can perform them. The second part of the argument may be sound according to the definition of automatic in Panel 3.1, but describing a system as automatic or not, based on its defensive use, is not. It is sensible to try to define Phalanx as automatic or autonomous based on published definitions to see if there is any consensus.

The definitions from JDN3/10 are given in Appendix A3.1. It defines an autonomous system as one which is *capable of understanding higher intent and direction . . . .* [9]. If the 'higher intent and direction' is taken to be initiating the system to detect nearby moving objects and then stop any weapon hitting the ship (or other defended asset), then Phalanx is autonomous even though its actions are deterministic. This is consistent with the definition of an autonomous system in Panel 3.1 (from BS8611:2016). The conclusion is that Phalanx, as originally built, can be described as an autonomous system using engineering definitions. As it is a weapon system, the logical next step is that it is an AWS.

The next question to be addressed is whether Phalanx is an AWS or an automatic system from a legal perspective. The critical issue is human responsibility for the consequences of its lethal decisions. The International Committee of the Red Cross and Red Crescent (ICRC), in Reference [3], suggests that (strong) AI in a system as one differentiator between automatic and autonomous:

> An autonomous weapon system is one that can learn or adapt its functioning in response to changing circumstances in the environment in which it is deployed. A truly autonomous system would have artificial intelligence that would have to be capable of implementing IHL.

The authors then argue that deployment of such systems is a paradigm shift which should only happen after addressing their fundamental ethical, legal and societal issues. Phalanx does not have strong AI so would not be an AWS using this definition.

More recent legal thinking recognises that systems are complex organisations of functions which make up the overall system. This is similar if not identical to the systems engineering approach of defining a system by identifying and specifying functions. The question is which functions are critical for a lethal decision and must be taken by a human. The ICRC have suggested in recent documents [10,11] that an AWS is one that can select (i.e. search for or detect, identify, track) and attack (i.e. intercept, use force against, neutralise, damage or destroy) targets without human intervention. The functions of detect, identify, track and attack are the same as those identified in Chapter 10 from an engineering perspective.

It can be concluded that legal and engineering approaches to understanding, and defining AWS are in broad agreement. They are both consistent with the definition in Panel 3.1 which is based on the ICRC proposals. The key point is that it may not be possible to make a precise prediction of the autonomous system's decisions whilst performing its task. Despite this, the overall activity must lie within a range set by a combination of its design and use if it is to meet IHL. Later chapters (mainly Chapters 10 and 11) give guidance on achieving this aim by design.

3.5   Autonomy levels

3.5.1   The need for autonomy levels

The engineering profession has responded to the problem of distinguishing between automatic and autonomous systems by introducing the concept of autonomy levels. They can be illustrated with an example:

> Early aircraft autopilots simply held course and speed, assuming that the pilot would immediately take control if there were any unexpected events (Automatic). Modern autopilots in civil airliners calculate and fly routes based on a wide range of variables such as weather and fuel consumption to meet time and air traffic control criteria. Pilots act as supervisors, monitoring their aircraft's route and ensuring it complies with Air Traffic Control Instructions. The aircraft can be said to be autonomous.
>
> If the aircraft is flown manually between waypoints given by the pilot's electronic flight bag[3] the aircraft's autonomy lies at some level between autonomous and automatic. The concept of autonomy level is introduced to quantify the level in a defined structure.

---

[3]An electronic flight bag is a specialist laptop loaded with all the data needed by a pilot for a flight and which can be connected to the aircraft's avionics system. It is certified to high levels of safety and reliability.

At no point is the pilot concerned with the detailed flight control instructions which have replaced those from the traditional control stick. These come from an autonomous subsystem, which can be advisory or supervisory depending on the specific flight control system.

Every autonomy level includes numerous decisions made by the system which are never reported to the human controller. These are considered to be too trivial or too rapid to need human approval. With increasing autonomy level, the authority level of these unreported decisions also rises. In control system terminology, they move from the inner control loops to the outer loop.

Autonomy-level definitions must take into account the interactions between humans and the autonomous systems. These include remote control (automatic); refer decisions to a human before acting (advisory); act on a decision unless stopped (supervisory); act on the decision and then report the action to a human at a later stage of the mission (fully autonomous). The nature of the autonomous system will give both constraints on, and requirements for the minimal levels of autonomy required for that level. One example illustrates this: delays in the communication channel of a few milliseconds may be irrelevant in some space applications, but the Earth-vehicle delay of several seconds or minutes will be an important driver for the system design of a deep-space probe.

The use of autonomy levels as a concept is now ubiquitous, but each industry sector is developing its own set of levels and definitions. This is probably the only way to use them as their definition must relate to the human–machine interface and the division of tasks and decisions between humans and machines. Appendix A3.1 gives ten different and sometimes mutually inconsistent sets of autonomy levels.

Two different approaches to defining autonomy levels in a consistent way have emerged. The definition can be either for a specific type of system, or be generic and then applied for the specific type of system. A report for the US Army Corps of Engineers calls these contextual and non-contextual [12] and shows the short-comings in the current lack of standardisation of definitions. The report does emphasise their potential value in testing and assessing progress in developing a new autonomous system and the importance of the division between human and machine tasks.

Autonomy levels in defence were originally seen as analogous to Technology Readiness Levels[4] (TRLs) which have proved invaluable in planning research and development (R&D) programmes, giving a readily understood way to plan technical developments. TRLs are given in Section 5.2.4 and Appendix A5.

Increased understanding of the benefits and limitations of autonomy levels and TRLs has led to a recent proposal to combine them for Unmanned Air Vehicles (UAVs) to give a unified framework for Unmanned Air Systems (UASs) Autonomy and Technology Readiness Assessment (ATRA) [13]. The aim is to give a quantifiable method to assess status and progress in technology planning for specific and actual environments. Reference [13] draws together a set of terminology definitions

---

[4]TRLs are covered in Section 5.2.4 and Appendix A5.

for UAS and determines autonomy levels based on the degrees of the UAS involvement and efforts in performing autonomy-enabling functions or guidance, navigation and control functions.

Again, this emphasises the need for autonomy levels to be defined for the specific industry sector or application area. When this is recognised, they become a powerful tool as a metric for discussions about autonomous systems.

3.5.2   Autonomy levels for non-military systems

Early papers on autonomy levels treated the increased capabilities required to go from automatic to autonomous as a spectrum and divided that spectrum into levels. These are analogous to colours in the visible spectrum and like colours there is no precise boundary except by a generally agreed definition using relevant criteria. The early thinking was also dominated by human interactions with computers rather than interactions with autonomous machines.

One early paper presenting an autonomy spectrum and autonomy levels was by Barber and Martin [14]. They were attempting to derive a quantitative measure of autonomy, deducing the following insights:

- Any definition of autonomy must be set in the context of the system's goal;
- If there is an autonomy measure, a, it must be a variable with multiple levels; They took values between 0 and 1;
- Autonomy is a spectrum.

They use the term *agent* rather than system as they also considered AI systems.

Many industries now use autonomy levels for common understanding of autonomy between the various communities within that sector. Details of representative definitions are given in Appendix A3.1. They include space, road vehicles and marine as well as defence.

3.5.3   Autonomy levels for military systems

Military planning for UAV developments in the early 2000s assumed that there would be a steady development starting with remote control, progressing through nearly autonomous single air vehicles to groups or swarms of UAVs. These would be capable of self-organisation and transfer of local command of the swarm around its members. Applications were driven by the aim of reducing risk, with the main ones being referred to as the 'three Ds' – dull, dirty or dangerous; examples are surveillance of areas over a long time, operating in contaminated areas and destruction of Improvised Explosive Devices (IEDs).

Developments in the use of UAVs led to changes in predictions about their development. This can be seen in the US DOD UAV development roadmaps, published and updated every few years from 2000 by the US DOD. The first and most recent are given as references [15] and [16]. The first had the premise that UAS development will progress through levels from 1, remote control, to 10, full swarm autonomy. These are called Autonomous Control Levels (ACLs) and presented as a time-based plot in Figure 4.4-2 in Reference [15] and as a table in

Appendix A3. Although they can be considered to be contextual, as defined in Reference [14], they became obsolete.

Reference [16] published in 2011 simply states four 'commonly referenced description of the levels of autonomy that takes into account the interaction between human control and the machine motions'. This change from ten closely defined levels is probably due to the operational necessity for UAVs capable of long-term surveillance, some of which eventually became weaponised. These produced a change in the philosophy of use for UAVs from the replacement of manned aircraft to new roles exploiting their ability to loiter over areas which may have legitimate targets in them.

The increase in surveillance data for large areas of territory also brings requirements for extra personnel to interpret the data, with improved quality of decision-making by the UAV pilot if he/she can be given timely access to this interpretation. The extra contextual information can make lethal decisions that must comply with the API (See footnote in Section 3.4) 'proportionality' consideration more complex[5]. The social and political pressures to keep a human clearly in the loop also had a significant effect on the implementation of autonomy. Article 36 Reviews[6] had always been applied to UAVs but the increased information available for real-time lethal decisions has brought increased scrutiny for decision aids and the need for up-to-date information by the weapon commander.

Despite the reduction in 'roadmap' autonomy levels, they still provide a useful consideration for autonomy in critical functions in the targeting process. One example comes from the US Air Force Research Laboratory (AFRL) [17]. It expresses the levels in terms that are generally applicable to UAVs and then as specific capabilities in the OODA loop[7]. The table of autonomy levels and OODA functions is given in Appendix A3.1. Shortcomings in the OODA loop have been explained in Section 2.7 and the more modern targeting cycles introduced in Section 8.4. However, the idea of considering autonomy levels at different stages of the targeting control processes is one of the best ways to approach the problem of making autonomy in critical function compatible with IHL. This approach will be refined later in this book, using criteria for authorised power as definitions of autonomy level for specific functions at each stage of the four-part process.

The concept of UAV swarms in Reference [17] was based on low numbers of capable UAVs operating in the same way as a flight or wing of manned combat aircraft operating together. As UAV operations evolved, it became less important relative to other military needs. Parallel civilian developments of micro-UAVs lead naturally to the idea of large numbers of small UAVs.

There have been several attempts to define autonomy levels unambiguously. An agreed and widely accepted set of definitions would provide a framework for

---

[5]Proportionality decisions are described in Sections 7.6 and 10.3.4.
[6]Article 36 Reviews are the process of checking that a new 'means or method of warfare' is legal under international law. They are described in Chapters 7, 8, 11 and 12.
[7]The OODA loop is a common military description of the targeting process. It is described in Sections 2.7, 8.4 and 8.5.

comparing systems and generally accepted test criteria. One widely referenced framework is the Autonomous Levels for Unmanned Systems (ALFUS) [18] and is discussed in Appendix A3.1. This concludes that there is no single definition of autonomy level. Instead it has four reference levels of autonomy and then gives a methodology for assessing and expanding them in specific applications by examining the level of complexity of the requirement. The number and definitions are chosen by the team drawing up the unmanned system requirements.

3.6   Autonomy, trust and work-sharing

Widespread acceptance of autonomous systems will only come about if there is trust of the autonomous system both by the operator who uses it, and the societal acceptance that autonomous systems technologies can, in general, be trusted. The latter is partly met by a recognised regulatory regime and other factors discussed in Chapters 6 and 9. The associated question is whether there will be public acceptance that the use of systems relying on autonomous functions is ethical. This can only come about if there is public debate about the aim of regulations and laws for sensitive applications of autonomous systems which engineers must join.

There is considerable anecdotal evidence that autonomous systems, with a clear boundary and interface with the human operator or commander, are trusted by them. This applies even when use of the autonomous system is not essential. Eventually, the human may transfer all decisions that they can to the autonomous system, and certainly will do so under conditions of high workload. Authority to act on these decisions and the allocation of responsibility for the consequences for the results of those decisions will be an issue for debate. User education, human–machine interface design and regulation will all play a part.

Transfer of decisions to an autonomous system for high workload conditions is not new. The example of the Missile Attack Warning System (MAWS) given in Panel 3.2 is one. The first MAWS in the 1950s were very simple systems that deployed flares or chaff as soon as a threat was detected by a scanning Infra-Red (IR) sensor. Modern ones are much more complex, and it is clear that there is scope to use more sophisticated software in the internal decision-making process. Examples could be comparing the IR signature and approach trajectory of an air vehicle in the vicinity with a database of known threats and using AI to estimate its potential threat based on its behaviour over time. This then brings database reliability to make the match and the search process used in the behaviour verification process. This is an example of the type of problems that AI will bring into weapon reviews.

Integrated human and machine decision-making is not a problem in principle, but there will be a dynamic transfer of work between the human and their automated support systems. Under conditions of high workload, the system may well be designed to take over lower priority tasks that are under human control in less-demanding circumstances. It is absolutely essential that the responsibility for every decision is unambiguously allocated to one node in the control chain, whether human or machine. Changes in responsibility over time must be clear to both the human and the machine. This is a normal requirement for humans in any military C2 chain. The general requirement for civil and military autonomous systems is that the automation process must include a process for reallocation of responsibility between human and machine. It must include a check that the automated system has the authority, possibly defined as authorised power, to make the decision and implement the consequent command.

Panel 3.2:   A simple example: A Missile Attack Warning System (MAWS)

An MAWS on an aircraft analyses missile parameters such as its IR signature, speed and bearing to decide if the incoming object represents a threat. When activated, it decides if a collision is probable, it will not wait for the human to react, but will respond by firing the aircraft's Defensive Aids Suite (DAS). This example illustrates one of the many differences in approach between the designer and the user. The engineer will calculate success probabilities and false alarm rates for the following steps in the operational use:

1. Detection of an object that may be a missile, e.g., by the detection of its Ultra-Violet (UV) launch flash
2. Measurement of kinetic properties by its sensor suite, e.g., by radar or measuring increased optical or IR size over time
3. Identification of that object as a threat to the aircraft, e.g., by IR signature of plume or Radar Cross Section (RCS); then by estimating its change of bearing; no change means a collision is certain
4. Choose best subsystem to deploy from the DAS, deciding if the missile is radar, IR or optically controlled
5. Deploy DAS subsystem. These may include one or more of: optical or IR flares; chaff; a towed decoy; and an anti-radiation missile.

The pilot only has two questions before the mission:

1. Will the MAWS protect me against all missiles in the opponent's inventory?
2. Will the limited stock of DAS countermeasures be wasted on missiles that are not fired at me, or false alarms like birds?

The debate conducted by the procurement agency and their technical staff is to translate probabilities into system performance in likely campaigns. The balance to be struck is between the rate of consumption of countermeasures against false alarms, the false alarm rate and the successful identification of a lethal threat.

The above discussion does not include any questions about any legal restrictions on firing DAS subsystems such as IR flares or anti-radiation missiles over populated areas. When agreed, system parameters are the basis of a contract, and actual delivered performance must be verified before final payment.

The pilot always has only one question during the mission: Is that object really a hostile missile aimed at me and will the countermeasures deal with it? Only a Yes or No answer is allowed.

The level of trust and the transfer of authority to the machine will need to be carefully defined for any new system or upgrades to existing ones. The design will have to ensure that the highest level of automation does not automate an assessment that was assumed to be made by the operator when the system was originally specified and delivered. This is important for weapon systems and emphasises the need for consideration of Article 36 reviews for an upgrade to any weapon control system.

Given the multitude of definitions of autonomy level and the changes to a human's workload, it is worthwhile asking if there can be a consistent definition of levels that allows workload or trust to be equated with autonomy level. There is no clear answer at present, but the Autonomous Systems Technology Related Airborne Evaluation and Assessment (ASTRAEA) programme[8] developed these ideas for UAS from those in the UK MOD 'Cognitive Cockpit' project.

The Cognitive Cockpit used a concept of 'Contractual Autonomy' to give a manned aircraft pilot the ability to decide during a mission which tasks could be delegated to the avionic system and which were kept under his or her direct control [19]. Two sets of tasks were not included, the first is those where authority to act is always delegated to the avionics; the second is those where authority is never delegated.

The set of tasks considered were where the pilot may want to retain authority in some circumstances, and in others pass it to the avionics suite, perhaps because of workload problems. Each task in this set has four levels of authority between machine and pilot, designated 1 to 4. The details of the split for each task are defined and fixed by design. The pilot then chooses which of these levels is used at any given time for each task. At mission start, the pilot chooses default levels for them. This choice can be changed during the mission so that the pilot prioritises tasks to keep an acceptable workload. This is known as Pilot Authorisation and Control of Tasks (PACT).

Table 3.1 takes the PACT levels from Reference [19] defining the levels and commander's authority in the first three columns, the unmanned system authority from the ASTRAEA programme in column 4, assigning it to an AI decision-maker in the unmanned system and adds limitations to full autonomy. Column 5 is the authority for an AI decision aid and shows that it is limited to a maximum of giving advice based on fairly extensive analysis. The difference between levels 3 and 4 is that at level 4 there is on-board autonomy in that function.

Non-military applications of work-sharing systems are under development, for example AI-based medical diagnostic systems are used to provide an additional opinion for a diagnosis. Moving them to trigger a treatment plan would be a large ethical step even if it was a simple technical one. This emphasises the need for the division between decision-aids and autonomous action-initiators to be kept clear. This division becomes less clear when the automated system is, for example, a sensor with an alarm measuring a patient and remedial action is possible by an

---

[8]ASTRAEA is described in Panel 6.1 and the accompanying text in Section 6.3.3.

Table 3.1   PACT levels for a unmanned system with transfer of authority between commander and unmanned system with AI

Levels	Operational relationship	Commander authority	Unmanned system with AI decision-maker – authority for action	AI decision aid authority
0	Under command	Full	None	None
1	At call	Full	Advice only if requested	Present data from one type of sensor after initial processing
2	Advisory	Acceptance of advice	Advice	Present fully processed information
3	In support	Acceptance of advice and authorising action	Advice, and if authorised, action	Present options to pilot, based on off- and on-board data analysis
4	Direct support	Revoking action	Advised action unless revoked	Chooses option and presents it to AI decision-maker
5	Automatic	Interrupt	Full, with limitations	As above

automatic system. The requirement for human judgement which can consider other wider factors must form an integral part of setting requirements for the system of systems of patient, medical staff and equipment.

A mixture of human and machine judgement will be a difficult problem for military decision-support systems if used for the weapon-release decision. It is likely that non-military decision-aid technologies will gradually be transferred into military use as they become widespread. Their trust, reliability and repeatability issues will need to be understood and acceptability criteria set for them. There will be an impact on Article 36 Reviews and possibly international treaties.

3.7 Control system developments

The preceding sections of this chapter have shown that the key issues for ethical acceptance of an autonomous system is that its behaviour will always be within defined limits, and that the autonomous system's operation will always be controlled by responsible humans. The engineering discipline of control theory is now well developed with an extensive literature and countless applications. Good introductions are to be found, for example, in two books also published by the IET [20] and [21].

The wider aspects of controlling the human operator are the subject of laws, regulations and local processes. These are covered in Chapters 6 and 7. This

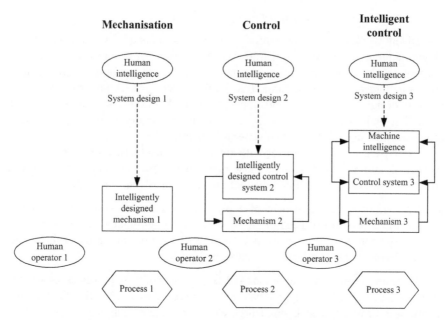

Figure 3.1 Development of control systems. (Based on Figure 17.3 in Reference [20])

chapter is restricted to developments in control theory that relate to the effectiveness of the human operator when using them.

3.7.1 Intelligently designed mechanisms

Figure 3.1 illustrates the development of control systems since the early stages of the Industrial Revolution. Humans developed controlled processes with predictable outputs, usually involving feedback loops. The automated processes were, or are, under close human supervision even with the addition of electric power in the late nineteenth and twentieth centuries. They were, and still are, designed using the process shown under the heading 'Mechanisation' in Figure 3.1.

Generally the early control systems were analogue, based on a few feedback loops and were well understood. However, even when they work correctly there are still accidents or unforeseen events due to either a human or equipment failure, or circumstances arising which were not foreseen by the designer or user.

User safety became an important issue both for cost and humanitarian reasons. Gradually legislation evolved to provide this and continues to evolve with developments in technology and the concept of duty of care. The aviation and oil industries provide exemplars of the symbiotic relationship between legislators, commercial drives and technologists. It can be anticipated that this will continue even with the advent of AI.

3.7.2   Intelligently designed control systems

The advent of digital techniques produced changes in control system engineering, playing a large part in moving it from 'Mechanisation' to 'Control' in Figure 3.1. Large computers gave the design engineer the ability to perform complex calculations in an acceptable timeframe. Eventually the design techniques evolved into modern design philosophies and concepts with the extensive use of complex modelling and simulation methods, as well as the ability to control nonlinear processes. The use of models is discussed in Section 3.8.

Clearly the control system, as with any system must be specified correctly and the actual system, as delivered, must meet these specifications and meet the customer's needs. There are two distinct processes involved, V&V [22]:

- Validation is the assessment of a planned or delivered system to meet the sponsor's operational need in the most realistic environment achievable.
- Verification is the process for determining whether or not a product fulfils the requirements or specifications established for it.

There are several other definitions and explanations of these processes, especially for software design. Basically, validation is to increase confidence that the requirements, specifications, model, simulation, etc. do represent what is actually needed, whilst verification is the detailed confirmation, usually through thorough testing, that the final product meets its specification. Remember that the specification is the formal description in engineering terminology of the delivered product, validation having shown that it represents the needs of who or what is the recipient of the product.

V&V are carried out at all stages of the design processes described in Chapters 4 and 5. This can lead to the verification at one level in the work breakdown structure becoming the validation of the specification for the next one down. Although the V&V processes may then be carried out at the same time, their aims should be kept separate.

Complementing improved design methods, microprocessors were developed from the 1970s onwards and used for dedicated signal processing and control purposes. The best-known examples for control purposes are probably the Motorola Z68000 series and the IBM/Motorola Power PC. There has been a steady development from 8-bit central processing units (CPU) running at kHz speeds to 64- or 128-bit CPUs running at GHz speeds.

Microprocessors brought the advantages over preceding analogue systems of greater robustness and stability. Limited system upgrades are available by reprogramming the chip instead of replacing components or circuit boards as would be the case with an analogue system.[9]

---

[9]The control system is often part of an asset that has a lifetime of many years if not decades, leading to requirements for increased performance beyond the capability of the in-built microprocessor even with software upgrades. A further problem is that the control software may be written in a specialist language with very few practitioners. The equivalent problem for an analogue system also can only be solved by a complete redesign, even for an upgrade which could be implemented as a software upgrade for the digital system.

Since the 1990s, increased speed and chip complexity has led to sophisticated processors, such as those from Intel, now being the CPU of personal computers and the ARM microprocessors being widely used in smart phones. This enables system designers to incorporate sophisticated algorithms in control systems and signal-processing applications giving levels of automation which are impossible for unaided humans.

An example of this level of complexity in a control system is resource management in a modern military radar system. One or more microprocessors are used to analyse return signals, optimise beam pointing and decide the next transmitted signals for improved radar performance. They may also provide the guidance parameters for the associated weapon system.

The cost of developing algorithms and implementing them in code can be high and leads to the practice of reusing the software from previous successful systems if possible. This practice usually succeeds, but great care must be taken in checking the detailed compatibility of the two systems. Panel 3.3 describes how a failure can occur even using well-tested software.

Panel 3.3:   The loss of an Arianne 5 satellite launcher

The crash of the Ariane 5 satellite launcher about 40 seconds into its first flight is a well-documented catastrophic failure due to the reuse of software. Ariane 4 software developed for a 16-bit processor was used in Ariane 5's 64 bit floating point system; rounding errors that would have been truncated in 16 bits were not, and built up beyond anticipated values due to differences in trajectories between Ariane 4 and Ariane 5 rockets. Full details are in the ESA Board of Enquiry's report [23]. The ESA press release summarises it as [24]:

> A chain of events, their inter-relations and causes have been established, starting with the destruction of the launcher and tracing back in time towards the primary cause. These provide the technical explanations for the failure of the 501 flight, which lay in the flight control and guidance system. A detailed account is given in the report, which concludes:
> 'The failure of Ariane 501 was caused by the complete loss of guidance and attitude information 37 seconds after start of the main engine ignition sequence (30 seconds after lift-off). This loss of information was due to specification and design errors in the software of the inertial reference system. The extensive reviews and tests carried out during the Ariane 5 development programme did not include adequate analysis and testing of the inertial reference system or of the complete flight control system, which could have detected the potential failure.'

The Arianne failure in Panel 3.3 occurred despite the rigorous testing and analysis required before software is used in a safety-critical system. One logical next step to avoid a repetition of this type of accident is to mandate that every control system has to be designed for the specific application, starting from first principles. Besides being unacceptable on cost grounds, it ignores the advantages that can come from bringing in algorithms based on well-proven techniques and implementation.

Increasing public reliance on automatic control of complex systems is an accepted feature of the modern society. Most passengers are unconcerned that railways such as London's Dockland Light Railway and the Paris Metro have no driver, or that during most of an aircraft's flight the pilot is only acting as a supervisor to the autopilot. An even more extreme example of a highly complex automated system with total public acceptance is Air Traffic Control (ATC). All these systems do, however, have qualified humans in very close proximity for emergencies. A failure of an ATC computer, such as happened at London Heathrow in December 2014, and the subsequent flight delays illustrates this clearly [25]. Despite, or perhaps because of the rare occurrence of this type of incident, there is wide public acceptance of these complex systems. Software upgrades to complex in-service systems are a relatively common source of failures, again despite extensive analysis and testing prior to the upgrade.

Increased system complexity means that a control system can no longer be considered as isolated, but is part of a larger system of systems, and interacts with other parts of it. A human may be considered to be part of the larger system of systems, making human–machine interface design an essential consideration for complex systems. Human–machine interfaces have been discussed in Chapter 2 in the context of decision-making and situational awareness.

3.7.3   Intelligent control systems

Humans are being replaced in many areas due to the development of AI. It is a rapidly developing field with a widening range of applications. Section 2.5 introduced the concepts of weak and strong AI. Weak AI is human rational decision-making, following fixed rules and processes. Strong AI is analogous to Recognition-Primed Decision-Making (RPDM). The control systems of Section 3.2.2 include weak AI whilst those in this section and on the right of Figure 3.1 (Intelligent control) are based on strong AI.

A system or process controller using the output from the 'machine intelligence', 'control system 3' and 'mechanism 3' blocks in Figure 3.1 is possible and will be developed if not already implemented.

One of the most, if not the most, important advantage of strong AI systems is that they can be presented with a relatively unstructured problem and respond with a solution. They achieve this by assessing a wide range of information, and weighting it against criteria. The method of assessment and the criteria used may have been developed by the AI system using a combination of design and learning from training and experience.

Numerous comparisons are made between the results obtained from strong AI systems and human reasoning. Sometimes the AI result is a surprise to the recipients as it is valid and may have been missed by the humans. However, the internal logic is not always understood and is almost impossible to deduce because the logical argument comes from the system's experience since its delivery by the design team. This cannot occur with the conventional control systems of Section 3.7.2 as they have defined inputs and responses to them which result from known algorithms implemented in code which is only changed at known times.

We have to consider whether the advantages of AI-based systems can outweigh the precision of a dedicated control system. The answer is not clear and depends on the application. This book deals with autonomous systems that will be regulated by a legal or statutory authority so the comparison must be based on the approaches taken by the regulators. The regulators will be asking the following types of question:

1. How repeatable is the system output when presented with the same inputs?
2. How reliable and repeatable is the output over ranges of input parameters?
3. Is the input data clearly defined?
4. Are there limits on the range of output values over all potential input values and combinations of them?
5. Do the actions of the system meet the criteria set by the relevant regulations under all operational conditions?
6. How stable is the system performance over its lifetime?

These questions are part of the processes used in Chapter 13 for autonomous systems designed to meet ethical values. They must determine whether the system under review, and each of its subsystems are AI-based or deterministic. The difference is that there are well-established regulatory processes for conventional deterministic systems but they are only starting to evolve for AI-based systems. There will be an additional set of questions for weapon systems which are discussed as part of Article 36 reviews in Chapters 11 and 12.

The problem we have to address is how the autonomous systems designer can use AI techniques to offer advanced capabilities, but ones that will meet the legal and ethical requirements for the user. Chapter 2 examined how human decisions are made, but introducing AI means we have to question whether we can trust the machine's conclusions, their reasoning and reliability. Essentially this is to examine the differences between a system designed for a human using a system based on classical control theory and one that has AI in the control chain. This problem is discussed in Section 6.9.

3.8 Models and control systems

3.8.1 Models of the process under control

All control systems are designed on the basis that the process which is to be controlled can be represented by a mathematical model. There are three ways to determine the equations to be used in the model:

1. The designer knows the behaviour of the subsystems in the process and their limits. If they have linear responses, it is then possible to make a mathematically

linear model of the process. The control circuits can then be designed using standard textbook methods. If the process is non-linear, there are techniques for designing a suitable control system; examples are given in Reference [21].

Validation of the process model can be carried out by extensive laboratory testing of each subsystem and the successively integrated subsystems until it is validated. Production tests of the finished control system then only need a few carefully chosen test points to demonstrate compliance.

The more difficult problem is verifying that the controlled process or product does behave in the manner expected, both for changes over the whole range of input (control) parameters and for combinations of them. If it is possible to gather a large amount of reliable data under controlled conditions, then verification will be tedious, but in principle is straightforward, however, in the majority of cases where this is not possible.

2. The process to be controlled cannot be represented with sufficient accuracy by a set of equations based on a theoretical understanding of it. This is the case for many processes. Chapter 6 in Reference [20] describes modelling and the methods which can be used in these circumstances.

    An aerospace example illustrates the problem. An aircraft flight control system may be highly linear by design with accurate mathematical models of the subsystems which are validated to a very high degree of precision. Aerodynamic forces on an aircraft are rarely linear, especially when the flight path is being changed by the use of the control surfaces in the wings and tail (ailerons). Flight trials are extremely expensive as well as high risk for a prototype design. The solution is extensive simulation coupled with wind tunnel tests prior to construction of the first aircraft. A demonstration aircraft is then built with minimal internal equipment, but representative of the aerodynamically critical shapes. Increasingly more complex manoeuvres are undertaken and the flight control system modified if necessary and validated.

3. A combination of the two. This is the most likely case, but designs are very specific to the particular problem. In general, the designer looks to identify as many parts of the system as possible that can be modelled mathematically, and determines empirical relationships for the remaining ones, as with the flight control system example above. It is important that the maximum range of anticipated inputs is determined and the maximum allowable output excursions are specified and limiting systems put in place.

Once there is a model of the process, the control system for it can be designed, but unless the process model is simple, testing the control system will be complex as with the flight control system example above.

3.8.2   *Models of the control system*

Modern control systems are usually implemented digitally, bringing the freedom to be able to incorporate complex equations and algorithms in them, but basic principles still apply.

The control system designer needs to have some basic information available before the design can start. If they are not all available at the beginning, assumptions will have to be made and updated as the data becomes available:

1. An aim for the future state or states of the process under control for a given set of inputs;
2. A possible set of actions that when implemented bring the process to a state nearer to the desired one;
3. A method of choosing the action to be implemented to achieve the desired aim.

It is interesting to note that these are similar to the four conditions identified by the Brehmer and Thumholm paper referenced in Chapter 2 for a military commander handling dynamic tasks [26]:

1. We must know the goal;
2. It must be possible to ascertain the state of the system that we want to control;
3. It must be possible to change the state of that system;
4. There must be a model of the system, a model that tells us how the system will react when we do things to it, including the case when we do nothing.

In the military case, the goal is mission success; the model may be maps, plans and intelligence information but the commander only has control over a limited part of the assets and process. Point two is simple for a known process with sensors, but there will be many unknowns about the enemy's systems and intentions in the military case; hence its absence in the first case and prominence in the second. The similarity between the two sets of conditions is not a coincidence; both are attempting to control a process, but under very different circumstances.

An engineer designing a control system will develop equations describing at least some of the subsystems in it. He or she will almost certainly build a model of both it and the process, and combine the two in a simulation. The complexity of the models and the resolution of the simulation will depend on the complexity of the problem and the sensitivity and precision needed for the output response to a given change in the input parameters.

A common design technique is to start with a simulation of the process and controller and steadily replace each part of the simulation with the real equivalent parts of the built system as they become available. This was once only feasible if all the real components could be physically located in one location, but high-bandwidth datalinks now remove this constraint. At the largest scale, modern military exercises have a mixture of simulators and actual military control rooms and platforms, such as ships, in all domains which execute the operations demanded by the simulators.

3.9 Control and the targeting process

The targeting process is at the core of all lethal decisions and, as has been made clear, all stages must be under control. The shortcomings of the OODA loop have already been identified, so it is better to consider targeting as the system shown in

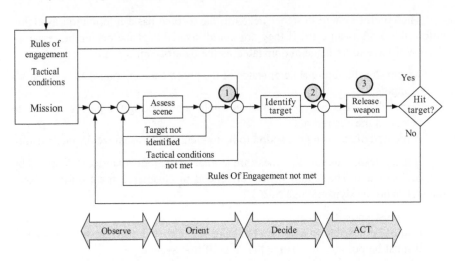

Figure 3.2   Targeting as a control system. (The numbered circles represent key locations which are discussed in the text)

Figure 3.2 and discuss its elements. The following precepts are made before consideration of Figure 3.2:

Precept 1 Scene

The target scene is an area larger than the target and its immediate area as it will be dynamic, probably with multiple objects moving around, into and out of it. These will need to be assessed in real time, but previously gathered data may reduce the size of this task.

The target area is the core of the targeting process. It must be regarded as the source of the critical information in any attack, the data being supplied by the commander's sensors. There are at least two types of information which must be extracted from this data: the nature of the potential target; and the type of object within the area which will be affected by the weapon when it strikes, allowance being made for target motion and system error budgets.

Precept 2 Sensors

It is often assumed that the main sensors will be on the weapon-carrying platform, which is usually the case. Battlespace networks enable remote sensing data to be passed to the weapon's commander in near real time. The use of Forward Air Controllers (FACs) passing information by voice or data link directly to a pilot to guide air strikes on to difficult targets is a well-established tactic. The principle is that the FAC is located near the target and has direct line-of-sight to it. Their task is to ensure that rules of engagement are met so that the pilot does not have to make their own rules of engagement decision. In some circumstances, the rules of engagement will guide a laser-guided bomb onto the target. The FAC usually has the legal responsibility and authority for weapon release, not the pilot.

There will also be information taken at earlier times which can supplement the current data. Digital maps are one example of this. Pre-existing information may be essential if the target area is obscured at optical and infra-red wavelengths so that the attack relies on radar and Electronic Warfare (EW) information.

Precept 3 Rules of engagement

There are rules-of-engagement which limit the actions of the attacking forces. These are set before the targeting process begins so they can be considered to be the reference state for the control loop.

Rules of engagement are normally issued as text so that a human can interpret them with minimal or no ambiguity. Expressing them in a digital form suitable for use in an automated system and still remain unambiguous for a human will be difficult, but essential for any form of automation in the 'Rules of engagement met' decision.

The comparators (empty circles) in Figure 3.2 derive their inputs from sensors with the comparison of the current target scene with the mission criteria needing complex functions, many of which will involve judgement. The figure has the appearance of a feedback loop with two feed-forward disturbance-reduction loops. In this case the disturbance-reduction signals produce one of only two states after the comparisons: 'Tactical conditions' are met; and 'Rules Of Engagement' are met. Both conditions must be met for weapon-release to occur. If either is not met, then the weapon is not released. The numbered circles in the figure represent actual or potential locations for human commanders.

The main elements in Figure 3.2 are:

Mission

This is a statement of the part the system will play in that part of the operation. There will be a statement of the aim, an initial plan with the criteria for the weapon-system to take an active part, the rules of engagement for that phase of the operation, and any other essential information for the commanders.

Decision: Tactical conditions met?

It is tacitly assumed in targeting that the weapon-carrying asset is in a suitable location for weapon release and a strike is still desired by the relevant superior commander. Conditions such as being in range and the release not triggering an immediate lethal response from the opponent must be satisfied. These are outside the scope of this book so it is assumed that the decision is that the conditions are met and the weapon can be released if the rules of engagement are met. If they are not met, tactical changes will be necessary which are not part of the targeting process.

Decision: Rules of engagement met?

It is shown in Chapter 7 that there are four criteria that must be met by any weapon system: necessity, distinction, humanity and proportionality. Assessment of these criteria is currently considered to be a matter for human judgement, but it has been shown that some aspects of them will be amenable to meaningful representation for an automated process [27].

The final decision will be made up of many intermediate decisions which will be approached in a logical structure. The possibility of making some of these decisions autonomously is discussed in Sections 8.5 and 8.6.

Weapon systems have developed so that they can make lethal decisions according to built-in criteria. Land and sea mines have been in use for decades and may be designed to be detonated by only one set of targets and not another. (Anti-tank mines will not be set off by a human walking on them; tethered sea mines will only be released when a particular sound signature is detected.) These systems have a small subset of the decisions in the 'Rules of engagement met' decision in Figure 3.2 delegated to the automated part of the system. The 'Release weapon' decision is completely automatic so there is no human at Location 3.

Process: Assess scene

This is essential for any targeting process: the target must be identified and confirmed; its surroundings must be monitored for excessive collateral damage; weapon guidance may require continued observations; the weapon's effect on the target will need to be assessed; there may be previously undetected defences, etc.

Assessment covers many operations and technologies. The operations as well as the sensors and databases may be distributed over several coalition assets with the results fed to the relevant commanders in the C2 chain. The processes will have automated functions which may be complex and have sophisticated algorithms. These decision aids will need to be thoroughly tested before use, and the commander will need to be aware of their strengths and weaknesses.

The time delay between scene activity and the commander making the lethal decision is a critical factor for both the 'Rules of engagement' and 'tactical conditions' decisions. This is the time delay from the inner feedback loop and may be minutes. It is generally assumed that if the target area is likely to change in this time so that collateral damage is likely to become unacceptable, then the weapon will not be released.

Process: Identify target

Target identification is closely related to the assessment of the scene, but will be a clearly separated stage. The criteria for the identification must be strictly defined and applied using rigorously tested procedures and algorithms. There are likely to be other objects in the target area and the scene which will be relevant to tactical and rules of engagement conditions. These must be identified with appropriate accuracy for the weapon and collateral damage decisions.

Action: Release weapon

Weapon release follows target identification if two sets of conditions are met: tactical and rules of engagement. There may be several steps in the guidance process from release to strike, but these will be automatic after initiation for most weapons. The time at which it is initiated after the conditions are met will be a matter of judgement by the commander. It will be necessary to continue with the 'observe scene' and 'scene analysis' during this time to ensure that rules of engagement are met at weapon release time rather than some time before. Acceptable time delays and the precautions to be taken will be agreed tactics and may well be part of them.

The situation is more complex if the commander has control of the weapon for some time after release, for example a laser-guided bomb, as the commander will have the potential to guide the weapon away from the target if it becomes clear that it is no longer legitimate to strike it. There will always be a time after release when it is not possible to change the weapon's aim point. Normally, there are agreed processes for handling this type of situation.

Weapon guidance system

This will be a straightforward control loop based on the difference between the weapon's estimated impact point and the target's location at that time. It will have a form such as that shown in Figure 3.3.

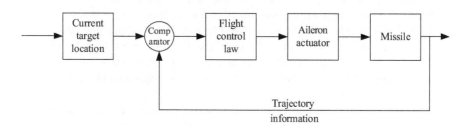

Figure 3.3   Missile guidance control loop

The role of the human commander's decisions in Figure 3.2 is critical if he or she must be responsible for the weapon-release decision. Three locations are shown as numbered circles. Location 1 does not directly affect the legality of the decision. It is a tactical decision which can overrule or delay the release of the weapon. It is reasonable to assume that it will be a human for the foreseeable future, even if the weapon-carrying asset and weapon have a high level of autonomy in their critical functions.

Location 2 is the most important one from an IHL perspective as it is where the qualitative judgements are made. As mentioned above, Reference [27] discusses the possibility of automating the comparisons of the scene and target status with the rules of engagement, but it does state that there will need to be changes in the way that rules of engagement are expressed. Some rules are easy to implement automatically, for example, a protected area can be expressed in map coordinates and identified on the scene images with a flag giving an electronic stay on any weapon aim-point locations in that area. Other rules are more difficult as engineering tolerances are quantitative whilst legal ones are qualitative. Consideration of a pair of questions illustrates this:

• Ask a lawyer whether a 75% probability of an object being identified as a known target is reasonable grounds for weapon release;
• Ask an engineer what performance he or she will specify for an object classifier to give reasonable assurance that the object belongs to one class of known targets and not to another class of objects including 'unknowns'.

It can be seen that in most circumstances human judgement cannot be removed from targeting decisions with current and imminently foreseeable technology. Therefore, it is essential to have a human at location 2. Location 2 is clearly inside the loops in Figure 3.2. When this is the case, the targeting process can be described as Man In The Loop (MITL).

There may be circumstances when the 'Rules of engagement met' decision can be made by an automated system. The naval Phalanx close in weapon system is one example but it is only operated under very restricted conditions such as it only being active when the ship is at sea in a high-risk area so the only objects in the protected zone will be hostile.

A human at location 3 will have the ability (if not the time) to overrule an automated 'Rules of engagement met' decision and weapon-release initiation. This can be described as Man On The Loop (MOTL) as they are 'above' the inner loop.

MITL, MOTL and 'man out of the loop' can be emotive terms in discussions about the legality and ethics of autonomous systems. It is the author's opinion that widespread use of Figure 3.2 can bring some common understanding to the terms enabling rational discussions.

The term 'meaningful human control' is sometimes used as an alternative to debates about the position of the man in, on or out of the loop. Figure 3.2 is still relevant, and is discussed in Section 10.9 with the concept of 'authorised power' defined in Panel 1.1 and Panel 3.1.

Appendix A3   Definitions of autonomy and autonomy levels

A3.1   Autonomy-related definitions from published civilian standards

The two civilian standards referred to, and used[10], in Chapters 1 and 3 are:

1. *ISO 8373, Robots and robotic devices — Vocabulary* This gives an extensive vocabulary which is called up by BS 8611:2016. The definition of autonomy given in Panels 1.1 and 3.1 comes from here. (The ability to perform intended tasks based on current state and sensing, without human intervention.)

---

[10]Permission to reproduce extracts from British Standards is granted by BSI Standards Limited (BSI). No other use of this material is permitted. British Standards can be obtained in PDF or hard copy formats from the BSI online shop: http://www.bsigroup.com/Shop.

2. *British Standard BS 8611:2016, Robots and robotic devices – Guide to the ethical design and application of robots and robotic systems.* This builds on ISO 8373 and gives guidance on ethics applied to robots.

Selected definitions from these standards are given in Tables A3.1 and A3.2, with their paragraph numbers from the standard. Neither standard defines autonomy levels.

Table A3.1   *Autonomy level definitions from ISO 8373*[*]

Para no.	Word	Definition
5.5	Automatic operation	State in which the robot (2.6) executes its task programme (5.1.1) as intended.
2.2	Autonomy	Ability to perform intended tasks based on current state and sensing, without human intervention.
2.7	Control system	Set of logic control and power functions which allows monitoring and control of the mechanical structure of the robot (2.6) and communication with the environment (equipment and users).
2.29	Human–robot interaction (HRI)	Information and action exchanges between human and robot (2.6) to perform a task by means of a user interface (5.12). Plus examples and note.
2.22	Integration	Act of combining a robot (2.6) with other equipment or another machine (including additional robots) to form a machine system capable of performing useful work such as the production of parts. Plus note.
2.28	Intelligent robot	Robot (2.6) capable of performing tasks by sensing its environment and/or interacting with external sources and adapting its behaviour.
2.6	Robot	Actuated mechanism programmable in two or more axes (4.3) with a degree of autonomy (2.2), moving within its environment, to perform intended tasks. Plus notes 1 and 2.
2.27	Robot cooperation	Information and action exchanges between multiple robots (2.6) to ensure that their motions work effectively together to accomplish the task.
2.14	Robot system	System comprising robot(s) (2.6), end effector(s) (3.11) and any machinery, equipment, devices, or sensors supporting the robot performing its task.
2.16	Robotics	Science and practice of designing, manufacturing and applying robots (2.6).
7.9	Senor fusion	Process to obtain improved information by merging information from multiple sensors.
2.30	Validation	Confirmation by examination and provision of objective evidence that the particular requirements for a specific intended use have been fulfilled. Plus note.
2.31	Verification	Confirmation by examination and provision of objective evidence that the requirements have been fulfilled. Plus note.

[*]Permission to reproduce extracts from British Standards is granted by BSI Standards Limited (BSI). No other use of this material is permitted. British Standards can be obtained in PDF or hard copy formats from the BSI online shop. www.bsigroup.com/Shop.

Table A3.2   Autonomy-related definitions from BS 8611:2016[*]

Para no.	Word	Definition
3.1	Autonomous system	System which has the ability to perform intended tasks based on current state, knowledge and sensing, without human intervention.
3.2	Ethical harm	Anything likely to compromise psychological and/or societal and environmental well-being. Plus note.
3.4	Ethical risk	Probability of ethical harm occurring from the frequency and severity of exposure to a hazard.
3.5	Ethics	Common understanding of principles that constrain and guide human behaviour.
3.6	Harm	Injury or damage to health.
3.9	System	Set of parts which, when interoperating, has behaviour that is not present in any of the parts themselves.

[*]Permission to reproduce extracts from British Standards is granted by BSI Standards Limited (BSI). No other use of this material is permitted. British Standards can be obtained in PDF or hard copy formats from the BSI online shop. www.bsigroup.com/Shop.

A3.2   Military definitions of autonomy

A3.2.1   NATO discussion of terms

There is an extensive discussion by Williams of the meaning of many words related to autonomy, robots, automatic, autonomy levels and targeting using OODA [28]. Williams shows that there are ambiguities between various definitions but concludes with a definition of *Autonomous functioning in a system*, with supporting arguments for each phrase or part in an accompanying table. Williams' definition is given here:

> **Autonomous functioning in a system:** Autonomous functioning refers to the ability of a system, platform, or software to complete a task without human intervention, using behaviours resulting from the interaction of computer programming with the external environment. Tasks or functions executed by a platform, or distributed between a platform and other parts of the system, may be performed using a variety of behaviours which may include reasoning and problem solving, adaption to unexpected situations, self-direction, and learning. Which functions are autonomous – and the extent to which human operators can direct, control, or cancel functions – is determined by system design trade-offs, mission complexity, external operating environment conditions, and legal or policy constraints. This can be contrasted with automated functions, which (although they require no human intervention) operate using a fixed set of inputs, rules, and outputs, the behaviour of which is deterministic and largely predictable. Automatic functions do not permit the dynamic adaptation of inputs, rules or outputs.

This definition is not used in this book. The first sentence is consistent with the definitions in Panels 1.1 and 3.1 as it relies on operation with no human intervention. The rest of the definition is a set of discussions about that first sentence which are expounded elsewhere in this book in the context of systems engineering.

A3.2.2 UK Joint Doctrine Note JDN 3/10

UK developments of doctrine and technology for autonomous systems led to early definitions of terminology. JDN 3/10 [29] gives the two definitions below as well as others for aircraft. Although it has been superseded by JDN 2/11 and so is no longer definitive, these definitions are not redefined in so they should still be extant.

Automated system

> In the unmanned aircraft context, an automated or automatic system is one that, in response to inputs from one or more sensors, is programmed to logically follow a pre-defined set of rules in order to provide an outcome. Knowing the set of rules under which it is operating means that its output is predictable.

Autonomous system

> An autonomous system is capable of understanding higher level intent and direction. From this understanding and its perception of its environment, such a system is able to take appropriate action to bring about a desired state. It is capable of deciding a course of action, from a number of alternatives, without depending on human oversight and control, although these may still be present. Although the overall activity of an autonomous unmanned aircraft will be predictable, individual actions may not be.

A3.2.3 US DOD Directive of 2012

The US DOD published Directive 3000.09 in 2012 with the definitions given below [30]. It introduced the concept of a semi-autonomous weapon system. Although this term is not used widely, it shows that defining autonomy in a weapon system is difficult.

Autonomous weapon system

> A weapon system that, once activated, can select and engage targets without further intervention by a human operator. This includes human-supervised autonomous weapon systems that are designed to allow human operators to override operation of the weapon system, but can select and engage targets without further human input after activation.

Human-supervised autonomous weapon system

> An autonomous weapon system that is designed to provide human operators with the ability to intervene and terminate engagements, including in the event of a weapon system failure, before unacceptable levels of damage occur.

Semi-autonomous weapon system

> A weapon system that, once activated, is intended to only engage individual targets or specific target groups that have been selected by a human operator. This includes:
> Semi-autonomous weapon systems that employ autonomy for engagement-related functions including, but not limited to, acquiring, tracking, and identifying potential targets; cueing potential targets to human operators;

prioritizing selected targets; timing of when to fire; or providing terminal guidance to home in on selected targets, provided that human control is retained over the decision to select individual targets and specific target groups for engagement.

'Fire and forget' or lock-on-after-launch homing munitions that rely on tactics, techniques, and procedures (TTPs) to maximize the probability that the only targets within the seeker's acquisition basket when the seeker activates are those individual targets or specific target groups that have been selected by a human operator.

Unmanned platform

It is an air, land, surface, subsurface, or space platform that does not have the human operator physically on board the platform.

A3.3   Non-military definitions of autonomy levels

A3.3.1   Barber and Martin [31]

This early paper discusses autonomy of agents at some length, finally giving two definitions:

- Autonomy is an agent's active use of its capabilities to pursue some goal, without intervention by any other agent in the decision-making process used to determine how that goal should be pursued
- An agent's degree of autonomy, with respect to some goal that it actively uses its capabilities to pursue, is the degree to which the decision making process, used to determine how that goal should be pursued, is free from intervention by any other agent

The authors then develop a concept of a spectrum of autonomy levels:

Agent autonomy can be described as a spectrum. An agent's autonomy (a) increases from 0 to 1 across it. Three discrete autonomy level categories are identified, defining salient points along the spectrum.

Command-driven ($a = 0$) – The agent does not make any decisions about how to pursue its goal and must obey orders given by some other agent(s).

True consensus (a around 0.5) – The agent works as a team member, sharing decision-making control equally with all other decision-making agents.

Locally autonomous/master ($a = 1$) – The agent makes decisions alone and may or may not give orders to other agents.

Supervised autonomy levels exist between the command-driven and consensus levels, and *supervisory* autonomy levels exist between the consensus and locally autonomous/master levels.

It is important to realize that these autonomy levels also apply only to a particular goal.

Barber and Martin's paper takes a theoretical approach and does not give definitions of autonomy levels, instead it looks on the degree of autonomy, *a*, as a continuum. This is consistent with the concept developed in this book of defining autonomy levels through authorised power and the criteria for a specific function. Authorised power itself is defined in Panels 1.1 and 3.1.

A3.3.2 Parasuraman *et al.* [32]

This paper takes an earlier set of autonomy levels from Sheridan [33], and considers them in relation to a four-part model of human decision-making based on: (i) information acquisition; (ii) information analysis; (iii) decision selection; and (iv) action implementation. They show that each of these four parts can be at different autonomy level in one system.

The authors make some small changes to Sheridan's levels and use the definitions in Table A3.3 for autonomy levels.

Table A3.3   Autonomy level definitions from Parasuraman et al.

Level	Definition
1	The computer offers no assistance: humans must take all decisions and actions.
2	The computer offers a complete set of decision/action alternatives, or
3	narrows the selection down to a few, or
4	suggests one alternative
5	executes that suggestion if the human approves, or
6	allows the human a restricted time to veto before automatic execution, or
7	executes automatically, then necessarily informs the humans and
8	informs the human only if asked, or
9	informs the human only if it, the computer, decides to
10	The computer decides everything, acts autonomously, ignoring the human.

A3.3.3 Space industry

ESA developed the levels given in Table A3.4 and they have been published under joint work by the ESA and NASA [34], although they come from a previous report which does not seem to be currently available [35]. It can be seen that these have the specific context of space operations with the lack of human proximity to the automatic or autonomous systems.

Table A3.4   ESA's mission-execution autonomy levels

Level	Description	Functions
E1	Mission execution underground control.	Real-time control from ground for normal operations.
	Limited on-board capability for safety issues.	Execution of time-tagged commands for safety issues.
E2	Execution of pre-planned, ground-defined, mission operations on board	Capability to store time-based command in an on-board scheduler
E3	Execution of adaptive mission operations on board	Event-based autonomous operations. Execution of on-board operations control procedures
E4	Execution of goal-oriented missions operations on board	Goal-oriented mission re-planning

A3.3.4   Road vehicle industry

SAE International standard J3016 was published in 2014 to simplify communication and technical collaboration across the road vehicle industry. It defines five autonomy levels from no autonomy in level 1 to full autonomy at level 5. It is available at https://www.sae.org/standards/content/j3016_201806/

A3.3.5   Maritime industry

The maritime industry is concerned about the problems of reduced crew numbers and increasing automation when at sea, with consequent increased human and financial risk. Lloyds Register originally defined autonomy levels, but found that these were inadequate and replaced them with cyber-accessibility levels. These are both given as Tables A3.5a and A3.5b [36].

Table A3.5a   Lloyds Register autonomy levels for shipping (© Lloyds Register, reproduced from Reference 36 with their permission)

Level	Definition
AL 0	Manual – no autonomous function.
AL 1	On-ship decision support
AL 2	On and off-ship decision support
AL 3	'Active' human in the loop
AL 4	Human on the loop – operator/supervisory
AL 5	Fully autonomous – (Unsupervised or rarely supervised operation.)
AL 6	Fully autonomous – (Unsupervised operation.)

Table A3.5b   Lloyds Register cyber accessibility levels for shipping (© Lloyds Register, reproduced from Reference 36 with their permission)

Cyber Accessibility Level	Definition
A 0	No access – no assessment – no descriptive note – for information only.
A 1	No access – no assessment – no descriptive note – for information only.
A 2	Cyber access for autonomous/remote monitoring.
A 3	Cyber access for autonomous/remote monitoring and control (onboard permission is required, onboard override is possible)
A 4	Cyber access for autonomous/remote monitoring and control (onboard permission is not required, onboard override is possible)
AL 5	Cyber access for autonomous/remote monitoring and control (onboard permission is not required, onboard override is not possible)

A3.4 Military definitions of autonomy level

A3.4.1 NATO-suggested definitions

NATO Working Group AVT-146 stated the definitions given in Table A3.6 [37] which are taken from an earlier report.

Table A3.6 NATO autonomy levels quoted in Reference [37]

Level	Description
1	Remotely controlled system – system reactions and behaviour depend on operator input
2	Automated system – reactions and behaviour depend on fixed built-in functionality (pre-programmed)
3	Autonomous non-learning system – behaviour depends upon fixed built-in functionality or upon a fixed set of rules that dictate system behaviour (goal-directed reaction and behaviour)
4	Autonomous learning system with the ability to modify rules defining behaviours – behaviour depends upon a set of rules that can be modified for continuously improving goal directed reactions and behaviours within an overarching set of inviolate rules/behaviours

A3.4.2 The US Department of Defense levels

The US DOD has published several unmanned system roadmaps since 2000. The first [38] discussed autonomy levels developed by the Air Force Research Laboratory (AFRL). It was referred to in the DOD Roadmap of 2000 and published in 2001 as an Autonomous Control Level (ACL) chart. It is given in Table A3.7.

Table A3.7 Autonomous Capability Levels (ACL) from Figure 4.4-2 of Reference [38]

ACL	Description
10	Fully autonomous swarms
9	Group strategic goals
8	Distributed control
7	Group tactical goals
6	Group tactical re-plan
5	Group coordination
4	On-board route re-plan
3	Adapt to failures and flight conditions
2	Real-time health diagnostics
1	Remotely guided

This evolved over time and a later, more definitive version was published in 2007 [39] with an interpretation of the levels in the phases of the OODA loop. This is given in Table A3.8.

Table A3.8 The Air Force Research Laboratory (AFRL) Autonomous Control Logic (ACL) chart from Reference [39]

Level	Level descriptor	Observe	Orient	Decide	Act
		Perception/Situational awareness	Analysis/Coordination	Decision making	Capability
10	Fully autonomous	Cognisant of all within battlespace	Coordinates as necessary	Capable of total independence	Requires little guidance of any sort
9	Battlespace swarm cognisance	Knows intent of self and others (friendly and threat) in a complex/intense environment; on-board tracking	Group strategic missions assigned, threat tactics inferred	Distributed tactical group planning; individual mission decision-making; chooses targets	Group executes mission with minimal supervisory assistance
8	Battlespace cognizance	Proximity inference – intent of self and others (friendly and threat); reduced dependence upon off-board data	Strategic group goals assigned; threat tactics inferred; aided target recognition	Coordinated tactical group planning; individual task planning and execution; choose targets of opportunity	Group executes mission with minimal supervisory assistance
7	Battlespace knowledge	Short track awareness – history and predicted battlespace data in limited range, time-frame and numbers; Limited inference supplemented by off-board data	Tactical group goals assigned; enemy intent estimated	Individual task planning/ execution to meet goals	Group accomplishment of tactical goal with minimal supervisory assistance
6	Real-time multi-vehicle cooperation	Ranged awareness – on-board sensing for long range, supplemented by off-board data	All below plus enemy location sensed/ estimated	Coordinated trajectory planning and execution to meet goals – group optimization	Group accomplishment of tactical goal with minimal supervisory assistance Possible close air space separation (1–100 yds)

(Continues)

Table A3.8 (Continued)

Level	Level descriptor	Observe	Orient	Decide	Act
5	Real time multi-vehicle coordination	Sensed awareness – Local sensors to detect external targets (friendly and threat), fused with off-board data	All below with prognostic health management; group diagnosis and resource management	On-board trajectory re-planning – optimizes for current and predictive conditions; collision avoidance	Group accomplishment of tactical plan as externally assigned; air collision avoidance; possible close air space separation (1–100 yds) formation in non-threat conditions
4	Fault/Event adaptive vehicle	Off-board awareness – friendly systems communicate data	All below plus: Rules of engagement assigned; inner loop changes reflected in outer loop performance	On-board trajectory re-planning – event driven; self-resource management; deconfliction	Self-accomplishment of tactical plan as externally assigned
3	Robust response to real-time faults/events	Health/status history and models	Tactical plan assigned; RT health diagnosis; compensate for most control failures and flight conditions (i.e. adaptive inner-loop control)	Evaluate status vs. required mission capabilities; Abort/ return to base if insufficient	Self-accomplishment of tactical plan as externally assigned
2	Changeable mission	Health/status sensors	RT health diagnosis (Does UAV have problem(s)?); off-board re-plan (as required)	Executes pre-programmed or uploaded plans in response to mission and health conditions	Self-accomplishment of tactical plan as externally assigned
1	Execute pre-planned mission	Preloaded mission data; flight control and navigation sensing	Pre/post flight built-in-test; report status	Pre-programmed mission and abort plans	Wide airspace separation requirements (kilometres)
0	Remotely piloted vehicle	Flight control (altitude, rates) sensing; 0n-board camera	Telemetered data; remote pilot commands	None, off-board pilot	Control by remote pilot

The 2011 DOD Roadmap [40] reduced the number of autonomy levels to those given in Table A3.9 and describes them as 'the most commonly referenced description of the levels of autonomy that takes into account the interaction between human control and the machine motions'.

Table A3.9   Commonly referenced levels of autonomy (from DOD, Reference [40])

Level	Name	Description
1	Human operated	A human operator makes all decisions. The system has no autonomous control of its environment although it may have information-only responses to sensed data.
2	Human delegated	The vehicle can perform many functions independently of human control when delegated to do so. This level encompasses automatic controls, engine controls, and other low-level automation that must be activated or deactivated by human input and must act in mutual exclusion of human operation.
3	Human supervised	The system can perform a wide variety of activities when given top-level permissions or direction by a human. Both the human and the system can initiate behaviours based on sensed data, but the system can do so only if within the scope of its currently directed tasks.
4	Fully autonomous	The system receives goals from humans and translates them into tasks to be performed without human interaction. A human could still enter the loop in an emergency or change the goals, although in practice there may be significant time delays before human intervention occurs.

A3.4.3   Autonomy Levels For Unmanned Systems (ALFUS)

The ALFUS aim is to provide a framework for characterising and articulating autonomy for unmanned systems (UMS). It does this by having standard terms for requirements analysis and definition. The principle is that autonomy levels can be defined following a generic process, but autonomy itself must be defined for the particular system under consideration as:

> A UMS's own ability of integrated sensing, perceiving, analysing, com-municating, planning, decision-making, and acting/executing, to achieve its goals as assigned by its human operator(s) through designed Human-Robot Interface (HRI) or by another system that the UMS communicates with. UMS's Autonomy is characterized into levels from the perspective of Human Independence (HI), the inverse of HRI. Autonomy is further characterized in terms of Contextual Autonomous Capability (CAC).

The group that developed ALFUS recognised that it is not possible to have one generic set of definitions for autonomy levels. They opted for a generic ontology for discussing levels and supplementary ontologies for defence and Search And

Rescue (SAR) operations with proposals for manufacturing. It was published in two main publications [41,42] then and has been further developed as SAE 6128 [43].

The required autonomous capabilities which together give the overall system autonomy are each called a Root Autonomous Capability (RAC) which is defined to be:

> The collective sensing, perceiving, analysing, communicating, planning, decision-making, and acting/executing capabilities as specified in the Autonomy definition.

ALFUS also does not have a linear scale of autonomy level. Each level is defined by the summaries of the metrics which themselves may not be linear. The scale can be from 0 to 10, but this is not mandated.

The ALFUS terminology in Reference [42] defines four modes of operation which are given in Table A3.10. Assessment of autonomy level starts by examining three levels of complexity in an operation:

- Environmental complexity
- Mission complexity
- Human independence.

Table A3.10   ALFUS modes of operation

Name	Definition
Fully autonomous	A mode of UMS operation wherein the UMS accomplishes its assigned mission, within a defined scope, without human intervention while adapting to operational and environmental conditions.
Semi-autonomous	A mode of UMS operation wherein the human operator and/or the UMS plan(s) and conduct(s) a mission and requires various levels of HRI. The UMS is capable of autonomous operation in between the human interactions.
Tele-operation	A mode of UMS operation wherein the human operator, using sensory feedback, either directly controls the actuators or assigns incremental goals on a continuous basis, from a location off the UMS.
Remote control	A mode of UMS operation wherein the human operator controls the UMS on a continuous basis, from a location off the UMS via only her/his direct observation. In this mode, the UMS takes no initiative and relies on continuous or nearly continuous input from the human operator.

Reference [41] gives a methodology to assess these complexities to arrive at an autonomy level. It starts (Paragraph 7.6.2) by defining four reference levels of autonomy. These are given in Table A3.11. The methodology then allows expansion of these by individual programmes using the ALFUS framework. Reference [41] gives two examples of further expansion to nine and eleven levels.

Table A3.11   The ALFUS reference levels of autonomy

Name	Description
Remote Control/ Tele-operation	Human, while off-board the UMS, directly controls all of the RACs.
Human lead	Human directly controls more than the UMS does of the RACs.
Shared	Human and the UMS directly control the RACs at equivalent levels of effort.
UMS lead	The UMS controls more than human does.
Fully autonomous	UMS performs all of the RACs.

References

[1]   NATO AAP-6 NATO Glossary Of Terms And Definitions, 2015 Edition.

[2]   Wallach W., and Allen C. *Moral machines, teaching robots right from wrong*. Oxford: Oxford University Press; 2009.

[3]   31st international conference of the Red Cross and red Crescent; Geneva 2011: Report reference 31/C/11/5.1.2, 2011, pp. 39–40.

[4]   UK Ministry of Defence Joint Doctrine Note, JDN 3/10, 2010, *Unmanned aircraft systems: terminology definitions and classification*. This has been superseded and is only available as an archived report at https://www.gov.uk/government/uploads/system/uploads/attachment_data/file/432646/20150427-DCDC_JDN_3_10_Archived.pdf

[5]   British Standard BS 8611:2016, *Robots and Robotic Devices, Guide to the ethical design and application of robots and robotic systems*, British Standards Institute, 2016.

[6]   en.oxforddictionaries.com [accessed on 18 September 2018].

[7]   Raytheon, *Phalanx close-in weapon system – Last line of defense for air, land and sea*, http://www.raytheon.com/capabilities/products/phalanx/ accessed on 29 April 2017.

[8]   http://icrac.net/2014/06/banning-lethal-autonomous-weapon-systems-laws-the-way-forward/ accessed on 20 February 2015.

[9]   UK Ministry of Defence. *Joint Doctrine Note 3/10, 2010, Unmanned aircraft systems: terminology definitions and classification*, 2010. Available at www.mod.uk/dcdc

[10]   Statement of the ICRC at the Convention on Certain Conventional Weapons (CCW), Meeting of Experts on Lethal Autonomous Weapons Systems (LAWS), 13–17 April 2015, Geneva.

[11]   International Committee of the Red Cross and Red Crescent. *International humanitarian law and the challenge of contemporary armed conflicts*. Report 32IC/15/11, Geneva, October 2015. pp. 44–45.

[12]   Durst P., and Gray W. *Levels of autonomy and autonomous system performance assessment for intelligent unmanned systems*; ERDC/GSL Report no SR-14-1, 2014.

[13]   Kendoul F. *Towards a unified framework for UAS autonomy and technology readiness assessment (ATRA)*. In: Autonomous Control Systems and Vehicles 2013. pp. 55–71, Springer, Japan.

[14]   Barber K.S., and Martin C. E. 'Agent autonomy: Specification, measurement, and dynamic adjustment'. *Proceedings of the Autonomy Control Software Workshop at Autonomous Agents* May 1999 (Agents'99), Seattle, WA, pp. 8–15, 1999.

[15]   US Office of the Secretary of Defense. *Unmanned aerial vehicles roadmap 2000–2025*. Washington; 2001.

[16]   US Office of the Secretary of Defense. *Unmanned Systems Integrated Roadmap FY2011-2036*. Washington; 2011.

[17]   Sholes E. 'Evolution of a UAV autonomy classification taxonomy'. *IEEE Aerospace Conference, Big Sky*; March 2007. IEEE 2007, pp. 1–16.

[18]   National Institute of Standards and Technology. *NIST Special Publication 1011-I-2.0 Autonomy Levels for Unmanned Systems (ALFUS) Framework, Volume I: Terminology*, Version 2.0. 2008.

[19]   Taylor R.M., Brown L., and Dixon B., *From Safety Net to Augmented Cognition: Using Flexible Autonomy Levels for On-Line Cognitive Assistance and Automation* Paper presented at the RTO HFM Symposium on 'Spatial Disorientation in Military Vehicles: Causes, Consequences and Cures', held in La Coruña, Spain, 15–17 April 2002, and published in RTO-MP-086.

[20]   Leigh J.R. *Control Theory A Guided Tour*. Control Engineering Series, no. 72. London IET 2012.

[21]   Ding, Zhengtao, *Nonlinear and Adaptive Control Systems*. Control Engineering Series, no. 84. London 2013.

[22]   *Systems Engineering Guide*. Bedford MA: Mitre Corporation; 2014; p. 419. Downloaded from: https://www.mitre.org/publications/technical-papers/the-mitre-systems-engineering-guide [Accessed on 7 March 2018].

[23]   Ariane 501 Inquiry Board Report. Available at: http://esamultimedia.esa.int/docs/esa-x-1819eng.pdf [Accessed on 23 September 2018].

[24]   ESA Plain Text Press Release No. 33–1996: Ariane 501 – Presentation of Inquiry Board report. Available at: http://www.esa.int/For_Media/Press_Releases/Ariane_501_-_Presentation_of_Inquiry_Board_report [Accessed on 23 September 2018].

[25]   One typical report for this incident can be seen at: http://www.telegraph.co.uk/news/aviation/11290412/Flights-grounded-at-all-London-airports-in-air-traffic-control-computer-meltdown.html [Accessed on 23 September 2018].

[26]   Brehmer B, and Thunholm P. 'C2 after contact with the adversary: Execution of military operations as dynamic decision making'. *Proceedings of the*

16th International Command and Control Research and Technology Symposium; Quebec, Canada, June 2011.

[27]   Gillespie, T., and West, R. 'Requirements for autonomous unmanned air systems set by legal issues', *International Command and Control Journal*. 2010, Vol. 4(2): pp. 1–32.

[28]   Williams A. 'Defining autonomy in systems: Challenges and solutions' in Scharre P. (ed.) *Autonomous systems; issues for defence policymakers*. NATO, Capability Engineering and Innovation Division, Norfolk, USA. 2013, Chapter 2, pp. 27–62.

[29]   UK Ministry of Defence. *Joint Doctrine Note, JDN 3/10, 2010, Unmanned aircraft systems: terminology definitions and classification*. This has been superseded and is only available as an archived report at https://www.gov.uk/government/uploads/system/uploads/attachment_data/file/432646/20150427-DCDC_JDN_3_10_Archived.pdf [Accessed on 15 September 2018].

[30]   Department of Defense. *Directive no 3000.09, Autonomy in weapon systems*, November 2012.

[31]   Barber K. S., and Martin C.E. 'Agent autonomy: Specification, measurement, and dynamic adjustment'. *Proceedings of the Autonomy Control Software Workshop at Autonomous Agents* May 1999 (Agents'99), Seattle, WA, pp. 8–15, 1999.

[32]   Parasuraman R., Sheridan T.B., and Wickens C.D. 'A model for types and levels of human interaction with automation' *IEEE Transactions on Systems, Man and Cybernetics-Part A: Systems and Humans*. 2000, vol. 30(3): pp. 286–297, 2000.

[33]   Sheridan T.B., and Verplank W.L. 'Human and Computer Control of Undersea Teleoperators' MIT Man-Machine Systems Laboratory, Cambridge, MA, Tech. Rep., 1978.

[34]   Vassev E., and Hinchey M. *Autonomy requirements engineering for space missions*. NASA Monographs in Systems and Software Engineering (Springer 2014).

[35]   CSS Secretariat. *Space engineering: space segment operability*, Technical report, ESA-ESTEC, Requirements and Standards Division, ECSS-E-70-11C, Noordwijk, The Netherlands, 2008.

[36]   *ShipRight Procedure – Autonomous ships 1st edition 2016*, Lloyds Register, Southampton 2016. Now replaced by: *Cyber-enabled ships – ShipRight procedure assignment for cyber descriptive notes for autonomous & remote access ships Version 2.0*, December 2017.

[37]   Protti M., and Barza R. *UAV Autonomy –Which level is desirable? – Which level is acceptable? Alenia Aeronautica Viewpoint*. NATO Research & Technology Organisation report AVT-146; 2007. (This is taken from an earlier NATO report: NIAG (SG/75) "Pre-feasibility Study on UAV Autonomous Operations", 2004.)

[38]   US Office of the Secretary of Defense. *Unmanned aerial vehicles roadmap 2000–2015*. Washington; 2001.

[39]   Sholes, E. 'Evolution of a UAV autonomy classification taxonomy'. *IEEE Aerospace Conference, Big Sky*; March 2007. IEEE 2007. pp. 1–16.

[40]   US Office of the Secretary of Defense Unmanned Systems Integrated Roadmap FY2011-2036. Washington; 2011.

[41]   Huang, Hui-Min, Messina E., and Albus J. *Autonomy levels for unmanned systems (ALFUS) framework, Volume II: Framework Models Version 1.0.* NIST Special Publication 1011-II-1.0, 2007.

[42]   Huang, Hui-Min, *Autonomy levels for unmanned systems (ALFUS) framework, Volume I: Terminology.* NIST Special Publication 1011-I-2.0, 2008.

[43]   SAE 6128 Unmanned Systems Terminology Based on the ALFUS Framework.

Chapter 4

Operational analysis to systems engineering

4.1 Introduction

All significant expenditure has carefully thought out and documented processes for financial and technical activities. These vary from sector to sector and nation to nation, usually with significant regulation through national or international bodies. The company, institution or government makes its financial case using a business plan. This is usually prepared when the engineering deliverables are based on design proposals and associated financial and risk estimates. The technical proposals should be based on an initial 'scoping' study which will have delivered a design concept, broken down into its principal component systems. Designs will be sufficiently detailed for cost and time estimates to be given to the business team.

This chapter covers the analysis and engineering processes which are used to move from the first idea, through the initial concept to a firm design which can be made by one or more companies, tested, delivered and meet the customer's expectations. Detailed design of the products making up the system is described in Chapter 5. The process is similar for the bid, development and design stages, but the time invested and detail involved will differ significantly. Detailed design, test and integration work will not be carried out during the scoping study but their costs and timescales must be estimated accurately.

Moving from concept to firm requirements with details of customer expectations and methods of use is known as operational research for civil applications and operational analysis for military ones. It was originally called operational research, and this name persists for non-military applications. The stages from customer requirement to a tested and delivered system, including post-delivery support, probably including upgrades are known as systems engineering[1]. There is not a firm division between operational analysis or research and systems engineering. Modelling is an essential part of operational analysis and systems engineering, with simulation as a natural adjunct. It is then possible and practical to study the effectiveness of new equipment and capabilities for a variety of environments and uses. Operational research methods and their applications have an extensive

---

[1]Systems engineering does not usually include product design which is covered in Chapter 5. Note that the term 'system' is not precise. One company may consider its product to be a system but it could be a module to another. (An aircraft engine is a complex system, but may only be a major module to the airframe manufacturer.)

literature (see, for example, [1,2]) and have been suggested for complex civilian applications [3].

Operational research is also used for the analysis of current and completed campaigns so that lessons can be learnt in a methodical and scientific way (see, for example, [4]).

Every project should move seamlessly from operational analysis or research to systems engineering. In practice, many of the processes are iterative as ideas firm up and design issues arise. The balance between analysis, systems engineering and product design will depend on the type of autonomous systems required, its anticipated life, and the market requirements. However, there are contractual payments at significant points in the programme. These points are usually major milestones in the design and production programmes. The formal transition from analysis to systems engineering processes and the validation of the system design are often taken as suitable milestones. These are marked by major design reviews. Successful completion of the reviews, and major actions from them, are often used as payment criteria. Successful project and technical management teams place reviews at dates in the programme which allow some iteration across the boundaries before the system design freezes which occur at payment milestones.

Companies making an investment in a new product have design and cost aims set by their perception of the potential market. Once these are set, there is usually tight cost control, but plans should include a funded risk-management plan, and flexibility in response to problems. The autonomous systems covered in this book have physical components which interact with humans and so are part of a larger system-of-systems (sometimes called SOS). Autonomous systems for the consumer market will probably be self-contained units with clearly defined interfaces with their users. The design may be driven by the appeal of new facilities that attract new users and create new markets or expand old ones. An example is the small remotely piloted aircraft or drone. Alternatively, the autonomous system may deliver an improved process or capability that increases the efficiency of an existing process. An example is capable robots in manufacturing processes. In many cases, the product will be one that will have a life of only a few months or years before being replaced by a superior one.

Autonomous systems for government organisations are likely to be part of a larger programme, giving it critical capabilities that make the investment viable. Government investment costs are set by national budgets which aim to balance income to national strategic priorities. Government priorities change over periods of several years, often less than the timescale from concept to in-service for a major infrastructure programme or a military platform. As a result, funding is phased with release of major blocks of funding at major milestones called main gates. As discussed above, these generally coincide with significant contractual deliveries and design reviews. Balancing an iterative design process with the financial and contractual conditions imposed by milestones is one of the more difficult aspects of programme management. The problems are compounded when a government customer is forced to introduce breaks in the programme of months or even a financial year due to wider national constraints.

This chapter discusses the operational analysis and engineering processes that are used when the autonomous system is part of a larger investment programme. The examples are for a military system, but the principles apply to all programmes. The terminology may be different for infrastructure projects, and the processes will be much simplified for a consumer product, but must cover the same ground to give a successful product. The combination of investors/budget-holders, potential users and technical advisers must define what is affordable and ensure that it meets strategic requirements.

4.2 Terms and tools

4.2.1 Systems of systems

'Systems of systems' is a widely used term to describe complex systems comprising many smaller systems. System of systems, system, subsystem and module are defined for the purposes of this book in Section 5.1 and Figure 5.1.

System of systems engineering describes all the engineering associated with these complex systems. It can be argued that all systems, by definition, comprise sub-systems which may or may not have complex interactions. This shows that the difference between a system of systems and a system is mainly one of scale, so there is no fundamental difference between system-of-systems engineering and systems engineering. This is the view taken in this book. The military and engineering aspects are then part of either or both capability and systems engineering.

'System of systems' used as a noun is useful as it has developed a meaning in defence circles to describe a defence capability made up of several major items or platforms which can be used independently, but when operated as a closely coupled system deliver a far greater capability than the simple sum of the parts. The coupling is usually delivered through networks and a unified Command and Control (C2) system. An Integrated Air Defence System (IADS), such as those discussed in Chapters 11 and 12, could be described as a system of systems. A slightly more pragmatic and tongue-in-cheek definition of a system of systems is a capability that exhibits multiple wicked problems as soon as the previously well-behaved systems are coupled together.

Systems of systems usually have good connectivity with the problems that arise from multiple interactions between many systems. Therefore, it is to be anticipated that highly automated and autonomous systems will exhibit multiple wicked problems as well, especially if there is any level of Artificial Intelligence (AI) and machine learning involved.

4.2.2 Emergent behaviours and wicked problems

A complex system of systems will have many, possibly many hundreds of interactions, both internal and external. Even in an ideal test environment or test facility, there will be unexpected behaviours which are usually reproducible. These need not be detrimental and sometimes give an unexpected extra capability if utilised reliably. These are called emergent behaviours and occur in most complex systems, especially if they are software-intensive. They should be analysed and their cause explained and documented in an appropriate way.

The complexity of systems of systems and their multiple internal and external interactions mean that it can be difficult, if not impossible, to return a system of systems to the state it was in when a problem occurs. Making a small change to try and understand the problem usually changes the symptoms and often the problem. This makes it difficult to diagnose the fault and fix it. This type of problem is known as a wicked problem. They have been defined by several authors and organisations, but the most appropriate definition for us is one ascribed to Horst Rittel by the CogNexus Institute [5]:

> A wicked problem is one for which each attempt to create a solution changes the understanding of the problem. Wicked problems cannot be solved in a traditional linear fashion, because the problem definition evolves as new possible solutions are considered and/or implemented.

Although these are widely discussed and many examples given, there are no specific solutions. Rather, there is recognition that they can only be solved by making proposals, debating them and reaching a consensus on an acceptable solution. It is possible that a wicked problem is, in fact, a complex emergent behaviour that only occurs in a small subset of cases.

4.2.3   Architectures

Large organisations, systems and programmes can be complex and difficult to describe in a way that supports efficient management and support. A consistent description is called an architecture. It can be achieved by the use of a set of diagrams, textual descriptions and connecting links. These are called views in an architecture. When drawn up to describe an architecture from a particular per-spective, such as that of the system designer, or the communication network, they are called viewpoints.

The UK Chapter of the International Council On Systems Engineering (INCOSE UK) have drawn on a variety of interpretations of what an architecture is, to give a summary which covers the common ground [6]:

> The architecture of a system is its fundamental structure – which may include principles applying to the structure as well as specific structures.

One of the landmark papers describing how a system with a complex archi-tecture can be described with a set of useful and consistent views was by Zachman [7]. This gave a table of analogies between a building's architecture, the design of an aircraft and then the design of an IT system.

Table 4.1 is based on Zachman's table, but it treats the aircraft as a system to make it more relevant to autonomous system design. The table's analogues are only illustrative, but each row shows that there is a different way of describing the same product for different groups working on or using the same item. It should be noted that the system architecture cannot be separated from its operating and commercial environments.

Architectures are drawn up for a purpose, generally to answer a specific question. The scope of an architecture can be defined for the specific problem in

Table 4.1   Analogies between architectures for a building, an aircraft and an IT system (Based on Table 2 from Zachman 1999 and reprinted courtesy of International Business Machines Corporation © 1999 International Business Machines Corporation)

Generic	Building	Aircraft as system	IT system
Rough order of magnitude	Bubble chart	Concept (Range, load)	Scope/objective
Owner's requirements	Architect's drawings	System of systems of air traffic control, airports and support	Model of business or IT system description
Designer's concept	Architects plans	Aircraft subsystems (Airframe, engine, avionics, etc.)	Model or description of the information system
Builder's representative	Contractor's plans	Subsystem specifications	Technology model (or technology-constrained description)
Out-of-context representation	Shop plans	Contracts on subsystem suppliers	Detailed description
Machine language representation	–	Detailed subsystem designs	Machine language description (or object code)
Product	Building	Aircraft	Information system

hand and will be different to the scope for another problem. A well-planned large project will define the architecture framework and ontology to be used. All contractors will have to conform to this. An existing large system with multiple contractors may not have had one framework mandated for it, so there is the potential for confusion. The financial risk from this confusion will have to be balanced against the cost of producing common architectural viewpoints over all subsystems and contractors. An intermediate solution may be to agree a common ontology and meta-model[2] for the system of systems with each contractor using their own internally consistent viewpoints.

4.2.4   Architecture frameworks

An architecture framework is a logical structure which is used to give consistent descriptions of a system for its development, implementation and sustainment. There are several recognised frameworks with different scopes. The scope may be restricted to the system of systems which includes the system under consideration. More commonly, it includes the business processes using the system, the personnel involved and support structures. The international standard ISO/IEC 42010 addresses architecture descriptions and architecture frameworks.

There are a wide range of architecture frameworks, some of which are given in Table 4.2. The first three military ones DODAF, MODAF and NAF are closely

---

[2]The meta-model is a precise definition of what can be described in the architecture through each of the viewpoints.

Table 4.2   Architecture frameworks

Name	Website	Application
The Open Group's Architecture Framework (TOGAF)	www.opengroup.org/subjectareas/ enterprise/togaf	Civil
Zachman Enterprise Architecture	www.zachman.com	Civil
US Department of Defense's Architecture Framework (DODAF)	http://cio-nii.defense.gov/sites/dodaf20	Military
The UK Ministry of Defence's Architecture Framework (MODAF)	www.mod.uk/modaf	Military
NATO Architecture Framework (NAF)	http://nafdocs.org/introduction/	Military
Unified Architecture Framework (UAF)	https://www.omg.org/spec/UAF/About-UAF/	Mainly military

related but distinct. There is agreement that these will evolve into one Unified Architecture Framework (UAF). Reference [8] gives details of progress and the relationships between the frameworks. Most large systems companies will have a preferred framework and the necessary infrastructure, tools and training to support work with it. The choice of framework is a strategic one and will depend on the company's business sector, their customers and collaborative work with peer companies and subcontractors.

The NAF, MODAF, DODAF and UAF all give a set of views that make up each viewpoint. Producing every one for a project or capability is a daunting and, in fact, unnecessary task. Before any architectural work is undertaken, the architect should establish the reason for producing it, and the questions it is to answer. Architecting is not a process that should be undertaken by a newcomer without training or advice and supervision from an experienced architect. There are numerous software tools available for the architecting process and the choice depends on a combination of the user's organisational preferences and the purpose of the architecture. In some cases, the architectural questions can be answered by using Microsoft Office programmes (PowerPoint, Word and Excel) in a logical and rigorous process with good configuration control. The Systems Modelling Language (SysML) and Unified Modelling Language (UML) are generally accepted as suitable languages for use with most architecture frameworks.

System definition and design using an architecture framework is not trivial and it is possible to complete a design without using one. However, the discipline and consistency it brings to the design saves cost and reduces the risk of redesigns significantly. In the author's experience, it is always sensible to adopt the discipline of a framework and produce a set of views using that framework's definitions and viewpoints. This results in the designer deciding which views are needed and considering many key issues at an early stage, saving time later.

4.2.5 UML and SysML

It is possible to develop architectures for some purposes using basic office tools such as PowerPoint for diagrams, Word for text and Excel for data and links between system modules. However, more complex systems need descriptions that use models of information flows and the interactions between modules and the system's environment.

A large number of modelling languages were developed in the 1980s and 1990s to describe complex software products. Eventually, the UML was developed as a non-proprietary language. It is now widely used to describe systems as well as software, but is not necessarily optimised for systems. The SysML is an extension of UML specifically for systems applications. Neither language is used in this book. Readers who want to learn more about them and use them can refer to the Object Management Group[3] (OMG) website (www.omg.org) and two books by Holt [9] and Holt and Perry [10].

UML and SysML can be, and are, used as the mechanism to implement the architecture frameworks. They are both based on a set of thirteen diagrams with information flows and other links between the objects in them. Many architecture views in the frameworks are given as UML and SysML diagrams.

4.3 Contradictions in technology developments for military use

There is a generally accepted view that technology is progressing rapidly and probably accelerating. The advent of cheap fast processors whose power follows Moore's Law and their consequent capabilities supports this view. Modern business and consumer applications give capabilities undreamed of a few decades ago. The result is that processor-based applications are driven by highly competitive, rapidly evolving civil mass markets. Their lifetime is short, probably much less than a decade, often due to the appearance of better products.

Armed forces utilise the new advances if possible, and do when their application is in what may be called a civilian environment such as stock control and administration. They struggle to use them for combat operations and exercises. Why? The answer is a combination of factors such as: arduous military environmental requirements; small production numbers for a manufacturer to meet these requirements; the need for support over many years; and support in difficult parts of the world.

Military-specific technologies have developed for different reasons, giving new capabilities. The driving requirements have been to outperform other nations' equipment. The combination of a fighter aircraft and its missiles, for example, must win in combat with a hostile aircraft. The result is a steady improvement in the combat performance of weapons and their platforms. Speed, manoeuvrability and

---

[3]The Object Management Group® (OMG®) is an international, open membership, not-for-profit technology standards consortium, founded in 1989.

Table 4.3   Lifetimes for sample military aircraft

	Hawker Hunter	Panavia Tornado	Eurofighter Typhoon	F35, Joint Strike Fighter	B2 Spirit bomber
Start design	1948	~1970	1986	1997	1970s
In service	1953	1980	2003	2016	1993
Out of service	~1970	>2020	2050?	2070	?
Lifetime (years)	17	>40	>47?	>54?	>50?

stealth are good examples and apply at platform level. Their cost is high, giving very expensive products with few applications in mass markets.

Table 4.3 illustrates the changes in development time and lifetime for aircraft, but similar timescales apply to other large platforms and missiles. Technology development times of one or more decades are set partly by budgetary constraints which come from political decisions, not investment decisions based on potential markets. In-service lifetimes are set by assessments of likely opponents' capability developments and the time to develop the replacement capability.

The contradiction in technology developments is the decade-long development time for high-performance military platforms with decades of lifetime but which must have additional capabilities based on civil processor-based products with short development times and lifetimes.

Military shortcomings arising from these contradictions have been made even more obvious due to engineering successes over the last three decades in producing highly capable cheap consumer products, but with a comparatively short life, and the increased reliability of more expensive products such as cars where expensive development costs can be amortised over hundreds of thousands or millions of units rather than the tens or low hundreds for military products.

4.4   Defining and delivering military requirements

Initial identification of a requirement can arise in many ways and the first steps require relatively small budgets. A nation usually carries out regular capability audits which test their existing systems and infrastructure against a set of likely scenarios. These may be defensive, such as an attack by a hostile nation on an ally, or offensive such as landing a large force on a defended coastline. Capability gaps are identified and analysis carried out to propose possible solutions to the problem. This is known as threat-based planning.

Requirements may arise from new technologies. The huge investment by companies in civil markets gives a continual flow of products which offer new military capabilities. Autonomous technologies of all types and autonomous systems, in particular, are examples that have already changed military operations and will continue to do so. The inverse of this is the danger to a nation if an opponent develops a technology that they are likely to use against the first nation.

The military planners must also include potential threats that their own nation may consider illegal under IHL such as those used by some non-state organisations.

A common first step in defining a need is to plan conceptual operations using the new technology or countering the hostile threat. The concepts are developed to the level of detail thought necessary by the planners. The approach which evolved over the last century is to think through, in detail, how a likely war will be fought and the operations required to win it or to counter success by the opposing forces. This is known as war-gaming although the term can give rise to confusion with the popular war games played by many people on their computers and dedicated consoles. Strategic plans have the least detail, but lead to detailed considerations of the critical operations. It should always be remembered that one of the certainties of warfare is that the enemy will always do something you do not expect. There are many related maxims, for example, 'no plan survives contact with the enemy'.

Scientific methodology was applied to plan complex operations during World War II and became known as operational analysis. The invasion of France in 1944 (D-Day) was one of its earliest and biggest successes. Operational analysis showed that the speed of operations required extensive fuel supplies and a harbour where none existed. The solution was rapid deployment of an oil pipeline from the Isle of Wight in southern England to the invasion area in Normandy and the assembly of a pre-fabricated harbour which was towed in sections from England to a site near the invasion beaches. These two components provided sufficient fuel for initial movements, and the heavy equipment which could not be supplied over beaches.

The engineering, as well as military lesson learnt from the above example, and numerous others, is that it is essential to think through an operation or design in great detail. It is only when this thought process is completed do the real problems emerge. 'The devil is in the detail' is a very apt proverb. Although even the best plans need changes, it is the author's and other colleagues' experience that when problems emerge, an initial, well-thought-out plan has enough detail and/or resources in it to provide a good chance of success even if extensive re-planning becomes necessary due to changes in circumstances.

How does this relate to new autonomous systems which have potentially game-changing capabilities? The processes of capability-based planning (Section 4.5), operational analysis and systems engineering can be used to identify the likely military requirement for their desired capabilities, using methodologies that are understood by both military user, procurement authority and materiel supplier. (Materiel is used here rather than material, as the latter word ignores the human user and support needs.) Detailed analysis using possible autonomous systems should reveal the opportunities and disadvantages of automation when faced with practical dilemmas and uncertainty. Operational analysis gives context, so legal assessments of automated and autonomous weapons can be made in realistic and relevant scenarios, and can provide guidance to operational commanders for their planning and use. The processes are described in the next sections.

4.5   Capability-based planning

Capability-based planning was developed in a Canadian technology demonstrator programme with an aim of breaking down the problems of threat-based planning [11]. The traditional approach was to look at the major threat, and develop forces and equipment to counter or defeat it using operational analysis as part of threat-based planning. This leads to 'stove-piped' procurement where a state develops a countermeasure to the most threatening materiel of their opponent. (Stove-piping also means developing expertise or equipment in isolation from related areas.) Examples include the development of new fast-jets and new submarines with performance tuned to be superior to a potential adversary's new equipment.

Threat-based planning is highly effective at assessing the value of a new piece of equipment when used against a range of threats, but does not take into account wider needs. Long lifetimes for platforms such as ships, tanks and aircraft necessitate their incremental upgrades as technology develops and likely threats evolve. It also leads to a lack of common standards or interpretation of standards giving interoperability problems even within one state's military forces.

The capability-based approach to the affordability problem of stove-piped procurement is to recognise that a state must have the capability to defeat its potential opponents' forces, not just specific weapons or systems. Success can come from a combination of systems rather than a niche one. It may be possible to negate the value of an opponent's new fast-jet attack aircraft by building an IADS and continuously upgrading its component systems as technology evolves. This provides a counter to the new aircraft as effective as a new specialist air-combat fighter. The IADS will have the added advantage that it can monitor all the state's airspace and be capable against multiple threats with deployed Surface to Air Missiles (SAMs).

The capability-based planning approach is described in a TTCP[4] document [12]. That paper summarises the difference in approach as the question 'what options are there for new artillery?' is replaced by 'how do we provide fire support to land forces?'

Capability-based management was included in the Canadian programme and the results published in 2005 [13]. That paper discusses the problems of integrating all aspects of a project and gives recommendations on the management strategies and the engineering tools required for successful capability management. Most nations now adopt some if not all of the recommendations given in that paper and integrate engineering effort and data across their organisations.

There are several significant aspects of starting with capability.

● A capability can be delivered by one or more of several options;
● The capability will be required for use over a long, possibly indefinite period so it will be defined at a strategic level;

---

[4]The Technical Cooperation Programme (TTCP) is an agreement between Australia, Canada, New Zealand, the United Kingdom and the United States for cooperation on military technologies.

- Technology changes for both sides of a potential conflict, but the capability requirement will remain, so the materiel will have to evolve over time. This has led to the concept of through-life capability management explained in Section 4.11;
- Maintaining a capability incudes logistics, training and a host of other issues. These are defined in different ways by different nations; see Appendix A4 for more details;
- A capability not to be able to do something is not a usual requirement, but may be necessary for autonomous systems due to the legal restrictions discussed in Section 2 Chapter 1.

4.6 The capability-based Vee diagram

The military market is a combination of the senior officers and the political part of government. The market solution to the technological contradiction in Section 4.3 is to demand delivery of enduring capabilities which need not be product specific. Realisation that military capabilities are delivered by combinations of platforms and materiel which change over time led to the formal concept of military systems rather than platforms and weapons, and the development of MODAF, DODAF and NAF to describe them.

The engineering solution to the technological contradiction is the development of the disciplines of capability and systems engineering. Traditional engineering methods to produce a single product for any market can be called product engineering. The boundaries between the disciplines is blurred in the same way that a product of any size is usually designed as a system, even though it may form one subsystem in a larger system.

The management process to deliver capability is summarised in Figure 4.1 which is known as a Vee diagram with time as the horizontal axis. There is no agreed Vee diagram, every author and organisation have their own variant. This one emphasises capability and the systems engineering processes, with the outputs being architectures, requirements and plans. Another version is given in Section 4.12 which emphasises the detailed design aspects of systems engineering. The common theme of them all is a supply process down one side of the Vee, production in the lower section, delivery with successive stages of integration up the right-hand side. Incremental upgrades to maintain or improve capability over time are then a continuing on-going process. Figure 4.1 shows these common features and also shows the inputs and outputs of each step and the test philosophy. The figure sets the framework for later sections in this chapter.

The Vee diagram is a simplification and in practice most of the steps are iterative. This is illustrated in Figure 4.2 which is based on a diagram from the UK Defence Procurement Authority in ~2008, but not officially published. It shows how policy and budgets lead to products which are integrated to produce capabilities until they become obsolete. (The original figure was colloquially known as the 'egg diagram' after its first appearance.)

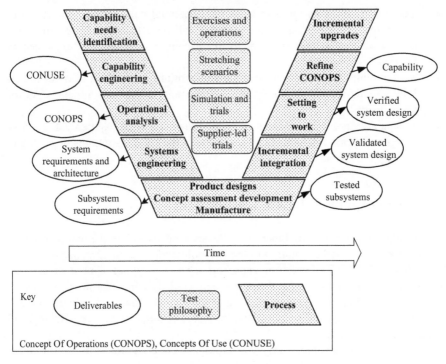

Figure 4.1   Capability-based Vee diagram

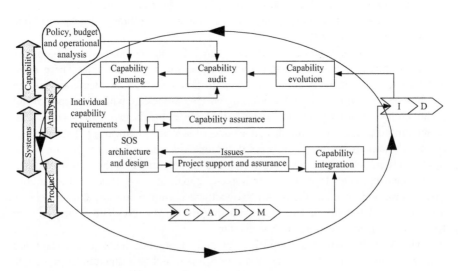

Figure 4.2   The 'egg' diagram for military systems
   [C = Concept, A = Assessment, M = Manufacture, I = In-service, D = Disposal SOS =
   System Of Systems]

4.7   Establishing solutions by operational analysis

Operational analysis is one of the main tools used to look at the cost-effectiveness of the various options open to the budget-holders. The aim is to achieve a balance of investment across the armed forces for their likely deployments on time-scales of decades. A good guide to operational analysis is the NATO Code Of Best Practice for Command & Control Assessment [14], although originally for one application, the code has widespread applicability and is used here. The NATO process is shown in Figure 4.3 (which is taken from it) and shows that the analysis is iterative.

The starting point (Sponsor problem in Figure 4.3) for a new product or capability may be a fairly vague concept that a new technology must bring significant benefits, but it is not clear what they will be. An example might be that an autonomous ground vehicle with detectors could provide an efficient way to locate and destroy mines in urban and rural areas. This concept can then be developed into a set of design concepts and assessed as potential products in business cases. Modelling and simulation is central to nearly all analyses of possible capability requirements and has an extensive literature (see, for example, [15]).

Operational analysis starts when the initial idea or concept can be expressed as a coherent concept or problem. This is the 'Sponsor Problem' in Figure 4.3 which is applied to autonomous systems in Section 4.8. A military sponsor may be a senior officer who is looking at the nature of warfare in ten years' time, or a more junior person looking at a specific problem with current equipment.

In the commercial world, the sponsor could be a marketing executive looking for a new market, or replacing a current product range with new concepts. In this section, the assumption is that the sponsor is a military one, but the approach is widely applicable.

Defining the problem is a critical problem and should be discussed in detail with the sponsor, with the likelihood of numerous iterations. There will be many unstated assumptions that must be identified during this stage. In addition, the sponsor and analyst may not be in direct communication if there is a contractual relationship between their organisations which has defined channels for communication. The sponsor problem will be formulated in a mutually agreed form before formal analysis can start.

The sponsor problem will almost certainly include the replacement or upgrade of one or more parts of the military inventory to give a different capability. There should be an existing architecture for the relevant part of the inventory. If so, it can be used to provide a set of architecture framework views which can act as a basis for further analysis. If not, a set should be drawn up. These views are sometimes called the 'as-is' views. The analysts will then prepare the probably smaller set showing the architecture with the required changes to the affected parts, sometimes called the 'to-be' views.

The sponsor problem may well be concerned with the use of weapons in difficult or complex situations. There will definitely be the Article 36 question: *is this*

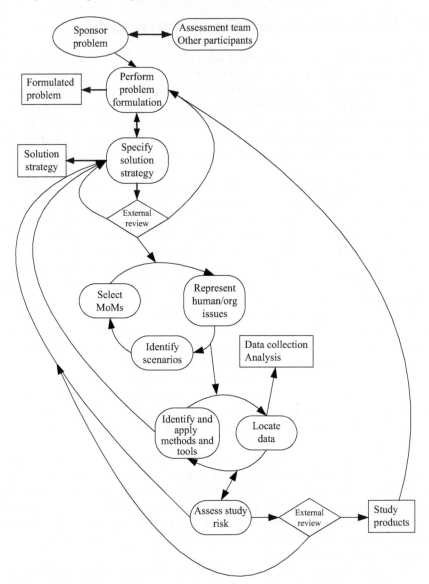

Figure 4.3   Operational analysis process
[Based on Figure 2.2 from the NATO Code Of Best Practice]

a new means or method of warfare?, explained in detail in Chapter 11. A negative answer will still need to be substantiated by a description which must represent the system's capabilities, i.e. a set of architecture views showing how the concept will be used and the differences from current capabilities. It should be noted that automated capabilities which are not an integral part of a weapon system may still give a positive answer. An example would be an automated object identification system with outputs to a human in a weapon control system but without any caveats or restrictions.

The analogy for civilian applications to the military problem on legality of use is that of meeting codes of practice for ethics and demonstrating adherence to regulations. These questions are discussed in detail in Chapter 6.

It is unlikely that a request for a more automated capability or service will be a like-for-like replacement. In general, the customer is assuming that there will be an improved capability as well as any cost saving. Even if the new equipment is replacing an earlier, obsolete one, there will be different human–machine interfaces and interactions. One big danger is false expectation based on 'hype' about robots and their capabilities. The inverse problem must also be included in planning, that is the emergence of new or improved capabilities which were not requested by the sponsor but may be highly desirable for them. It is unlikely that there will be little or no contact between the architect and sponsor pre-contract and the former will have some ideas about solutions. One practical solution is for the architect or a well-briefed representative to present draft selected views showing a possible solution at a very early stage, and elicit comments and questions.

Even if military planners have adequate resources, how do they deal with the uncertainty arising from enemy action or unforeseen circumstances? They cannot rely on doctrine, training and good luck when a war or operation starts, although the first two are essential and good luck helps. The answer is flexibility and the ability to analyse the problem accurately, re-plan rapidly and then act effectively. The 40-year-old Cold War with two power blocks of differing technologies, but comparable strength was a period of great military certainty and relative stability. Leaving aside the Vietnam War and peace keeping in the Balkans, the first major coalition war was against Iraq, following their unex-pected invasion of Kuwait. The multi-national military infrastructures had to modify equipment designed for Cold War use in Europe for desert fighting in a period of a few weeks. This was in parallel with the logistics of moving materiel to the Middle East and training coalition forces to work together. Flexibility in approach and interoperability between coalition partners became key. Rapid and rigorous analysis of the execution of the strategic plan led to detailed plans so that the coalition forces had the correct materiel in the right places during the campaign.

Military planners use a wide range of techniques to predict the course of a campaign. Operational analysis is the scientific approach to give effective advice for military planning and action. Threat-based planning is operational analysis when the scenarios assume a military operation against a specific threat. Opera-tional analysis takes one scenario at one point in time, but the process may be repeated for a range of scenarios and timeframes to test the robustness of a parti-cular solution. The extension to cover force evolution and the wider logistic and support problems is part of capability-based planning described in Section 4.5.

4.8   Operational analysis for autonomous systems

Most states will look at long-term trends to identify likely adversaries and areas of conflict. Much of the detail and probabilities are not published, but some idea of the nature of the broad future context can be seen from the two UK documents [16,17]. The first document has been superseded by the second, but it contains more detail and gives more specific detail about likely areas of conflict. The second document discusses trends in automated systems and their use both by the UK and by hostile states and organisations. The predictions for highly automated systems are for three types:

1. High-end general purpose systems which can carry out a variety of roles and are almost certainly high value and will need to have some self-defence capability;
2. Specialist systems which will probably be small, relatively cheap and only capable of specific types of operation;
3. Swarms of small systems which may overwhelm a traditional defence system.

This division is not proscriptive, but gives a starting point in posing and answering the question: what type of unmanned system is needed to meet the military requirement? It also gives a starting point in answering the inverse question: what military advantage comes from this new unmanned system technology that is now technically possible? In both cases, the process is to ask whether a specific niche is being addressed or is there a more widely applicable capability on offer? The answer to this question becomes the first of many assumptions in the analysis.

There will be an assumption that an autonomous system is to be assessed and that requirements and restrictions must be derived for the level of automation in the system, the targeting chain and the human-machine interactions. Alternatively, the main requirement may be to look at the offering by an industrial consortium of an existing or new technology demonstrator. Assumptions will then be made about the force mix that will be available to the state over the timescales of the new system. These will be based on existing equipment and items already in procurement with their lifetimes. The problem is then to assess whether the product offers a tangible military advantage and improves force effectiveness.

The next step is to postulate the various force-mixes and systems available to a commander and how these can be combined into a system of systems delivering a set of military capabilities. Their effectiveness is then assessed by modelling campaigns and operations in various scenarios. It is likely that a state's procurement authority and military planners will have a set of recognised scenarios derived from their predictions of the future. Use of these then allows an unbiased assessment of the benefits of the new autonomous systems compared with manned systems.

The main concern for highly automated systems is that every unmanned system must always be under control, with data or information transfer between the human and it. It is shown in Section 10.3 that if an unmanned system is not under the control of its commander then it is illegal and should not be procured or used in operations. This makes command and control systems one of the key areas for consideration. There are a range of tools available to analyse these systems. One example is those available in the UK MOD [18].

The entry point into the process was taken to be the 'Sponsor problem' in Section 4.6. However, it could be at another place. For example, Article 36 concerns about military use of a new civilian autonomous capability in operations could make the 'Human-organisational issues' the entry point.

A strawman specification will have to be drawn up and considered in a scenario. For example, if it is a long-range Unmanned Air Vehicle (UAV) operating in contested airspace with no radio emissions, there will be concerns about changes in the target area and the reliability of target identification algorithms. An architecture is created and analysed through the use of different views. The analysis then uses different concepts and models to assess the benefits of the different concepts through measures of merit.

There is no single measure of merit for any system. Instead there are several measures which are relevant for different purposes. The NATO Code Of Best Practice has adopted five levels of measure of merit which are given below with comments on their relevance to autonomous systems and legal reviews.

1. Measures of policy effectiveness, which focus on policy and societal outcomes;
    These are the high-level issues which would look at the effect on public opinion, whether domestic, belligerent or neutral, of using autonomous systems in a campaign or an operational area. It would be assumed that there is a method of legal use rather than ask what is acceptable.

2. Measures of force effectiveness, which focus on how a force performs its mission or the degree to which it meets its objectives;
    The introduction of high levels of automation and autonomous systems into a force mix should increase its effectiveness, so measures of force effectiveness will be used to decide the mix of autonomous and manned vehicles.

3. Measures of C2 Effectiveness, which focus on the impact of C2 systems within the operational context;
    These will be used to assess the effectiveness of the autonomous systems in specific scenarios. They will almost certainly be used to assess different rules of engagement for the proposed or actual autonomous system. Analysis of the measures will lead to requirements for specific aspects of an autonomous system's performance. An inability to operate under effective and acceptable rules of engagement will remove a proposed autonomous system from the campaign order-of-battle or raise questions about the cost-effectiveness of procuring it.

4. Measures of performance, which focus on internal system structure, characteristics and behaviour;
    These are measures of effectiveness in the physical, electronic, network and cognitive domains at system and system of system level. This is the level which will decide if an autonomous system can be used legally. The measures

of performance needed to test for legality have to be defined. These will include the autonomous system's ability to meet the four principles of Additional Protocol I (API) to the 1949 Geneva Conventions as explained in Chapter 7.

5. Dimensional parameters, which focus on the properties or characteristics inherent in the physical command and control systems.
    These are the physical parameters of the system and include the details of the architectures describing the system of systems. Methods for data collection and dissemination and their use as information by humans and machines will be quantified as far as practical to give the dimensional parameters for use in identifying the higher measures of merit. Specifying the autonomous capabilities of the autonomous system as dimensional parameters and their assessment in deriving measures of merit is a difficult problem that needs to be solved. Authorised power, defined in Panel 3.1 and discussed in Chapters 12 and 13, gives one possible approach.

Defining the problem is a key issue and is also iterative. However, drawing up tables of the different type of measure of merit gives an effective way of highlighting key issues in introducing higher levels of automation into any military materiel.
The process should end with:

1. A description of the military capabilities that deliver the desired improvement in force effectiveness;
2. An outline technical requirement for the systems delivering the capabilities;
3. Tables of each of the above measures of merit, quantified if possible or scored for comparison in various scenarios.

It will be necessary to draw up a technical, system-level description of the system of system delivering the capability at an early stage in the process. This is carried out using capability and system engineering techniques. The results will then be part of the data located and collected in the central part of Figure 4.3.
    There is an underlying assumption that the delivered system will meet the legal requirements discussed in Chapters 7 and 8. The system cannot use any banned weapons such as cluster bombs or chemical weapons. This is easily checked at the first stage of the analysis. However, meeting the requirements of Article 36 is not so straightforward. Chapters 11 and 12 discuss how this is taken into account at this stage.

4.9   New types of engineering for military systems

There is an assumption that processor-based products will be in-service, or use, for a period of a few years and then go out-of-service or be withdrawn and replaced by a superior product. Backward compatibility of a new product with an existing, older system is a well-known problem, especially with software.

The problems arising from this complexity include:

- Upgrades with technologies that were not considered at the concept phase may result in major redesigns of major systems or modules;
- Civil products offering superior performance but not meeting full military requirements, with prohibitive costs to develop military equivalents;
- Supporting equipment supplied by companies that may have ceased trading during the platform's life;
- Life-extension programmes imposed late in the product's life due to budgetary pressures;
- Variations between in-service products due to changing circumstances during its life and phased introduction of upgrades across a fleet of the same basic platform;
- Low production numbers giving high one-off costs which can only be spread over a few tens to hundreds, and rarely thousands, of units.

Recognition of the extra complexity and describing it in a way that brings out the capability need is shown by the widespread inclusion of through-life capability management techniques discussed in Section 4.11 and given in Appendix A4.

It should be noted that many of the perceived benefits come from integrating many services and equipment into more complex systems by the use of networks, common interface standards and protocols. Information can then be accessed from a variety of sources which may be of a provenance known only to the data generator. Internet and other network service providers do not concern themselves with the internal structures of users' equipment, its reliability and the quality of the data generated. This is unacceptable for decision-making in a military environment. There must be a clear command chain based on direct human-to-human links which relies on an understanding of the reliability of the available data and information from the network.

Military commanders at all levels must always have a clear command chain with actions based on reliable information and authorised decisions. It is not acceptable for weapons and the weapon control system, to have an indeterminate command structure, operable solely due to the use of common standards and protocols for data exchange. This could arise if commanders at a low level interpret information and make decisions without authorisation from their superiors to make both the interpretation and the decision. Human commanders may be able to use their initiative and make lethal decisions under some limited circumstances, but this would not be acceptable for an automated system.

Engineers have responded to these challenges for both military and civilian applications by developing the techniques of system architecting, systems engineering and, more recently, capability engineering described in this chapter. It is important to understand the extra complexities addressed by capability engineering compared with systems engineering. However, the boundaries between these engineering disciplines are not fixed. Pragmatically, they should be regarded collectively as a set of techniques offering a range of approaches to solving a particular problem; engineering judgement is used to decide the most appropriate approach.

A proposal to increase the level of automation in an existing platform or service is likely to have to consider all the complexity problems given above. A proposal for a highly automated system may offer a new type of solution based on the potential cost savings arising from the reduced numbers of personnel required to operate it, but this may not be the case. Originally, one predicted advantage of the use of UAVs was cost savings for this reason. In practice costs have escalated due to the vastly increased number of UAV hours flown compared with strike and surveillance aircraft, requiring many more UAV operators than aircrew and more ground-support staff, plus the improved imagery products requiring increased numbers of intelligence personnel to interpret them.

Chapters 11 and 12 show how an existing capability with current levels of autonomy could evolve into one which relies on high levels of automation, with consequent problems for Article 36 reviews.

4.10   Capability engineering

Capability engineering can be described as the process of taking a customer's description of a need in their terminology, understanding their wider requirements and turning this into one or more proposed systems that satisfy the need. The output can be at a diagrammatic level or detailed SysML models. It should have enough detail for systems engineers to derive procurement specifications and compliance matrices[5] to be drawn up. It is the top section of the 'egg' in Figure 4.2. The step between the Manufacture and In-service parts of CADMID[6] represents both a capability evaluation as the product enters service, and possible upgrades to give a different, required capability. The roles of capability, systems and product engineering should be apparent from this figure, with increasing engineering detail created as one goes down the figure.

The step from 'Manufacture' to 'In-service' is not necessarily a short one. A complex system, whether for military or civilian use, has to be thoroughly tested before release. The production numbers are low, possibly only one and rarely more than 100. The user-environment is also unique, already in place and difficult to simulate at the supplier's facilities. Software applications have extensive 'beta testing' where experienced users, who are usually not employees of the production company, test an advanced version and report back on problems and their general experience. The military equivalent is 'setting to work'. The product is first transferred from the manufacturer's facilities to military ones. It is usual for a joint supplier/military team to then use the product in increasingly complex and realistic scenarios. Setting to work is a key part of the right-hand side of Figure 4.1.

---

[5]A compliance matrix is a table with the customer's requirements in the first column and columns identifying how closely the requirement is met and the contractor's view of whether the requirement is met or not.
[6]The CADMID cycle is discussed in more detail in Section 5.4.

Capability engineering is often carried out by government employees such as those in military research laboratories or specialists in the procurement agencies. The usual alternative to this is a contract with a specialist company to carry out this work on the government's behalf. The output from the capability engineers usually forms the basis of the final procurement specification and contract.

Capability management has three fundamental aspects:

- Capability generation is the creation of a capability with all aspects necessary to maintain it over its predicted life which is typically 10–20 years using assets which have a life of up to 50 years with numerous upgrades;
- Capability sustainment and readiness covers the elements necessary to keep a current capability operational for the time-frame of current operational planning. This is typically 5 years, recognising that the capability must evolve to match likely opponent's capabilities;
- Capability employment is the planning for, and conduct of, operations using the capability in the immediate future.

All three will be included explicitly or implicitly in the statements of needs to fill a capability gap.

Military procurement is preceded by an assessment and audit to reveal capability gaps. The trigger for these can be one or more of: planning for the end of life of an equipment type (The D of CADMID); the 'lessons-learned' studies after a campaign; budget cuts rendering some facilities unaffordable; development of a new weapon type or threat by another state or non-state organisation; or the possibility of a campaign in a new area of the world with a different environment to those currently foreseen. If the audit shows a gap, it is followed by studies to identify possible solutions which will include re-use and upgrading of in-service equipment and services. It is also possible that the capability may be provided by an external service supplier such as a bespoke contract with a satellite communication service provider to supplement or replace most of a military comms satellite network.

The procurement process starts with the identification of the capability need and the results of the studies into possible solutions, the capability planning stage. Military and budgetary experts will have compiled statements of their desired capabilities. In practice, these statements can be grouped under headings representing different arms of the military complex who then effectively become a set of rival customers. In addition, each of the above aspects presents individually complex sets of problems which may have mutually incompatible solutions. The task for the capability engineering community is to understand the wide range of views and produce a set of requirements that can be developed into a system of systems design. At this stage it is possible to make rough order of magnitude[7] costings to fill the gap.

---

[7]The cynic might say that a supplier's initial rough-order-of-magnitude cost is one that must never be exceeded. This is because customers put that figure in their financial calculations and then do not have any more available for the project.

It is a truism that all engineering designs are a compromise and capability engineering is no exception. The difference with capability engineering is its assumption that there are multiple customers with changing requirements and that there may never be a single statement of customer need. Because of finite budgets, the cost to fill a gap will almost certainly be at the expense of funding to fill other gaps, bringing another level of compromise. Part of the advice which can be given by the engineering community is identification of other gaps which can be partially or completely filled by use of the proposed system of systems or by a (hopefully) small extension.

The first problem faced by capability engineers is that the multiple customers' problems may be stated very clearly and simply at their level but the solutions will involve several systems, some of them human, which change with time and go out of service, with support and logistics requirements in both peace and war. Additionally, a solution for one campaign type may be completely wrong for another. As an example, the reader may like to decide whether a carrier-borne low-altitude UAV providing excellent surveillance coverage for littoral operations can provide the same capability for an operation in jungle-covered mountains. Then the reader should consider if there is one type of UAV that can carry out both roles at an affordable cost.

It might be thought that capability engineering finishes when the system of systems is specified and the contracts placed on industry. In fact, it continues after this on the customer side. Design compromises by industry have to be made to contain costs, usually with capability compromises. Unless these are monitored closely and continuously, there is a high risk of the delivered product or products falling short of requirements in a critical capability that may not be a 'hard' contracted deliverable. Early identification of the risk can play a significant part in contract amendments during the development phase.

Technology demonstrator programmes can be considered to be one response to a potential capability gap due to worldwide technology developments. A decision is made to show that there is a militarily useful application of a technology, usually by building a demonstration of the key modules. Technology demonstrator programmes can be as large as a prototype ship or aircraft, or be more limited, such as the use of a commercial software suite for analysis of images from disparate sensors. They can be relevant to any part of the processes in Figure 4.1 or 4.2.

Levels of automation are increasing rapidly in the civil sector so there is scope for technology demonstrator programmes. These can play one of three roles:

1. Demonstrating the military utility of a civil autonomous systems development;
2. Demonstration of methods to raise the autonomy level of a uniquely military process. This category could be considered to include proof of safety-critical or mission-critical properties of the automated process;
3. Replacement of a human by an automated process.

Engineers are well placed to both identify military applications of the many developments in this fast-moving field, and help military staff make realistic assessments of their potential benefits. Ethical questions are likely to be a major part in the assessment of the application of more autonomy in weapon systems, so technology demonstrator programmes offer a way to evaluate the military application and assess it against legal and ethical criteria.

4.11 Wider capabilities

Through-life capability management was introduced in Section 4.5 in the context of technology upgrades over the lifetime of a platform to maintain a capability. This is one aspect but there are several others which different nations describe in slightly different ways. The two main ones are:

- The UK TEPIDOIL: Training, Equipment, Personnel, Information, concepts and Doctrine, Organisation, Infrastructure and Logistics. The UK has also added Interoperability as an overarching capability;
- The US DOTMLPF-P: Doctrine, Organization, Training, Materiel, Leader development, Personnel, Facilities and Policy

These are defined in Appendix A4. The US defines their terms as a required analysis whilst the UK defines them as capabilities which must be addressed during capability assessment. They are essentially the same. Increased automation will give changes in all these areas which are not necessarily within this book's scope. Some specific issues can be predicted which will affect future procurement and designs. These are discussed here under headings which encompass both the US and UK interpretations.

Policy, concepts and doctrine

> Unmanned systems and autonomous systems offer a relatively new way to conduct military operations. Therefore all these aspects will evolve over the next decades.

> There does not appear to be a policy by any state not to use autonomous systems but there are policies not to use Lethal Autonomous Weapon Systems (LAWS). There are calls to ban all research which could lead to autonomous weapons by organisations such as Human Rights Watch. See, for example, Reference [19] and subsequent papers referencing it, although there have been papers criticising them [20,21]. These form part of the international debate about types of weapons which could be banned under the CCW Protocol [22] which bans certain types of weapon. As agreement is reached, new state policies will emerge and must be checked at all significant stages in the introduction of increased levels of automation into weapon control systems.

> New applications of increased automation will be developed, giving rise to new military capabilities. Many will have support applications with no legal concerns, but some will be for weapon and weapon-delivery concepts. Proposals for these concepts will need to be assessed for credibility and legality under international law. Doctrines for their use will then need to be developed so that rules of engagement can be developed for specific campaigns. New

doctrines will need to become an inherent part of the engineering design process to turn any of these concepts into products.

Organisation

Responsibility for authorisation and implementation of a decision is a key factor for any organisation. Automation in decision-making will give engineering issues especially for decisions with uncertainty in the inputs. Reliability and accuracy become key issues and will need careful specification and demonstration.

Increased automation in support tools and equipment is likely to increase efficiency, which will lead to management changes, but not necessarily new engineering problems.

Equipment, personnel, materiel and logistics

Speed of development of autonomous systems technologies will lead to pressure for rapid introduction of new capabilities. Given the long lifetime of military platforms such as ships, aircraft and tanks, there will be incremental upgrades with the new capabilities. This takes time to complete for a whole inventory of platforms so there will be platforms of nominally the same type, but with different equipment fits and capabilities. Typical problems for commanders are around ensuring that they know the capabilities of their materiel and that they can all operate together effectively in a campaign. Logistic problems centre around ensuring that maintenance and support staff and supplies are correct for the equipment that require support.

An alternative approach is to adopt a 'throw-away' approach. The UK Watchkeeper programme assumes that ever-increasing performance will be required in the surveillance UAV and ground control station so there will not be a lifetime buy or a contract for the same equipment design over the decades of the programme. Instead, a modular approach is adopted with initial procurement for a low equipment numbers to meet initial demand, with development of more advanced UAV capabilities initiated. Then procurement is for next generation modules as the original buy are withdrawn from service [23].

4.12   Systems engineering

4.12.1   Overview

There are many courses and textbooks about systems engineering available with greater detail than is given here. A comprehensive overview is given in Reference [24]. The description given here is tailored to the systems aspects of autonomous systems, especially AWS.

Traditionally, systems engineering starts with a concept phase where new customer requirements are examined and a range of concepts tested against them until a likely solution is found. In practice, the procurement authority will have discussed the new capability with potential suppliers before issuing a formal request for proposals for concept development. This means that there is usually a fairly well-developed concept covering existing and planned materiel with the missing systems and interfaces identified. That part of traditional systems engineering will have been covered by the capability engineering processes discussed in Section 4.10. Effectively, the contract will be for work starting after a concept design review held by the customer.

It is likely that military requirements for autonomous systems will draw on developments from civil applications. In these cases there will be a concept such as: can we use autonomous car technologies in armoured vehicles to reduce the crew numbers? In which case we have a concept and the SE process effectively starts at the same place, i.e. at the end of capability engineering. There may not be any architectural design, and the process may be one of designing a set of subsystems which have to meet clear and unchangeable interface specifications. In this case, the design will go straight to the processes discussed in Chapter 5.

4.12.2   Inputs to the systems engineering process

The starting point for the following description is the information produced by the capability engineering process described in Section 4.10 and requirements for the wider capabilities discussed in Section 4.11. This is likely to be the contract on the supplier, so commercial considerations mandate that all engineering aspects are part of a project under tight project control from outside the engineering organisation.

We can assume that the systems engineers will have one or more possible system level solutions to the customers' needs and a set of requirements at different levels of quantification. It can be assumed that any performance trade-offs may have been very preliminary and subjective. They will almost certainly not include considerations of the strengths and weaknesses of the supplier's organisation. Project management should be aware of these, and will insist that the design accommodates, if not exploits, them and this will influence the many choices made throughout the whole supply chain.

There are several reasons why a military requirement is not always described completely to a supplier which include:

- Operational constraints that are so obvious to someone with battle experience that they do not state them;
- The product will interface with an undelivered system that is itself evolving due to trade-offs at that supplier;
- Commercial or national sensitivities about releasing information beyond an interface specification about another part of the system of systems;
- The customer may not be aware which parameters are difficult to achieve and has put tight tolerances on all of them, but is in fact able to negotiate;
- The customer may want to leave the supplier a degree of freedom for ease of supply.

If at all possible, the systems engineers should establish contact with the customer as the quantified systems trade-offs will give results which affect the utility of the product. Without this contact, the result may be an irrelevant product as a parameter which was 'traded-out' as it seemed minor to the company may be critical to the customer. Often these discussions can lead to mutually beneficial changes to the specification, especially if the engineering solution offers a new or improved capability. Contact is especially important with autonomous systems and the associated human-machine interface issues.

The example of a Missile Attack Warning System (MAWS) in Panel 3.2 shows how a 'simple' aircrew requirement for a system which tells them that a missile is coming towards them represents a complex set of engineering trade-offs and statistical thresholds. Multiple steps in the communication chain between aircrew and designers will almost certainly lead to mistranslations of the actual need into the performance of contractual deliverables. Without direct contact, there is a high chance of rework late in the programme or even rejection of the product and poor relations between customer and supplier.

There is always an internal tension between the supplier's engineers who want to deliver something that meets or exceeds the customer's expectations and the programme controllers who have to keep costs within budgets and keep schedules which may be disrupted by redesigns due to changes in the possible solutions. A practical risk is that an engineer with direct contact with the customer will tell them that a solution to a problem can be implemented; this can inadvertently become a contractual obligation that has not been costed or approved by the commercial team. Careful and sensitive management is required to minimise this risk.

4.12.3   The systems engineering process

Noting all the above provisos, systems engineering proceeds in a logical process familiar to all engineers. There are three main parts, not including subsystem production:

1. Understand the customer requirements;
2. Develop a system design and derive subsystem requirements;
3. Subsystem design, build and test is generally considered to be part of product design covered in Chapter 5;
4. Integrating the subsystems into the final product.

The chain of processes to carry out the first two steps is illustrated in Figure 4.4. Diagrams like this which show linear, logical steps in the design process are often called waterfall diagrams. They show a flow down from high-level requirement through reviews and trade-offs to a set of subsystem specifications. In practice, the initial design will have some shortcomings so the process is iterative. The iterations are indicated by the looping arrows in the figure. As a general rule, it is cheaper and quicker in the long run to change the concept at this stage than change the detailed design of, or specification for a subsystem even if it has not been built. However, the system engineers must be cognisant of the problems for the subsystem designers. If this systems engineering work is part of the operational

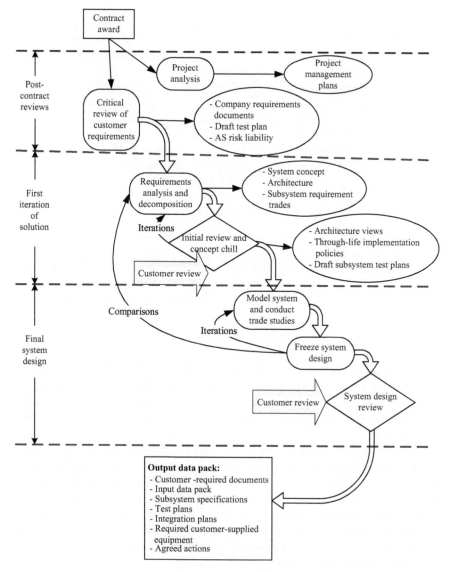

Figure 4.4   The waterfall diagram for systems engineering

analysis phase, then the 'output data pack' in Figure 4.4 forms part of the 'study products' in Figure 4.3.

It is important to realise that Figure 4.4 is only one side of a systems engineering Vee diagram. This is similar to the capability-based Vee diagram in Figure 4.1 and is shown in Figure 4.5. It assumes that the subsystems are made up

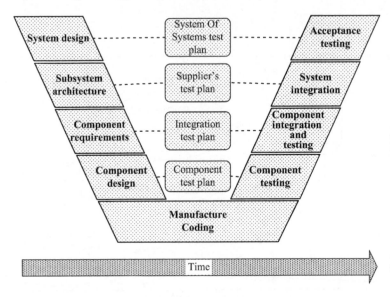

Figure 4.5   Systems engineering Vee diagram for subsystems

of modules which are themselves smaller subsystems. (This is the structure shown in Figure 5.1.) This version emphasises the design and production aspects of an autonomous systems programme. It includes the system integration part of the programme on the right-hand side and is more representative of the whole system task.

Similar diagrams can show the derivation of requirements down to module or software sub-routine level. Every company will have its own proscribed processes for all stages so only the principles are discussed here. Many contractors, including defence ones, must have programme, project and engineering design processes approved by government procurement authorities before they can be awarded a contract. Contract negotiation includes the access government has to the design and test processes. It is usual for customers, or their representatives, to be present at design reviews and critical tests.

Programme timescale pressures usually dictate that the design stages overlap with some work starting on the next stage before formal completion of the pre-ceding one. Sometimes this is forced on the team by the lead times for delivery of critical items. There is a financial risk involved and part of good systems engi-neering is minimising the technical risk and preparing mitigating contingency plans. The need for authorisation of work prior to completion of the previous phase is acknowledged by good programme and project managers, but they may not authorise it if the risk is too high. Formal review points and payment milestones must be met fully as failing to complete is unacceptable.

The steps in the process shown in Figure 4.4 are:

4.12.3.1   Post-contract reviews

- *Project analysis* – This is an essential check that the contract is exactly as expected. It results in the company's project plan which covers all aspects of delivery and includes the risk register;
- *Critical review of customer requirements* – This is the first major decision point after the company has accepted the contract. Financial commitments must be made and budgets set with associated timescales. The engineering part is to check that the full set of requirements can be met. This should have been carried out during the bid phase, but must be confirmed especially if there were rapid negotiations on detail prior to the contract being placed;

An important aspect of an autonomous system is the legal liability of the supplier for any unexpected behaviour or failure in the autonomous functions in the system when used by the customer or their customers. This is a new legal and technical area for both military and civil autonomous systems and their different legal regimes. It is different to the normal product liability as it relates to the consequences of the automated decisions and how those decisions are specified and made. Decision-making is discussed in Chapter 2. The extra requirements for integration and test are discussed in Section 5.4 as the liabilities potentially flow down the contractual chain, the cut-off point being an important if, as yet, unresolved issue.

The customer requirements will usually have been written in their terminology expressed in a way that can be interpreted by all companies bidding for the work. Some of the terminology and assumptions may not be directly transferred into the winning contractor's processes. Consequently the requirements must be expressed in a form suitable for them, i.e. they have to be 'translated' from the customer's 'language' into the company's 'language'. This may appear nugatory, but in fact it is a way of checking full comprehension and ensuring that there is no ambiguity in the use of critical words. A well-known example is that 'shall' and 'will' have very specific meanings in contracts. ('Shall' means mandatory, and 'will' does not mean that the requirement is mandatory.)

Companies relate requirements to their internal division or section that is best suited to deal with them, so the customer requirements will be re-ordered to reflect the company structure. It is important to check that none are missed, or extra ones inadvertently added, and that the sum of the 'company' requirements delivers the correct capability. Leaving requirements in customer language may give problems with the requirements flow down into the subsystems and parts of the company or subcontractors that have no contact with the customer or their needs. However, trials and tests on the right-hand side of Figure 4.5 will almost certainly be conducted against the customer's written requirements.

It is essential to start planning for the demonstrations that will be required at every payment milestone. There will be some mandatory tests, but the company must be sure that they minimise risk for the remainder of the contract.

If subsystems are supplied by subcontractors, the tests must test compliance fully. A subsystem failure to operate in the system is a major risk to the system integrator and/or supplier. Subsystem suppliers are usually contracted to provide support during system integration and tests but are not present at the customer acceptance trials unless there is a specific need.

4.12.3.2   First iteration of solution

- *Requirements analysis and decomposition* – This is the detailed analysis of the deliverable requirements and the generation of requirements for the sub-systems in the conceptual design. Compromises must be made to meet the usual size, weight and power constraints with trade-offs to meet available facilities. There will almost certainly be modelling involved with different models required for different aspects (e.g. mechanical Computer-Aided Design (CAD); network loading; Open Systems Interconnect (OSI) layer models). Regulatory standards will bring constraints as will requirements for the 'ilities' and through-life capability management;
- *Initial review and concept chill* – This is the second major decision point with agreement that the derived system concept is an acceptable basis for a final product. Acceptable definitions of 'Chill' will depend on the company and project. One definition could be that it is a design which is close enough to the final one for the programme managers to make financial commitments for the next stages of work and confirm delivery dates. It is accepted that there will be some minor design changes, but it should be possible to set accurate risk budgets to cover them.

It is good practice to hold discussions with the customer at the 'chill' stage even if they are not mandated. The customer sees progress and has the opportunity to express their views of a more detailed design and compare it with their desired capability. Commercial support to engineering is essential here as any changes will have financial implications.

4.12.3.3   Final system design

- *Model system and conduct trade studies* – This is the bulk of the systems engineering work and is one requiring considerable judgement as well as the use of formal assessment methods. It is similar to the processes in operational analysis described in Section 4.2, except that the trade-offs will be based on different criteria. The measures of merit will include measures of performance and dimensional parameters but will have others such as ease of design, pro-duction cost, delivery schedules, integration cost and risk of all types.

It is essential to recognise that there may be several iterations of the design before an acceptable 'final' one is produced. (See also Section 4.12.4 Spiral development.) The programme plan should include a first design and at least one iteration of it. However, it is foolhardy to make a quick start, producing a poor design on the basis that it will be improved by an iterative approach. It is essential to make a well thought-out first design with a comprehensive analysis of it; without this, it will be difficult if not impossible to understand its shortcomings when tested[8]. This is particularly important when hardware is involved as the time taken to produce and test it is usually a significant fraction of the project time.

Modelling of all types will be used. This will probably include simulations, but mathematical models should underpin most aspects of this work. This latter point is particularly important for the autonomous system's control systems due to the legal requirement to show not only who is in control, but also that the system provides adequate control under all reasonably predictable circumstances. It will be financially prohibitive to build a single detailed model for all aspects of the design trades so it is essential to maintain the models under rigid configuration control. The users must have a clear understanding of the assumptions in their models and how they differ from other models. Without this there will be confusion as all the models will evolve during this stage of the analysis.

One well-known tool from operational analysis that is used in systems engineering is the Red, Amber, Green (RAG) chart. This takes key measures of merit and rates them as unacceptable (Red), acceptable but may not fully meet requirements (Amber), and fully meets requirements (Green). Sometimes other colours are used such as blue to indicate exceeds expectations.

The idea is to have one senior person who understands and controls the versions of models used in each trade-study so that fair comparisons are made. Ease of design and build in the company will be given considerable weight in the studies which is not unreasonable, especially if the customer is known to expect certain 'signature' aspects of a given companies products. However, there is a danger that the final system choice will be based on internal company criteria with only a confirmatory comparison with the exact conditions of the contract. Although commercially acceptable it can be short-sighted and miss an opportunity of exceeding requirements in some aspects. These can be used either to offer extra capability for an extra cost or as trade-offs when product shortfalls elsewhere are detected later in the programme.

- *Freeze system design* – This is the date at which the final system design is chosen and comes under full company, and sometimes customer, configuration control. All models and documentation used in support of the decision must be lodged in a library. All changes necessitated by later developments will have to be formally recorded, with strict configuration control of all aspects.

It is often good practice to have a documented informal but thorough design review before design freeze so that any necessary changes can be made before full configuration control is imposed. This gives an opportunity to sweep up and review all suggested changes by different team members during the trade-offs. It also gives

---

[8]There is often an expectation to have every design 'right first time'. This is unrealistic for advanced products. The author, when a technical manager, ran a 'right second time, or an unacceptable design team'. If the team understood their design, then the failure could be explained and corrected for the second version. The team rarely had to make a third attempt.

the opportunity to seek cost-savings during the next stages. It may be the first opportunity to hold a detailed review of test and integration plans with the 'final' system design and compare it with the customer's acceptance criteria. This really is an area where 'the devil is in the detail'. The devil in this case is wicked problems causing long and expensive tests and trials and associated redesign work.

- *System design review* – The formal design review is a meeting which may last several days depending on the system complexity. Completion and formal issue of the minutes are usually a significant payment milestone. Company and customer policy will dictate the inputs and outputs, which take the form of data packs. It is virtually certain that there will be an action list to clear up issues that arose in the design review. Agreement must be sought as to whether any uncleared actions will prevent payment.

The customer will almost certainly be present at the design review with their technical advisers. A good design review has technically competent customers with a good grasp of the intended system use and through-life capability management issues. They will question contractor staff with a detailed knowledge of the system and the reasons for the decisions made during the trade studies. Specialised experts should be available if required to answer detailed questions. The role of the Chair is crucial. They must ask penetrating questions and understand the answers so that they can explore any issues and identify problems. Design reviews may spend excessive time on details and miss significant problems, so an important role of the Chair is to define less-important actions that can be cleared outside the meeting and identify methods to clear them.

4.12.4   Spiral development

Early versions of the waterfall diagram shown in Figure 4.4 did not explicitly include design iterations. When applied to software design, there was an assumption that detailed requirements and specifications with careful reviews and testing would lead to correct code being produced with very few changes needed. Despite this, iterations of all stages were found to be necessary. This could lead to cost and time overruns in projects which did not allow for them. Consequently, software engineers started to replace the waterfall process with spiral development [25].

Spiral development evolved and was also applied to some systems projects, but had to be refined, Its originator and a collaborator explained the reasons for this as well as some misunderstandings about the original concept in a report in 2000 [26]. They gave the following overview definition:

> The spiral development model is a **risk**-driven **process model** generator. It is used to guide multi-stakeholder concurrent engineering of software-intensive systems. It has two main distinguishing feature. One is a **cyclic** approach for incrementally growing a system's definition and implementation while decreasing its degree of risk. The other is a set of **anchor point milestones** for ensuring stakeholder commitment to feasible and mutually satisfactory system solutions.

The original spiral was fairly complex and a simplified version is shown in Figure 4.6. This shows its essence as starting with a requirements plan and lifecycle plan. The design then evolves with successive refinements, until the final, verified design is produced. Reference [26] changed the concept from the original by introducing wider stakeholder and lifecycle interests, the 'WinWin Spiral Model'. The intention being that the stakeholders and suppliers would both win something from the negotiations and trade-offs. This is similar in intent to 'Value-based design' in Section 13.3.2. The reader who wishes to use the spiral development model should consult the two references given here and conduct their own literature search for recent developments and applications.

One of the critical issues in all iterative processes, whether waterfall or spiral is that there must be rigidly timetabled project reviews for every iteration. Reviews of failures in testing must demonstrate that the cause is understood, and that there is a high chance of the proposed solution working. Reviews of successful tests must demonstrate that the test results were as predicted by previous analysis of the design, with any differences being understood and possible changes to the model or

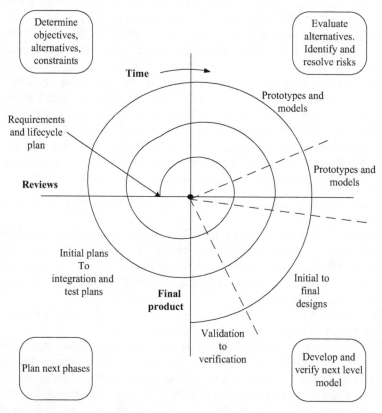

Figure 4.6   Spiral development

mathematical analysis documented. Without these outputs, a programme risk must be declared and close monitoring introduced.

4.13   Validation of the system design

In principle, validation of the proposed system design should be completely independent of the agencies and companies involved in the operational analysis described above. This is not possible in practice for complex systems for many reasons, including cost and time. It is possible to reach some level of independence by analysing the evidence with a different aim.

The aim of the operational analysis described above was to establish the 'big-picture' view of the campaign, such as the example used in Chapter 11 of achieving success in a littoral operation. The results reflect this, giving for example the extra number of each type of UAV required to achieve success despite finite loss rates. It is unlikely that dimensional parameters will have been developed in any detail or accurate measure of merit assessment weights estimated at this stage. Weighting measures of force effectiveness and measures of performance could show that electro-optical sensors give unacceptable surveillance UAV losses for day-night-all-weather operations. The operational analysis could not show that a synthetic aperture radar module can be fitted to the UAV, only that if fitted, an acceptable solution is possible. It requires detailed analysis of the dimensional parameters to specify the radar requirements. The level of autonomy of target identification will be a crucial issue as there is likely to be more advanced object identification algorithms for electro-optical sensors than for radar sensors. The consequence may be a requirement of off-board analysis of radar imagery which is not needed in the electro-optical case, reducing the autonomy level of the UAV.

The capability engineering process described in Sections 4.9 to 4.11 develops the detail that is needed to draw up procurement specifications. It can also be used as a method of validating the requirements as a parallel activity:

• Detailed architectures are developed using the preferred framework agreed with the customer;
• When a stable architecture is achieved, it can be simulated in as much detail as budgets allow. The simulation may have to comprise several types of model for the different nodes and subsystems. Ensuring consistency across the diverse models can be achieved by configuration control and adherence to one architecture meta-model;
• The measures of force effectiveness and measures of performance are revised in the light of feasibility studies using the models;
• The importance of the different measures of merit, and hence their weighting in deriving requirements, is assessed by analysing the simulation results. The robustness of the architecture can be tested by changing values of measures of merit and examining changes in system of systems performance;
• Dimensional parameters are derived. These are reviewed by design engineers for the feasibility of implementation. The use of weightings with qualitative or

better, quantitative trade-offs of the relevant measures of merit allows technical specifications to be derived which maximise the preferred military capabilities;

- The models and simulations can then be rerun with dimensional parameters based on the technical requirement specifications that will be put out to tender.

Repeating the original analysis and war-gaming using simulations based on the final architecture with detailed models of critical nodes and modules gives an acceptable validation of the procurement model.

4.14 Post-contract award changes

4.14.1 Pre-delivery changes

A simple product should meet its original requirements without any major changes in design concept between the Request For Quotation (RFQ) and delivery. With a more complex product, there will be differences between the concept at RFQ and the delivered product for a variety of reasons.

It is inevitable that although the concept in the quotation will be capable of meeting the specification, practical issues and details in the system and subsystem design will necessitate changes to it. In addition, development times of several years will inevitably lead to technology changes in the design, and necessitate further changes. The result is that there are numerous iterations of the design and specifications which necessitate changes to the procurement contract or contracts. This is understood by both customer and supplier, but cost and delivery time should not be relaxed[9]. The engineering implications of the contract changes must be continually assessed by the systems teams, the knock-on effects identified and project plans amended.

All design changes in the pre-delivery phase should be covered by contract changes, and the detail of the integration and test programmes must include them. Both sides will aim for delivery of a product that can be demonstrated to meet a 'final' issue of the requirement specification. The test philosophy however will have been set at an early stage of the design phase. The generic problem for the supplier is to demonstrate to the procurement authority that the delivered system meets specifications after all the design and specification changes have occurred.

The specific problems for autonomous systems concern both the speed of advance in AI and autonomy technologies and the lack of clarity in defining autonomy and related words. All military system of systems will have to meet legal requirements. The programme will have to generate test and trials evidence for successful legal reviews and to guide campaign planners in setting rules of engagement. The solution is to plan system demonstrations matched to requirements derived from the four underpinning tenets of IHL. Setting requirements from them is explained in Section 7.9 and Chapter 10.

[9]There is a 'conspiracy of optimism' where both customer and supplier rely on an increase in available budgets over the relevant time period. This can happen with major government procurement programmes where the project either has prestige or is too big to cancel.

Increased automation in specific subsystems and greater integration with other assets in the military inventory present a risk of inadvertent automation of both decisions and the production of the information they are based on. One specific question is whether the increased automation is in the direct decision chain for a lethal decision or not. The next question is whether increased automation affects any of the information presented to the human decision-maker in a way which could affect their decision. These two questions, which draw on the work in Chapter 10, can be used as starting points for devising test plans.

4.14.2   Setting-to-work changes for military systems

Setting-to-work is the process that follows customer acceptance of the product. The first delivery is likely to be an early model with limited capabilities. This is to give time for any necessary or desirable changes revealed during the process to be incorporated in later models. It is analogous to the alpha and beta testing of new software. Setting-to-work is likely to be a separate contract to the design and integration ones. The complexities of the deliverables and their interaction with the customer's system of systems usually necessitate that there is a dedicated team comprising supplier and customer staff that are co-located, with close collaboration on all issues. The systems engineers involved must have a deep knowledge of all parts of the delivered and customer system of systems.

Setting-to-work gives the ideal opportunity to generate the evidence for Article 36 reviews as they must be based on actual weapon system performance. The military user also wants to see how the actual, as opposed to specified, system behaves in an operational environment, operating with other assets. The human–machine interfaces will be almost certainly be in-service ones, so there is the opportunity to fully examine the combination of humans and machines making the decisions. The specific questions for the autonomous system and its system of systems concern the clarity of the division between the two and the authority level for each decision-making node in the C2 chain.

Appendix A4   Through-life-capability-management terminology

Section 4.7 discusses the wider capabilities of through-life capability management using the UK and US terms TEPIDOIL and DOTMLPF-P, respectively, for the Defence Lines Of Development DLODs:

- TEPIDOIL: Training, Equipment, Personnel, Information, concepts & Doctrine, Organisation, Infrastructure and Logistics. The UK has also added Interoperability as an overarching capability.
- DOTMLPF-P: Doctrine, Organization, Training, Materiel, Leader development, Personnel, Facilities and Policy.

These are defined in Tables A4.1 and A4.2, taken from the websites given in the tables, accessed on 2 November 2018.

Table A4.1   UK definitions of Defence Lines Of Development (UK MOD Crown Copyright)

UK (TEPIDOIL)	Definition from https://www.aof.mod.uk/aofcontent/general/sg_dlod.htm
Training	The provision of the means to practise, develop and validate, within constraints, the practical application of a common military doctrine to deliver a military capability
Equipment	The provision of military platforms, systems and weapons, (expendable and non-expendable, including updates to legacy systems) needed to outfit/equip an individual, group or organisation
Personnel	The timely provision of sufficient, capable and motivated personnel to deliver defence outputs, now and in the future
Information	The provision of a coherent development of data, information and knowledge requirements for capabilities and all processes designed to gather and handle data, information and knowledge. Data is defined as raw facts, without inherent meaning, used by humans and systems. Information is defined as data placed in context. Knowledge is Information applied to a particular situation
Concepts and Doctrine	A Concept is an expression of the capabilities that are likely to be used to accomplish an activity in the future. Doctrine is an expression of the principles by which military forces guide their actions and is a codification of how activity is conducted today. It is authoritative, but requires judgement in application
Organisation	Relates to the operational and non-operational organisational relationships of people. It typically includes military force structures, MOD civilian organisational structures and Defence contractors providing support
Infrastructure	The acquisition, development, management and disposal of all fixed, permanent buildings and structures, land, utilities and facility management services (both hard and soft facility management) in support of defence capabilities. It includes estate development and structures that support military and civilian personnel.
Logistics	Logistics is the science of planning and carrying out the operational movement and maintenance of forces. In its most comprehensive sense, it relates to the aspects of military operations which deal with; the design and development, acquisition, storage, transport, distribution, maintenance, evacuation and disposition of materiel; the transport of personnel; the acquisition, construction, maintenance, operation, and disposition of facilities; the acquisition or furnishing of services, medical and health service support.
Interoperability (overarching theme)	In addition to the DLODs, interoperability is included as an overarching theme that must be considered when any DLOD is being addressed. The ability of UK Forces and, when appropriate, forces of partner and other nations to train, exercise and operate effectively together in the execution of assigned missions and tasks. In the context of DLOD, interoperability also covers interaction between Services, UK Defence capabilities, Other Government Departments and the civil aspects of interoperability, including compatibility with Civil Regulations. Interoperability is used in the literal sense and is not a compromise lying somewhere between integration and de-confliction.

Table A4.2   US definitions of Defence Lines Of Development

US (DOTMLPF-P)	Definition http://www.acqnotes.com/acqnote/acquisitions/dotmlpf-analysis
Doctrine	The doctrine analysis examines the way the military fights its conflicts with emphasizes on maneuver warfare and combined air-ground campaigns to see if there is a better way that might solve a capability gap.
	• Is there existing doctrine that addresses or relates to the business need? Is it Joint? Service? Agency?
	• Are there operating procedures in place that are NOT being followed which contribute to the identified need?
Organizations	The organization analysis examines how we are organize to fight; divisions, air wings, Marine-Air Ground Task Forces and other. It looks to see if there is a better organizational structure or capability that can be developed to solve a capability gap.
	• Where is the problem occurring? What organizations is the problem occurring in?
	• Is the organization properly staffed and funded to deal with the issue?
Training	The training analysis examines how we prepare our forces to fight tactically from basic training, advanced individual training, various types of unit training, joint exercises, and other ways to see if improvement can be made to offset capability gaps.
	• Is the issue caused, at least in part, by a complete lack of or inadequate training?
	• Does training exist which addresses the issue?
Materiel	The materiel analysis examines all the necessary equipment and systems that are needed by our forces to fight and operate effectively and if new systems are needed to fill a capability gap.
	• Is the issue caused, at least in part, by inadequate systems or equipment?
Leader Development (or Leadership and Education)	The leadership and education analysis examines how we prepare our leaders to lead the fight from squad leader to 4-star general/admiral and their overall professional development.
	• Does leadership understand the scope of the problem?
	• Does leadership have resources at its disposal to correct the issue?
Personnel	The personnel analysis examines availability of qualified people for peacetime, wartime, and various contingency operations to support a capability gap by restructuring.
	• Is the issue caused, at least in part, by inability or decreased ability to place qualified and trained personnel in the correct occupational specialties?
	• Are the right personnel in the right positions (skill set match)?

(Continues)

Table A4.2 (*Continued*)

Facilities	The facilities analysis examines military property, installations and industrial facilities (e.g. government owned ammunition production facilities) that support our forces to see if they can be used to fill in a capability gap. • Is there a lack of operations and maintenance? • Is the problem caused, at least in part, by inadequate infrastructure?
Policy	Any DOD, interagency, or international policy issues that may prevent effective implementation of changes in the other seven DOTMLPF-P elemental areas

References

[1] Loerch A.G., and Rainey L.B. *Methods for conducting military operational analysis*; Publ Military Operational Research Society. 2007.

[2] Smith C.J., Oosthuizen R., Harris H., Venter J.P., Combrink C., and Roodt J.H.S. 'System of systems engineering – the link between operational needs and system requirements'. *South African Journal of Industrial Engineering*. 2012, vol. 23(2): pp. 47–60.

[3] Oxenham D. 'The next great challenges in systems thinking: A defence perspective'. *Civil Engineering and Environmental Systems*. 2010, vol. 27(3): pp. 231–241.

[4] Connable B., Perry W.L., Doll A., Lander N., and Madden D. *Modelling, Simulation, and Operations Analysis in Afghanistan and Iraq: Operational Vignettes, Lessons Learned, and a Survey of Selected Efforts*. RAND Corporation, 2014. Available at: https://www.rand.org/pubs/research_reports/RR382.html. [Accessed on 9 June 2018].

[5] Cognexus Institute website http://www.cognexus.org/id42.htm [Accessed on 9 June 2018].

[6] INCOSE Z-Guides, Sheet Z8 *System architecture*, V1. INCOSE UK 2010.

[7] Zachman J. 'A framework for information systems architecture'. *IBM Systems Journal*. 1999, vol. 26(3): pp. 454–470.

[8] Hause M., Bleakley G., and Morkevicius A. 'Technology update on the Unified Architecture Framework (UAF)'. *26th Annual INCOSE International Symposium (IS 2016) Edinburgh, Scotland, UK*. July 18–21, 2016.

[9] Holt J. *UML for systems engineering: Watching the wheels*. 2nd edn; Professional applications of computing series, no 4. London: IET; 2004.

[10] Holt J., and Perry S. *SysML for systems engineering*. Professional applications of computing series, no 7. London: IET; 2008.

[11] Pagotto J., and Walker R.S., 2004, in Suresh R. (ed) 'Capability engineering: transforming defence acquisition in Canada'. *Proc. SPIE 5441, Battlespace*

Digitization and Network-Centric Systems IV (SPIE, Bellingham, WA, 2004), pp. 89–100.

[12]   The Technical Cooperation Programme, Joint Systems and Analysis Group, Technical Panel 3. *Guide to Capability-Based Planning*. This was accessed on 24 March 2018 through the following link obtained by Googling TTCP and 'Guide to capability-based planning'. https://www.google.com/url?sa=t&rct=j&q=&esrc=s&source=web&cd=5&cad=rja&uact=8&ved=0ahUKEwigvszOiIXaAhVRlxQKHXJpBh8QFghOMAQ&url=http%3A%2F%2Fit4sec.org%2Farticle%2Fguide-capability-based-planning&usg=AOvVaw2CoaOXuOPtr9V2UrKUarxt

[13]   Lam S., Poursina S., and Spafford, T. *An Overview of the CapDEM Integrated Engineering Environment, Defence R&D Canada, Technical Memorandum No TM 2005-118*; 2005.

[14]   NATO Code Of Best Practice for command and control Assessment, 2002, available at http://internationalc2institute.org. This has two associated summary documents for analysts and decision-makers.

[15]   Tolk A. *Engineering principles of combat modelling and distributed simulation*. Hoboken, NJ: Wiley; 2012.

[16]   UK MOD Defence Concepts and Doctrine Centre (DCDC). *The future character of conflict* Part of the MOD Global Strategic Trends Programme. 2010.

[17]   UK MOD Defence Concepts and Doctrine Centre (DCDC) *The future operating environment 2035*. Part of the MOD Global Strategic Trends Programme. 2015.

[18]   Moffat J. 'Modelling human decision-making in simulation models of conflict'. *International Journal of Command and Control*. 2007, vol. 1(1): pp. 31–60.

[19]   Human Rights Watch. *Losing humanity. The case against killer robots.* November 2012. Available from https://www.hrw.org/report/2012/11/19/losing-humanity/case-against-killer-robots#page [Accessed on 26 March 2018].

[20]   Schmitt M., *Autonomous weapon systems and international humanitarian law a reply to the critics*. Harvard College Report 2012. Available at http://ssrn.com/abstract=2184826

[21]   Gillespie A.R. 'Humanity and lethal robots – An engineering perspective' in Verdirame G. *et al.* (Forthcoming) SNT Really Makes Reality: Technological Innovation, Non-Obvious Warfare and the Challenges to International Law, ESRC funded research project (ES/K011413/1) 2018.

[22]   UN convention on prohibitions or restrictions on the use of certain conventional weapons which may be deemed to be excessively injurious or have indiscriminate effects. 1980.

[23]   Thales document *Watchkeeper Unmanned Aircraft System (UAS)*; https://www.thalesgroup.com/sites/default/files/asset/document/3862_tha_watchkeeper_brochure_2013_pdf [Accessed on 26 March 2018].

[24]   The MITRE Corporation. *Systems engineering guide*. Bedford, MA; Mclean, VA, USA; 2014.

[25]   Boehm B.W., 'A spiral model of software development and enhancement'. *Computer*, 1988, vol. 21(5): pp. 61–72.

[26]   Boehm B. and Hansen W.J., *Spiral development: Experience, principles, and refinements* (No.). Carnegie-Mellon University, Pittsburgh PA Software Engineering Inst., Report no. CMU/SEI-2000-SR-008, 2000.

Chapter 5

Engineering design process

5.1 Introduction and overview

Chapter 4 showed how a market need or military capability is defined and turned into a set of functional requirements that can be delivered to a customer by a company. The systems engineering Waterfall Diagram in Figure 4.4 illustrates the process from contract award to a frozen system design. This chapter describes how a typical engineering company will turn the subsystem requirements in the Output Data Pack in Figure 4.4 into a final product that has been tested against those requirements and integrated into the larger system of systems.

This chapter, as with Chapter 4, gives the general principles common across the engineering sector. The application is for a subsystem which is a major part of a larger military infrastructure such as a major platform[1] or an improved communication system. Attention is restricted to subsystems with increased autonomy in one or more or their major functions in order to identify the type of problem which can be anticipated for autonomous systems. It is assumed that even if the ethical aspects of the design are encapsulated in law and regulations, they may not all flow down the traditional engineering design processes. Chapters 6, 7, 10 and 13 explain the legal issues in more detail for both military and civilian applications. Those chapters also indicate that the legal frameworks will almost certainly change in the future.

The meaning of the terms 'system of systems', 'system' and 'subsystem' depend on the context as it can be argued that any system is only a subsystem of a larger system. In this book, the terms are shown in Figure 5.1. Systems of systems are discussed in Section 4.2.1 and taken to be the customer's infrastructure, which the delivered system must fit into. The infrastructure taken here is the military one of a coalition conducting a military campaign against an opposing force. However, the design, test and integration issues are the same for any system fitting into a larger one. The subsystems are the major parts of one system such as an aircraft, ship or armoured vehicle. The next level of decomposition is called a module throughout this book for simplicity.

Systems, subsystems and modules should all have a set of performance requirements and interface definitions which form the basis of contracts for their

---

[1]The word 'platform' is use here in its generally accepted military meaning of a single large asset which carries other military systems or weapons. Examples are aircraft, ships, tanks and submarines.

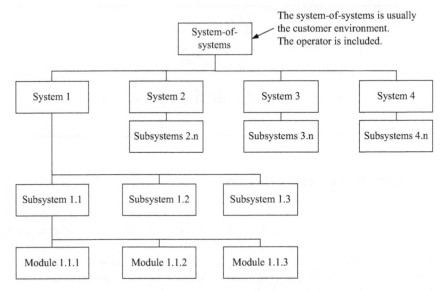

*Figure 5.1   Illustration of terminology: system of systems, system, subsystem and
module. The numbers are indicative of the nomenclature used to
identify a particular item*

delivery. Military system-level contracts are usually from the nation's procurement
agency, but they may be between companies or between divisions of a major
company. There will be contracts at every level shown in Figure 5.1. These may be
at any level from large subsystems down to subcontracts for small items, such as
consumables, and services.

The system contracts are placed on the basis of the need being satisfied by one
of the following three main processes:

1. A new system, platform or set of infrastructure;
2. Upgrades to old systems, platforms or infrastructure;
3. Increased integration of assets to give increased capability, but potentially with
   increased complexity as well.

All three will necessitate changes to some part of the system of systems, and
have implications for logistics and support. Costs for these should not be ignored at
any procurement stage, but design decisions can easily increase them. In principal,
inadvertent extra costs can be avoided by detailed specifications and/or the
designer having a good understanding of the wider aspects of the design and its
intended use. Some of these costs are readily foreseen such as the need to dredge a
harbour entrance for a new larger aircraft carrier but some may not, such as the
need for extra intelligence analysts arising from the large volumes of surveillance
data from unmanned aircraft. The last decade has seen increasing use of 'service'
contracts which are for the provision of a capability or performance specification
over a timescale which is a significant fraction of the lifetime of the system or

system of systems. The cost of support then falls on the supplier who imposes a requirement for low all-life costs on their design team. Often-quoted example is the 'power by the hour' contracts for aircraft engines or helicopter rotor blades.

This chapter makes the assumption that there is one contract for a stand-alone system, that it is either an autonomous system or has autonomy in at least one of its major functions, and that it can be tested before delivery. The tests may not be exhaustive but should be reasonably comprehensive. It is likely that an autonomous system will have a range of responses and will be operating in dynamic complex environments that cannot be fully replicated for tests. As a result, it will be very difficult, if not impossible, to define an exhaustive test set for all eventualities.

Although it is assumed that the design is a result of one large contract, the process will be similar for products developed in anticipation of a market need, whether for a military or civilian market. There will always be a requirement and design specifications[2], but these will be created within the company. These designs are often funded without customer support and are usually known as Private Venture (PV) funded projects. The source of funding may be by the company investing part of its profits, shareholder investment or venture capitalists. Given the speed of developments for autonomous systems, there will be many PV designs funded by a wide range of investors. There is a consequent need for a return on these investments and pressure to exploit the result, whether as a stand-alone product or as part of a larger system. Financial pressures can add to the tensions within a company. These are explained in Section 5.2 and can include a short-term need for a return to investors that prevents longer-term investment in the more advanced technologies needed by the strategic plans.

Immediately after, or sometimes before final customer acceptance, the product(s) have to be integrated into the customer's existing system. Military systems do not immediately go into service but have a 'setting to work' phase. This is where specialist units operate the system with the existing military infrastructure. The time taken for this phase depends on the complexity of the product and its priority. It can vary from a few days for an Urgent Operational Requirement or Need[3] (UOR or UON) for a small item to years for a major platform such as a new fast jet or first-in-class ship. This is analogous to an extended test period and initial operations for civilian applications.

Financial and design control are imposed through the management processes in the organisations with responsibilities for delivery. This chapter gives an outline of their general principles and the problems posed to them by autonomous systems. They are illustrated by the example of a littoral operation in Section 11.4.

The civilian ethical, legal and regulatory product requirements discussed in Chapter 6 are usually sufficiently well developed and understood to be an integral part of most companies' processes for current products. Autonomous systems are

---

[2]A requirement is what is wanted, the set of design specifications express what is delivered. The two should be the same, but it takes considerable effort to ensure this is the case.

[3]There is a minor difference in terminology here, the US use UON whilst the UK use UOR for a capability that is needed by the nation's forces whilst conducting an operation.

introducing new problems and are widely discussed even if adequate solutions are not yet in place. Chapter 13 suggests some approaches to solutions.

Military systems have the additional requirement to meet International Humanitarian Law (IHL), outlined in Chapter 7, and potentially face extensive 'Article 36' reviews discussed in Chapter 11. Design processes evolve over a long time and with experience from many products. The 'stove-pipe' philosophy explained in Section 4.5 led to the demonstration of adherence to IHL being carried out post-delivery, by military staff and lawyers. Consequently design processes to meet IHL requirements are a relatively recent development and are not necessarily fully embedded in processes, regulations and standards.

Unanticipated post-delivery requests for extra information or even new trials programmes to demonstrate adherence to IHL can be very expensive to satisfy and lead to delays to entry into service. This chapter describes widespread engineering processes and identifies areas where autonomous systems raise new problems. Design organisations should address these problems even if they are not explicitly included in contractual terms. The most critical application of IHL to system design is to systems and functions directly involved in targeting and weapon release, as shown in Chapter 8.

5.2 Management control

5.2.1 General principles

This section presents a synopsis of good engineering management principles following contract award. The intention is to show how the trade-off between time, cost, performance and quality of all aspects of the product is achieved by the hopefully constructive tension between the project and technical managers in the company. The simplistic view is that the programme or project manager is responsible for time and cost whilst performance is the technical manager's responsibility. The tension should arise because the project manager's aim is to deliver on time with minimum costs, whilst the Technical Manager wants to deliver the highest possible performance within the available budget. The role of the quality assurance (QA) organisation is to ensure that contractual and procedural processes are followed and regulatory standards are met. All three should understand the need to identify risks and agree mitigation plans. The company and its suppliers should be following ISO 31000 [1] for risk management. The main vehicle for managing risks in most engineering organisations is the risk register, which is discussed in Section 5.2.7.

Risks, their assessment and mitigation when they occur are often the source of conflict in an organisation. However, mutual agreement between managers on the mitigation plan when a risk occurs, and implementing it can be the source of great mutual satisfaction. The nature of technical risks will be a combination of technology problems and the company's structure. The latter is due to the fundamental problem that the project's Work Breakdown Structure must be a function of the company's organisation. The management structure of engineering, and other, companies is nearly always a matrix of departments and sections, with the management drive being project-based or skill-based[4], with cross-matrix efficiency-based coordination as necessary.

Autonomous systems add an extra level of complexity to the management responsibilities due to the current lack of clarity about legal responsibilities when the product fails in service. This is one area where the position for military autonomous systems is clearer than for civil systems as IHL and its existing international treaties ensure that new autonomous systems included in weapon systems must have 'Article 36' reviews at almost every major milestone from concept to entry into service.

It is not yet possible to give authoritative statements about the designers' legal responsibilities in most autonomous system applications. Chapter 12 gives some guidance for weapon systems where there are clear lines of authority and responsibility, whilst Chapter 13 makes proposals to clarify them for civilian systems. Those chapters should give a designer an idea of the type of questions that should be asked so that he or she can discharge their professional responsibilities. These are, in addition to, or may complicate application of existing standards and regulations for specific industries and environments.

5.2.2 *Project management*

Responsibility for the project costs and delivery time lie with the project manager.[5] He or she controls the overall budget, its division and allocation across the company. There are at least two key aspects to their role: liaison with the customer and/or programme manager and management of all aspects of the deliverable, including the design. They usually have considerable freedom of action, but are subject to scrutiny by the Quality Assurance (QA) organisation at all times, as is the rest of the company.

As stated earlier, there is usually some level of tension between the project manager and those with responsibilities for other aspects of the project such as the design quality. Autonomous systems are likely to be part of a strategic plan to expand into new markets or keep ahead of competitors in the existing ones. Therefore it is essential that the programme, project and technical teams have a clear understanding of the strategic issues. This includes having a mutually agreed understanding of basic terms such as autonomy and automated. Misunderstandings can easily occur as many terms can be used with different definitions such as the range given in Section 3.3 and Appendix A3 for autonomy and autonomy level.

The project manager will convene, but not necessarily chair the first review meeting discussed in Section 5.2.5 below. His or her input will be the wider aspects

---

[4]A company is said to be skill-based if the management hierarchy is based on technologies with departments organised by technical skills; projects are controlled from a specialist areas which load work onto the technical departments. Project-based companies have the management hierarchy based on major projects, with projects acting as departments and technical staff assigned to that project department for a period.
[5]A project has a defined output for delivery in a specified time within an allocated budget. A programme has wider aims, usually comprising several projects and other activities, such as marketing, with a strategic aim.

of the contract. The engineering team should identify the wider implications of their proposed work. The legal framework that applies to the autonomous system will have to be recognised and understood, whether military or civilian. Civilian subsystems and modules brought into use by the military will also have to meet IHL with weapon control systems and their associated decision support systems being especially critical. These issues are discussed in Chapters 6 and 7.

5.2.3   Technical management

Responsibility for the technical quality and performance of the company's products will lie with the technical managers who are responsible to a chief engineer or technical director. They will usually be supported and monitored by a QA organisation which must have some degree of independence from local management.

Design staff will usually want to exploit new technologies in their designs. A bid or new contract is an opportunity to do so. Local technical managers have to weigh the opportunities and benefits that will arise before completing a cost/benefit analysis with the project manager. Much of this process will be based on engineering experience, but all companies will have formal methods to carry this out in order to ensure that the process is comprehensive and as unbiased as practical. It is also essential that the technical managers understand the programme managers' problems and vice versa. Without this, the essential mutual respect may be absent and unnecessary problems arise.

Technical managers in all companies should ensure that they understand their design responsibilities under the relevant legal framework. These responsibilities are not necessarily clear for autonomous systems as the technologies and laws are evolving, so they should be addressed at the bid stage. If this was not possible, clarification should be sought during the contract review after award, but before its acceptance. A process such as that discussed in Chapters 10 and 13 could help the clarification, but however it is achieved, technical managers' responsibilities must be formally stated in the company's procedures. It is probable that the responsibility and consequent liabilities for system and system-of-systems performance problems can be resolved by an agreed set of tests and trials during integration and customer acceptance tests. This is especially true for autonomous weapon systems because of the need under IHL to be able to allocate responsibility for their performance when in operation, some of which will lie with the designers.

5.2.4   Technology Readiness Levels (TRLs)

Customer expectations can be unreasonably raised by claims that a particular technology is well developed and can be introduced into products with little risk in a short time. Public perceptions can also exacerbate these claims with little critical review, increasing military users' expectations. There may be an assumption that facilities in consumer products such as computer games or mobile phones can be implemented directly into military systems.

One of the best quantitative means of assessing these claims, and estimating the risks associated with introducing new technologies into products, is to use TRLs which were developed by the NASA in the 1990s. A retrospective view of their development is given in Reference [2]. Modified definitions of the TRLs have been defined for different purposes such as space, defence, oil and gas, and a more generic set by the European Union (EU). Some relevant definitions are given in Appendix A5 to this chapter. When used in a balanced way they provide a powerful management tool for all aspects of research, development, design and support activities.

There have been attempts to apply TRLs to software development with varying degrees of success. The report by Blanchette *et al.* [3] gives some of the problems due to the differences between software and hardware and the different risks in product development.

TRLs were rapidly followed by other readiness levels such as System, Integration and Manufacturing Readiness Levels (SRL, IRL, MRL). See, for example, Sauser *et al.* [4] for the early proposals and a mathematical approach to combining them. The majority have the least-developed level as 1 and the most developed as 9. Ross [5] gives a tutorial-type guide to some of the readiness levels and explains their use for products involving diverse technologies. The references in Ross' paper indicate the wide range of academic work developing readiness levels and the mathematical relationships between them. Austin and York [6] show a detailed mathematical application for assessing the readiness level of a system.

UK MOD uses one definition of TRLs across most projects, but takes a different approach to SRLs. They use an SRL matrix which is specific to an individual project or programme. There is a SRL assessment tool available from the UK MOD Defence Gateway [7]. This is an Excel spreadsheet which gives a useful way for assessing the overall readiness level of programmes without losing vital detail about risks from individual technologies. The summary sheet from the tool is shown in Table 5.1. Programme or project managers use eight systems disciplines plus one 'project-specific' discipline. These become the rows of the summary sheet which has 9 readiness level columns. Each matrix element is linked to a lower-level sheet which has detailed definitions for assessment and scoring of the project. Final assessment is by a RAG analysis[6].

The advantage of the multi-discipline method is that it gives a method of identifying system problems and their locations rather than one summary number which gives little or no management detail for corrective action. It is possible to use different definitions for SRLs, matched to that discipline in their project. It also recognises the practical problem that a large programme can have different SRL values in the different disciplines but still be ready to pass a major milestone or procurement gate.

Details of the various readiness levels and their relationships need not concern us here; what does matter is that the management organisation concerned can

---

[6]Red Amber Green (RAG) analysis is described in Section 4.12 as: *This takes key measures of merit and rates them as unacceptable (Red), acceptable but may not fully meet requirements (Amber), and fully meets requirements (Green). Sometimes other colours are used such as blue to indicate exceeds expectations.*

Table 5.1   System readiness level summary matrix for assessing a system (Taken from UK MOD Defence Gateway, UK MOD Crown Copyright)

Project plans need to be updated		1	2	3	4	5	6	7	8	9
System engineering drivers		☐	☐	☐	☐	☐	☐	☐	☐	☐
Systems disciplines	Training	☐	☐	☐	☐	☐	☐	☐	☐	☐
	Safety and environmental	☐	☐	☐	☐	☐	☐	☐	☐	☐
	Reliability and maintainability	☐	☐	☐	☐	☐	☐	☐	☐	☐
	Human factors issues	☐	☐	☐	☐	☐	☐	☐	☐	☐
	Software	☐	☐	☐	☐	☐	☐	☐	☐	☐
	Information systems	☐	☐	☐	☐	☐	☐	☐	☐	☐
	Airworthiness	☐	☐	☐	☐	☐	☐	☐	☐	☐
	Project specific	☐	☐	☐	☐	☐	☐	☐	☐	☐
	Maritime platforms	See the special guidance								

evaluate the maturity of a project and technology by assessing its current development stage in any application where it is in use and assess its maturity for their application. If the technology is in a Research and Development (R&D) establishment, it is probably at TRL4 or below and SRL1, which implies high programme risk and large development costs. A technology which is in production in one industry will be at TRL8, but if the environment is very different in the second industry, the level could be assessed as low as TRL4 or 5 and SRL of 2. Conversely it could be at TRL 4 but at SRL 5 or 6 if all interfaces use the same interface standards and drivers as the new system. Critical engineering reviews of technologies for new products provide a basis to plan the steps needed to introduce it into the second industry.

Well-established companies will have rules-of-thumb about the costs and typical time to progress through the TRLs from TRL2 to TRL8. For defence, as well as some other industries such as pharmaceuticals, the time for this is typically measurable in years or even decades. There are always challenges to technologists to shorten technology insertion times without increasing costs. Small- and medium-sized enterprises (SMEs) may not have rules-of-thumb. It is important that the TRL assessment is followed by detailed IRL and SRL assessments.

It is worthwhile noting that TRLs give a useful method to assess the credibility of some of the claims made for new technologies when applied to autonomous weapon systems. The reader should review the maturity of the demonstrated capability and its relevance to defence. They should then consider the developments needed to integrate it into the hard, soft and firmware of a practical weapon system and finally how it will operate in a defence infrastructure with its defined command and control structures and processes. A similar consideration of technologies for autonomous cars is also illuminating.

Ross, in his 'Background section', makes the important point that should never be forgotten in technology and system integration:

These approaches emphasize that the interfaces between subsystems are every bit as important as the subsystems themselves, and that no system can be deemed ready for deployment based on the module technologies alone.

Summarising, the various readiness levels should be used as management tools to identify problem areas in a complex programme involving disparate technologies and manufacturing processes at widely differing maturities. It is very easy to incorrectly read across a TRL from one application to another. They should not be used proscriptively, but rather as a useful tool to guide both project and technical managers to the problem and higher risk areas during product design and development. In the author's experience, the error in assessing any readiness level is at least ±0.5.

5.2.5 Initial critical review

The first step following contract award is a critical review of the customer's requirement and/or specifications, the proposed design, the costs and timescale to deliver the contracted deliverable. It is relatively common for the contracted specification to have some variations from the original invitation-to-tender. These should have been agreed by all relevant parts of the company in order to win the contract, but the detail may not have been considered in depth by the designers. There may also be some complications which were overlooked, or simplifying assumptions made during the bid, that must now be clarified and the risk assessed[7].

The initial critical review may take several days by a group of staff as it should have representatives from all relevant areas, not just those with direct responsibility for the deliverable outputs.

There are numerous aspects of every type of design which must be considered at the initial review. The translation of a customer contract into company requirements for all departments and manufacturing facilities is a complex issue. It must be thought through in detail and continuously monitored as the project progresses and the design evolves. Only those issues which are relevant to automated products in general, and weapon systems in particular, are considered here. The main areas reviewed are discussed under their headings in the following Sections 5.2.5.1–5.2.5.5.

5.2.5.1 Customer need

We assume that the customer is looking for a high level of autonomy from the product. This probably implies that they may not initially have known the autonomy level that is actually possible, resulting in detailed negotiations prior to contract award. If these did not occur, or were incomplete, the customer requirements may not be specified as a mutually consistent set of specifications. Any shortcomings must be identified at the initial review and methods of dealing with them agreed, preferably with flexibility in some aspects of the contractual arrangements.

---

[7]One Engineering Director used to inform the author, an engineering manager at the time, that a new contract was **NOT** an opportunity for the engineering organisation to make a massive loss for the company. ☺

It is also likely that the customer will want to incorporate more autonomy into the product, probably in short timescales and in response to both new operational needs and technology developments in other fields. This offers both sides the possibility of flexible management of problems provided the project demonstrates good progress and trust develops between all parties.

Typical questions which must be considered are:

1. Are all contracted design standards known and already in use? Currently there are very few standards for autonomy, so even definitions of the terms used in the specifications may be problematical. This makes the next question particularly important;
2. Is the requirement understood? This is a double-edged question. There is the requirement exactly as specified in the contract but this is the customer's interpretation of their need and so will have gaps and perhaps errors. The second factor is how the product will be used as part of their overall inventory and in use. The latter is difficult to specify and often has unwritten requirements. (The materiel used in a military action is known as the order of battle and has its own unique requirements.) Numerous design decisions will affect the way the product can be used and may prevent its use as envisaged by the military staff who wrote the original requirements;
3. How does the operator interact with the system? This is critical for any automated system and should be the subject of a special in-depth review;
4. What is the system of systems that the product will be used in, and where is that in the CADMID cycle[8]?
5. What post-delivery support is required and who is responsible for it? This is an area where the company's business policy is important as support is often more profitable than design. The trend is for customers' to procure a long-term capability instead of a simple product so the design company is also responsible for support and upgrades. A complex product, such as a highly autonomous system, will require in-service upgrades for changing requirements, obsolescence and technology advances offering new capabilities;
6. What supporting equipment or information will the customer supply during the contract duration? These are usually known as government or customer furnished equipment and government or customer furnished information;
7. Does the design team understand how the product will be used? It is this question that is most likely to identify legal concerns: civil liabilities for civilian applications, and IHL requirements for military applications.

5.2.5.2   Delivery criteria

This is arguably the most important consideration for any organisation. Unless achievable delivery criteria are agreed in advance, the supplier cannot claim payment, and the customer may never receive a satisfactory product. In practice, for large complex systems with long delivery timescales, there are many

---

[8]The Concept, Assessment, Development, Manufacture, In-service and Disposal (CADMID) cycle is described in detail in Section 5.4.2.

intermediate milestones and payments. In addition, changing requirements and budgets will give changes in the specification of the final deliverable, with associated changes to test and trials plans. Principle questions are:

1. What are the acceptance criteria in the contract? These could be trials, factory tests, successful integration into a larger system, simulation results with validation of the models used. Control systems with defined transfer functions should not be a problem. However, as both the control system and its operating environment become more complex, the acceptance criteria become more difficult to quantify and testing becomes more expensive. It is common for the acceptance criteria to be factory tests performed with customer representatives present. The tests will probably include the use of models, provided these are accepted as valid by the customer, or may be customer-supplied. Integration into the system of systems is then the subject of a different contract, agreed when the scale of the task can be assessed reasonably accurately. This is one area where the IHL requirements can be inserted through acceptance trials. The government procurement authority can specify that correct operation of the product during its 'Article 36' trials plan as an acceptance criterion;

2. Is there a risk of dispute about the product's performance when it is integrated into the system of systems? If so, is there a way to resolve them? This is a common problem area as there may be teething problems when any two new systems are brought together. Solutions may include a joint team of staff from several companies under the supervision of the end-customer. Very often the government will have its own technical experts who will make the decisions about the acceptability of the system's performance at the delivery date;

3. One solution is a phased delivery plan, with part payment following delivery of a system which has passed agreed factory tests. There is then a phase which could be the military setting to-work, or an equivalent civilian one with the system-of-systems integrator.

5.2.5.3 Design requirements

If the customer requires a high degree of autonomous behaviour from the product, it is likely that the design will have to incorporate novel techniques which represent a risk to a successful product and project. Risk is always converted into cost at the bid stage of a contract. Minimising these costs is an opportunity for flexibility and risk-sharing arrangements. Risk will be a key consideration in answering the following questions:

1. Are the interface specifications clear?
2. Will the design proposal which formed the basis of the bid actually meet the contracted requirements? If not, what must be changed?
3. Can the various functions making up the final design be unambiguously specified?
4. Which functions are deterministic? These functions will be amenable to standard design processes and have clearly defined inputs and resultant outputs. Therefore they can be tested rigorously and unambiguously;
5. Are there any non-deterministic functions? If there are, they represent a large risk both for specifying and testing them;

6. Which functions have a range of relatively undefined or unpredictable inputs? These are likely to be the systems that respond to the sensors or human inputs. The range of outputs must be defined by the designers if not already defined by the customer. It is likely that the test strategies for these functions will be the most difficult ones to agree;
7. Are any functions safety-critical? The level of criticality is an important criterion and will set the design procedures which in turn will affect costs;
8. How much of the design can use existing products, whether from within the company or from others?
9. Which parts of the design should have make-or-buy decisions? This is discussed in Section 5.2.6.

5.2.5.4 Project timescales

1. Are the contract timescales as envisioned when the bid was prepared?
2. What are the long-lead time components?
3. Is the level of automation clear with no further research or development needed? This is a critical question in view of the uncertainties about autonomy-related topics;
4. Are all suppliers known and have given believable assurances about their delivery times?

5.2.5.5 Project costs

1. Are the original cost estimates still applicable?
2. What changes have occurred since the bid that could affect costs?
3. What are the financial aspects of the risks on the risk register?
4. Is there an agreed method of budgeting for risk and contingencies?

5.2.6 *Make-or-buy decisions*

One early decision faced by the system supplier's managers is which subsystems of the product should be designed and/or built in-house and which should be contracted out. A platform supplier wishing to incorporate more autonomous capabilities may not have the AI skills to deliver these capabilities. They may have some research capability in the field but have made a business decision not to grow this into a design team.

A robust business strategy may be to form a strategic alliance between the platform supplier and a specialist Small or Medium-sized Enterprise (SME). This is a standard procedure for the design of Monolithic Microwave Integrated Circuits (MMICs). In that case, the subsystem supplier may have a small design team trained in the use of the circuit design tools for one particular semiconductor foundry who then produce the MMIC under strict conditions of commercial confidentiality.

Factors to be considered for make-or-buy decisions by a company are:

1. Does the company have the skills to design the subsystem? If not, does it want to grow them?
2. The supplier must allow full access to the company's QA authority;
3. Can the company manufacture the design?

4. Is there a company-preferred supplier who can deliver the subsystem?
5. Which subsystems have modules with long delivery times? Lead times due to limited resources, such as staff availability, often necessitate subcontracts even if the work could be carried out in-house;
6. Who owns the Intellectual Property Rights (IPR) and has rights to exploitation? Is there open-source software in the product?

Make-or-buy decisions for autonomous systems are a potential area of dispute and increased risk. This is because it is a highly dynamic field with companies of all sizes claiming remarkable advances, but not all delivering them. Results from a company's own research departments should be treated with the same degree of scepticism as those from anywhere else. In addition to the above questions, the following factors must be considered for military and other applications with critical outcomes[9]:

1. The formality and control of the supplier's software writing. This could be assessed using criteria such as the Software Engineering Institute's (SEI) Capability Maturity Model Integration (CMMI) levels. (More details are available at www.sei.cmu.edu. Accessed on 13 June 2018.)
2. The interfaces between the supplier's product and the rest of the company's system;
3. Mechanisms for long-term customer support of the supplier's product when integrated into the company's products;
4. How will the supplier's product be upgraded to give new or improved capabilities to the customer?
5. What are the consequences of the supplier's product failing in service?
6. The medium- and long-term prospects for the supply company have to be scrutinised carefully as they may go out of business or be bought up by one of the project company's competitors. The system house may make it a contractual requirement that they take the source code and IPR of the supplier if they go out of business or are taken over.

Considerations of the above points may lead to a decision not to incorporate the supplier's product in the company's project deliverable. Other alternatives are a buy-out of the supplier, or developing software in-house to deliver the same capabilities.

It should be clear from the above discussion that the use of advanced autonomous concepts represents a high risk for the project as there will be a need for the products or IPR of companies or research laboratories involved in autonomy R&D.

5.2.7 Risk registers

There will be an identified individual responsible for monitoring the risk and managing the mitigation actions, the Risk Manager. One of their essential tools is the risk register, mentioned in Section 5.2.1. The register is usually a table with

---

[9]Critical here is meant in the sense of the consequences of an incorrect output or system failure.

at least the following headings: Risk; Impact if it occurs; Probability of occurrence; Mitigation action.

Autonomous systems will have extra risk due to the uncertainties in their behaviours, especially if AI is involved. One way to mitigate design risks is to assign a small part of the overall budget to carry out a feasibility study of a different approach to the design of high-risk critical functions, analogous to taking out an insurance policy. This means that if the original approach fails, there is an alternative solution available, reducing the time overruns for the programme. The problem is establishing the budget for the insurance solution, as it is effectively introducing a budget reduction to the mainstream solution.

5.3   Project organisation

The organisational structure and monitoring processes in this section are standard custom and practice across many industries. Details and terminology may be different, but the basic aim is the same, i.e. to ensure that the delivered product meets the contractual requirements, on time, and that the costs incurred allow the company its anticipated profit margin. These aims are the same when the product is an autonomous system, but the risks may be different.

5.3.1   Work breakdown structure

The work breakdown structure is one of the fundamental parts of engineering management. It is how the company's contractual responsibilities are successively divided up until they can be placed across its organisation down to the lowest levels. The approach to deciding the work breakdown structure depends on the organisation of the company. As explained in Section 5.2.1, most engineering companies have some form of matrix management to deal with the problem of maintaining staff skills through line management across projects but still giving project managers the authority to demand deliveries on project timescales.

A work breakdown structure will have a structure such as that shown in Figure 5.2 with the solid black lines showing financial delegation and control for a 'skill-based' company. Solid black boxes in the figure are each a work package on a department; the dashed lines represent line management control. In this example, some work is carried out using subcontracts. Depending on their size and technical content, they may be placed by the design department or the project manager.

Work packages are each defined by a work statement, which is a contractual document, either within the company or on an external subcontractor for small tasks. These contain the exact specifications of the deliverable from that part of the company or the sub-contracting company. There will be a work statement defining the work at every level in the company. The work statement becomes increasingly detailed as the work is placed on successively lower organisational levels. Work statements should always be placed on named individuals who then have the responsibility for a deliverable and the authority to use the necessary resources. Even with small contracts, it is a good practice to produce work statements so that

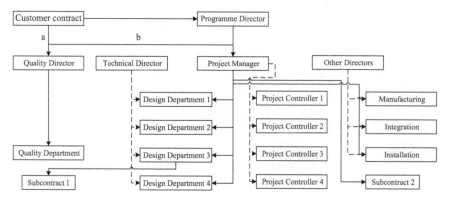

Figure 5.2   A typical work breakdown structure for a system
[*The solid lines show lines of financial authority. Dotted lines show line management control*]

responsibilities and authorities can be clearly identified in company terms rather than customer terms.

Work packages may also, directly or indirectly, form the basis for staff assessments so there is usually considerable negotiation between skill managers and project managers about them. It is vitally important that the result from all the work statements will be the company's contracted deliverable, and will generate payment.

Figure 5.2 shows the programme and project managers having direct control of the work content, but not line management authority over the design engineers. An alternative structure will have the large contract split between senior managers or directors who then divide the work between departments and so on down to small teams or individual employees. The project managers then have indirect control of the work content. The choice between these two, and other, types of control depends on the company management philosophy, the contract type and numerous other factors. However, it is usual, if not mandated, in defence contracts for the QA organisation to have its own work package directly from the original contract to ensure its independence from internal pressures (line 'a' rather than line 'b' in Figure 5.2).

5.3.2   Project monitoring

A well-run organisation will have systems in place to monitor spend and achievement on a regular, probably weekly, but at least monthly basis. Progress will be reported on individual work packages by the local project controllers shown in Figure 5.2. Their responsibilities may be simply monitoring spend and achievement, reporting it to the Project Manager who takes any necessary action. It is more likely that they will control and authorise all spending by the area which they are working with. The details of their authority and that of the line managers will be important in any post-delivery debate or legal enquiry into product failures in service.

Progress will be reported in different ways, the most common being the use of milestones. These are points in the mini-project defined by a work package and

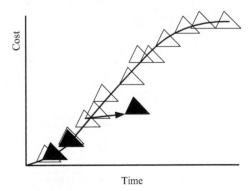

Figure 5.3   Cost-achievement plot
[The solid line shows budgeted expenditure as a function of time and milestone
achievement. Planned milestones are empty triangles. Actual cost and achievement are
solid triangles]

work statement where something has been achieved such as completing a circuit design, software routine, or starting the manufacture of a prototype item. The cost to reach this point will have been estimated and included in the budget. The Project Manager will have drawn up a cost-achievement curve at the beginning of the project. This is a spend-time plot such as the one shown in Figure 5.3. (They usually have this type of 'S' shape and are known as cost-achievement plots.) The various milestones are show as empty triangles at the start of the mini-project, and filled in as they are achieved and moved to the incurred cost and time coordinates.

When a milestone is achieved, the local project controller will confirm with QA that all contractual requirements for the milestone have been met on that date and can then establish the cost of achieving it through the company's financial control system. The date and cost will be added to the cost-achievement plot, indicated by the solid triangles in Figure 5.3. The vertical displacement gives the overspend and the horizontal displacement gives the time slip. In this case there is something seriously wrong that should have been spotted before. In practice the local project controllers and department managers will have been monitoring progress towards all milestones as a percentage of achievement. This gives the fine detail needed for close monitoring and detecting problems at an early stage. The overall project curve will be the summation of this data.

It is vital that the technical managers have access to the cost-achievement plots so that they can put corrective action in place well before a milestone and/or agree contingency plans with the project manager. In practice, this is often the source of most management activity in a project and exposes the conflict between the desire for technical innovation and the actual cost of achieving it. The preparatory action which gives a sound basis for the corrective action is a well-thought-out and costed risk register which includes technical and other risks to milestone achievement. (Other risks may include staff shortages, subcontractors delivering late and even incidents such as a cyber-attack.)

5.4   Autonomous (sub)system design

5.4.1   General principles for weapon systems

The design of the subsystems or modules of an autonomous system should follow the same principles as any other engineering design. The overall concept will have been analysed and decomposed into systems at a fairly high level such as platforms or networks. Design of most of the platform, system of systems or system will follow conventional processes and will not be discussed here. Increased automation in weapon systems however must take explicit account of IHL. This means that some specific guidelines should be followed in the early design stages in order to save costs during the Article 36 reviews.

It is shown in Sections 10.3 to 10.10.7 that the designers of weapon-systems should decompose the subsystems into decision-making functions and examine the requirements for these. This makes both engineering and legal sense as a recent paper from the ICRC [8] by international lawyers has examined the problem of autonomy and states that:

> Autonomy is sometimes discussed as a continuum; from remote controlled to automated and then to autonomous. However, this may not be a useful distinction when considering systems, including weapon systems, which may incorporate both remote controlled and autonomous operation for different functions. For example, although some armed unmanned air systems are often described as 'remote controlled' when in fact many of their functions are certainly automated and may even have a certain level of autonomy – such as in take-off and landing, navigation, pre-planned response to a specific event, and even some aspects of target acquisition.
>
> Therefore, for a discussion of autonomous weapon systems, it may be useful to focus on autonomy in critical functions rather than autonomy in the overall weapon system. Here the key factor will be the level of autonomy in functions required to select and attack targets (i.e. critical functions), namely the process of target acquisition, tracking, selection, and attack by a given weapon system. Indeed, as discussed in parts B and C of this paper, autonomy in these critical functions raises questions about the capability of using them in accordance with international humanitarian law (IHL) and raises concerns about the moral acceptability of allowing machines to identify and use force against targets without human involvement.

Engineering design using autonomous functions described in this book use the same principles, as discussed in Gillespie [9] for military applications and Gillespie and Hailes [10] for civilian applications.

The designers should have a recognised architecture of the command and control chains in control of the automated weapon system. The 4D/RCS model is used to describe autonomous systems in this book and is described in detail in Section 10.5.2. In the detailed design work described in this chapter, the important factor is the structure of a single node in the command and control hierarchy, and

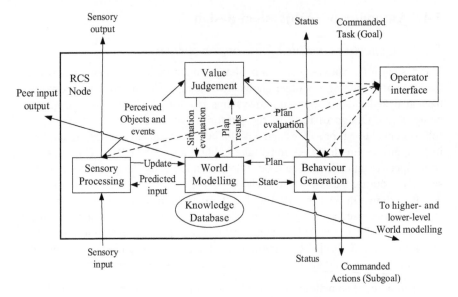

Figure 5.4   A single 4D/RCS node

the level of automation within it. A single node is shown in Figure 5.4. The elements of the node can be used to define the internal functions and write the detailed requirements for the specific application. Section 11.4 shows how this can be performed. The important point here is that the sensor processing, world modelling and behaviour generation functions are the critical functions of an Autonomous Weapon System (AWS). The subsystem design must ensure that these functions are clearly specified and can be scrutinised in an 'Article 36' review.

As discussed in earlier parts of this chapter, the designers should understand the wider context of their design. Part of this is where it fits in the CADMID cycle which is discussed in the next section.

5.4.2   The CADMID cycle

The development of a specific product has several phases which are common across all industries. These are usually described as the CADMID cycle which is mentioned in Section 4.10:

1. C Concept
2. A Assessment
3. D Development
4. M Manufacture
5. I In-service
6. D Disposal.

(Strictly speaking, it is not a cycle unless Disposal is preceded or followed by Concept or Assessment of a replacement item.)

This notation provides a convenient way to identify the main phases of a procurement programme. There can be a major design and financial review at the end of each stage which then triggers the start of the next phase. Payment milestones are usually associated with successful completion of a design review. 'Article 36' reviews must take place as part of, or in addition to these reviews and will draw on the technical results produced during the completed phase. As explained in Chapter 7, these are mandated under IHL. Engineering aspects of these reviews are described in detail in Chapter 11.

The CADMID programme phases for one product, even a complex one such as a new aircraft, can be readily identified. This is also true for a self-standing autonomous system such as Phalanx, which raised the autonomy level of close-in weapon systems. However, when the product is an incremental upgrade to systems within an existing platform or is a release of a more sophisticated software integration of existing systems, the CADMID phases for the overall programme may be difficult to identify clearly. Problems also arise when there are customer-requested changes necessitating rework of earlier, completed phases, or partial integration of early deliverables so that fundamental design decisions can be made for other deliverables in the programme.

The examples in the last paragraph show why the use of CADMID terminology has become less widespread in modern programmes. However, there must always be some equivalent programme break or decision points for financial and IHL reasons. Positioning the system being designed in a CADMID context does help identify the aims of the product and the issues that will be in the customer's mind when assessing it. For example, a product for the Concept stage may require a more innovative approach than a product for the Manufacture or In-service stages.

5.4.3 *Design validation and verification (V&V)*
5.4.3.1 **The reasons for V&V**
V&V are among the most difficult problems for the specifying, designing and accepting organisations of an autonomous system. There is no simple answer, so we will present the essential principles of V&V that apply to all systems, with a bias towards highly automated systems.

It is essential that a system, when delivered, meets the customers' requirements. In principle, the expectations are encapsulated in the requirements documents and the company offered a fully compliant bid, so the delivered product will fully meet the customer's expectations. When the system is simple, such as a nut and bolt, we can believe that this statement is true. However, it may not be possible to completely specify a complex system, even with many thousands of words. For example, it was said that NASA's Saturn rocket could not be launched until the weight of paper describing it was greater than the weight of the rocket. This has been repeated for many large products since then, generally with a multiplying factor on the weight of paper (☺). Another question is: whose expectations should the product meet? It certainly includes the procurement agency who are strictly-speaking the customer. However, the end-users will include the operators, but will also include the support and logistics organisations whose views are different to the operators about what constitutes a 'good piece of kit'.

Despite a system being tested and accepted by the customer, unexpected behaviours can happen. Even simple machines can behave in a way that is unexpected when it occurs, but may be explicable with hindsight. Unexpected behaviour in a word-processing software package is frustrating but not dangerous; in a highly automated weapon system, there could be severe consequences. Unexpected behaviours are almost certainly unacceptable and may lead to the weapon being withdrawn from service. Unexpected behaviour in a complex system is an example of the 'wicked problems' discussed in Sections 4.2.2. There is no reason to assume that a weapon system is any different. If we accept that it will not be possible to find and correct the cause of unexpected behaviours, the only solution is to ensure by design that the results from any that do occur are limited to be within an acceptable range and, if possible, a human such as the operator should be informed about the occurrence.

Summarising, there are two distinct questions that need to be answered:

1. Do the design requirements define the required system adequately?
2. Dies the performance of the delivered product meet the customer's expectations?

Expectations and stated requirements are likely to be different for an autonomous system as it may have non-determinate behaviour which cannot necessarily be predicted, even if it has limitations.

The design must include limitations on any unexpected behaviours which should be agreed with the customer, even if they are not in the contractual requirements. These are behaviours and properties that are seen only after the system has been integrated into the system of systems and is exposed to *in situ* test conditions. Clearly, planned behaviours due to the system that can only be realised by the complete system of systems are not a problem and are the desired ones. The problem comes when there are unexpected behaviours, often called emergent properties. The classic mechanical one is an oscillation due to an unplanned resonance.

A complex control system will have multiple and complex interactions with the environment. Sometimes the emergent property is irrelevant or beneficial, but should still be measured and recorded so that it can be characterised by an analysis of the design. The worst type of emergent property is one that is detrimental to some aspect of the required capability and is not easily explained, especially if its analysis turns out to be a 'wicked problem' as discussed in Section 4.9. Post-delivery unwanted emergent properties are a particular problem for any agreement that the system has been properly verified. These require careful analysis as well as removal with extensive further verification to ensure that their removal is complete. There should always be a risk budget assigned for their solution at the start of the project. They usually arise near the end of the project when most of the risk budget has been spent, potentially causing management problems.

Similar questions to the above two can be posed for upgrades to the existing systems.

The engineering processes to answer both sets of questions are:

- Validation checks that the defined requirements will meet expectations
- Verification checks that the final product meets its contracted specifications.

They are collectively known as V&V. Validation can be considered as a forward-looking activity whilst verification is testing that completed activities have been successful. They can, and usually are, used for subsystems and modules as well as the final system. The level of detail required will depend on the properties of the particular item considered. Verification, as defined above ensures that payment can be claimed from the customer. However, there are often some differences between the defined requirements, customer expectations and the contracted specifications. These should be identified as early as possible, taken into account in planning the verification process, and discussed with the customer. This may result in some financial adjustments to the project, either using the risk budget or a new contract.

It may be possible to represent a subsystem with one equation describing the output for the whole range of input signals. This becomes a simple model of the subsystem. A well-designed control loop is an example. A complex weapon system may need a complex simulation facility with a realistic human-machine interface, six Degree-Of-Freedom (6-DOF) models and 6-DOF Hardware-In-The-Loop (HWIL) simulators with a full-size missile on gimballed mounts in an anechoic chamber. Both these examples are deterministic systems so the validity of any tests can be assessed by analysis of the model's fidelity and the limitations of its environment.

Validation has been discussed in Section 4.13 in the context of testing the results of the operational analysis processes and the required system of systems. At the system and subsystem level, validation can be carried out by studying the requirements documents and comparing them directly with the predicted specification of the proposed design. This method can be used for a fairly simple system, or even a complex one if there are only a few key parameters to be considered. The output is a compliance matrix which lists the key parameters in the first column, the required values in the second and the predicted values in the third. There is usually a comments column and accompanying document(s) giving further explanations and assumptions.

In practice, the design proposed for the bid will not have been detailed or complete[10]. The bid team will have made many simplifying assumptions about the design. A good validation process should be part of the design process and is started immediately after the initial contract review. The basic approach is to build models of each subsystem and module and integrate their results to give the system's overall performance. Each subsystem and module model will be based on, and may be the same as the Computer Aided Design (CAD) models and simulations used by the designers. The predicted performance of the final system may not be a formal

---

[10]An exception to this can occur on a large programme, when there is often a separate feasibility contract placed before detailed design starts. Even in these cases, the design will not be complete unless feasibility is followed by a further design contract prior to manufacture and/or integration.

integration of all these models. They will have been produced for specific design reasons with many degrees of fidelity. They may also come from mutually incompatible design suites. However, there is a need to be an overall model which takes the outputs from the design models and integrates them into a prediction of overall performance. The costs arising from differences between the predicted and actual overall performance should be included in the project risk register and budget.

Verification of individual subsystems is usually an internal matter for the company, but overall verification is usually carried out with the customer. This can be through the agreed set of formal trials which form the contractual, delivery milestone. These may well be supplemented by further trials and tests by either the supplier, the customer, or both. They may continue well after delivery, extending into the setting-to-work phase or beyond.

5.4.4   V&V of autonomous systems

How do we approach V&V for an autonomous system? It has already been stated that it is a difficult problem. Validation will be predictive and will have its roots in the design tools used by the designers and suppliers. Verification will look at the actual performance regardless of how the system was designed. It should produce a refined understanding of the system, including both its desired and undesired features.

A first step is to compile a set of questions based on the currently foreseen, and likely, future uses of the system. All autonomous systems are there to provide assistance to humans by either providing refined information or undertaking tasks which the humans may not want to, or cannot, carry out themselves. This immediately raises a set of questions about the human-machine interface, how the system is controlled, and the limits on the system's freedom of action, among others. A system using AI to understand its environment, or to shape its response to its inputs is not deterministic. This makes it difficult to characterise and it may not be possible to model it with any degree of confidence in the results. If this is the case, there will be greater emphasis placed on verification than validation. Measured performance is then the critical issue, with a consequent increase in time and cost for tests and trials in realistic and operational scenarios.

The initial list of questions can be used as a basis for both the use of the outputs of the validation models and the approach to system verification. The list will have questions such as:

1. What is the aim of the system when in normal use?
2. Who is the customer?
3. Who is the system's user when in normal use?
4. How is the system controlled?
5. How does the system interact with its human controller?
6. How does the system interact with humans who are not its controller?
7. Are there test and training uses which will be different to normal use?
8. What are the limits on the system's actions when in use?
9. How will the system be maintained?

10. Will the system need fail-safe modes?
11. Will the system need to monitor its surroundings when not actually performing a task?
12. Is the system physically self-contained?, i.e. is it like a robot with a single physical instantiation?
13. Is the system electronically self-contained?, i.e. is all the functionality controlling its actions located within its physical boundaries?
14. Are all hardware, software and firmware interfaces defined?
15. What likely upgrades may be requested?
16. What technology changes are likely to occur during design and in the first years of service? Will these cause certification problems?

The reader may like to use the above list to draw up more detailed question lists for the following example autonomous systems:

• A bomb-disposal remotely controlled robot which is self-contained and amenable to unambiguous requirements and specifications;
• An unmanned submersible vehicle used for mine-hunting, initially released from one ship, but sending its results to any friendly ships in the locality and following the consequent instructions;
• A humanoid robot which is physically and electronically self-contained. The interfaces will also be defined, probably using optical and audio sensors. The problems come in defining its aims and how it interprets the human inputs such as voice data.

An automated weapon system is, or will be, part of a larger system of systems with one or more inputs which may not be fully specified. Additionally, it is highly unlikely that the design team will be starting with a blank piece of paper. There will have been extensive pre-contract negotiations. The contract will define government or customer furnished equipment and information. It is likely that the contract will be for a defined deliverable, but not for its integration into the military infrastructure. That is usually the subject of a further contract. The complexity of the system will probably necessitate a setting-to-work phase, followed by a formal introduction-into-service date, leading to an initial operating capability. Inevitably there will be support and upgrade contracts for the system until it reaches its full operating capability.

The architecture of the command and control system will be defined and should be available as government furnished information. If the automated system is fitted to an existing platform, there will be an interface specification and known combat-management-system. Many interfaces are defined by STANAGs and datalinks may be well-known ones such as Link-16. This information will have been used to create the system requirements document which forms part of the contract. The authors of the document have to balance prescriptive instructions to the supplier with allowing freedom for innovative design solutions. Some of this is achieved by having mandatory requirements and 'desirable' ones, allowing the supplier and customer to negotiate which 'desirable' requirements can be met within the cost and timescale constraints.

5.4.4.1   Validation of an autonomous system

The basic questions are:

1. What is being validated?
2. What are the criteria which it will be assessed against?

The basic aim of validation is to test that the product will meet its requirements when in operation. It is important to test that the set of specifications used within the company meet the customer's contractual requirements. Usually, there is an extensive use of models which become increasingly more complex as the design evolves, with gradual replacement as necessary with actual product hardware and software.

The supplier will have created a model of the system during the bid preparation stage. Its aim is to produce sales information for the bid. It will have been used to assess the system complexity (which usually means cost), project risks and outline sub-contract requirements. The sales and/or design engineers will have developed a design concept at an early stage of the bid process, if not before, and used that as the basis for the model. It is likely that they will have gone into considerable detail in some areas for various reasons during the bid, but not in all. The customers and their technical advisors will probably have seen demonstrations using the model and given some opinions about it. An early question is to ask if the bid model(s) are suitable as a starting point for design validation.

Following the initial review in Section 5.2.5, the design team will refine their concepts for the various subsystems and fill in many of the gaps in the necessarily simplistic design used in preparing the bid. At this point, it is essential to set a policy for the models to be used during the design process. It may not be practical to have one large system model as some subsystems may be adequately described by equations which can be solved using commercial simulation and modelling packages, whilst others, such as the MMICs mentioned in Section 5.2.6 will require high-fidelity complex simulations. It is highly probable that any company will have a range of design tools available, ranging from simulation and modelling packages to HWIL facilities.

5.4.4.2   Verification of an autonomous system

The basic questions are:

1. What is being verified?
2. What are the criteria which it will be assessed against?

The answers might appear to be straightforward – the deliverable from the contract is assessed against its specification. However, there are wider legal requirements for an autonomous system which may not have been explicitly stated in the specification. As stated before, these will come from IHL for military systems (Chapters 7, 8 and 10) and general ethical principles for civilian autonomous systems leading to product liability issues in the supply chain which are not yet fully encapsulated in law (Section 13.2).

Extra considerations come from the deliverable being used as part of a larger system of systems and will interact with other systems in it. The drivers on the verification programme, the programme plan and its costs may have to be agreed between several organisations. The legal and regulatory framework for autonomous systems is not yet clear so verification represents a significant risk to any programme.

Given the wider responsibilities of the operators of the system of systems, they will almost certainly demand a demonstration that the deliverable does not introduce unwanted or illegal behaviour in the system of systems. A military system has only one ultimate user, the state's military forces. It can be argued that the 'Article 36' reviews in Chapter 11 are simply a requirement that the state verifies that the 'new means or method of warfare' comply with IHL. The criteria then become those derived by processes such as those in Chapter 10, but may not have been called up directly from the contract documents.

The most cost-effective approach to verification is to make it part of the integration plan, i.e. to include elements of it in the right-hand sides of the Vee diagrams in Figure 4.1 and 4.5. The 'System requirements and architecture' in Figure 4.1 and the 'Subsystem requirements', shown in both Figures 4.1 and 4.5, will be inputs to the test and integration plan discussed in Section 5.4.5.

5.4.5   Integration and test

The various contracts and work packages in the work breakdown structure will deliver a system and subsystems which should all have been tested against their procurement specifications. In principle, if the design has been validated thoroughly against complete and rigorous specifications, the system will perform according to its specification and deliver the required capability. This is shown as the 'Incremental Integration' box in Figure 4.1, linking 'Tested subsystems' and 'Validated system design'. In practice, this is not achieved without considerable effort expended in integrating the subsystems. It is sensible for a company to have one person or small department identified as the integration authority. They must have defined authority for all aspects of integration and so may have an independent management reporting route. This ensures that they have a degree of independence from departments supplying the systems and subsystems.

The anticipated causes of undesired system performance may range from deviations due to changes during design, to harsh environmental conditions giving excessive noise in control loops. Additionally, the subsystems required to deliver a complex capability will themselves be complex, with difficulties in both specifying and testing them. There is also a danger of responses that are not easily understood such as the emergent behaviours and wicked problems discussed in Section 4.2.2.

The cost of single-step by single-step integration and complete testing at each stage is likely to be excessive as well as time consuming. This must be recognised as a fact of life and so procurement plans must minimise the predicted cost as well as mitigating the risk of cost overruns. One of the main elements of the solution is to draw up an integration and test plan prior to entering the bidding phase. This should then form one of the important influences in drawing up the structure and content of the contracts. The integration plan will deliver a system to the customer setting-to-work phase with performance that meets specification *as far as can be confirmed by the range of tests carried out*. The delivered system will be integrated into a complex infrastructure introducing more variables. Therefore, the setting-to-work team should have access to the information gathered during pre-delivery integration and testing. This knowledge is not necessarily documented as much will have come from extensive fault-finding and fault-fixing. IPR issues can cause complications, but can be solved if the setting-to-work team has members from all relevant companies. Usually the successful solution to a fault is recorded in detail but the unsuccessful ones are not although they may be relevant to faults which occur in the new environment.

Guidelines for integration and test plans include:

- Identify the key required system capabilities in its operational environment;
- Identify undesirable system characteristics that will reduce the system's capability or weaken protection against cyber or other attacks;
- Scrutinise the design validation results to find the key parameters and performance characteristics identified through measures of merit. This should indicate how critical their seamless integration is;
- Allow sufficient time for integration and debugging;
- Test facilities such as test ranges for military equipment may only allow very few trials. They may need to be booked several years in advance with little or no possibility to change dates due to programme slips;
- Civilian projects affecting infrastructure may only be allowed very limited times for access to the customer's system of systems. For example, only at weekends with the original configuration restored before Monday morning;
- The system is integrated in a 'bottom up' manner with the simplest components and modules combined first, steadily building up to complex subsystems which form the nodes in the architecture;
- Integrate as few subsystems as possible, ideally two, before conducting tests;
- Instead of immediately integrating several delivered complex subsystems, the best philosophy is to build test harnesses for each one. These are analogous to those used in software engineering, which provide controlled inputs and characterise the outputs. The unit under test can then be exposed to controlled conditions with known changes in input parameters and variables. It is possible to vary the inputs to force unacceptable behaviour and test its probability of occurrence in realistic situations. Specialist software can be purchased for this task;
- Have one company or person responsible for the integration of as many subsystems as possible. They must be able to draw on expertise from the suppliers of the subsystems in a contractual framework.

The above points apply to all types of system. An autonomous system requires extra considerations. It is essential that the system only responds to authorised commands and does not have undesired or unpredictable behaviour. The system-of-systems and system architectures become the foundation of the test plan and hence of the integration philosophy.

Command and control nodes may be human, or they may comprise one or more subsystem possibly from more than one supplier. The first step is to identify the node(s) that interface with the system and any nodes within it. The next step is to integrate the subsystems within a node. Each node will have specified capabilities and limits on its authorised power[11]. The authorised power comes from internal control systems which have specified inputs and outputs whether internal or external. The integration plan will therefore have a step which assembles the node from its previously tested subsystems, tests it, again with a test harness or in a simulation, and then integrates it with its surrounding nodes.

The example in Chapter 11 will have an integration plan for a system with relatively low levels of autonomy, with a different one needed for the autonomous capability in Chapter 12. The two are not detailed in those chapters, but they would follow the principles given above.

5.5   Module design

5.5.1   Setting the requirements

The final system design will be a compromise between the optimum designs of the various subsystems and their modules. It is likely that there will be periods of optimisation at various design stages. The factors affecting the final compromise include delivery dates, costs and company policies as well as technical risk, post-delivery support and potential upgrades, as well as the normal design compromises made during any design. These will not be discussed in this book as they are standard engineering practice, although an advanced design such as a highly autonomous system may need more compromises over a longer period than is normal for other products.

It is assumed that the systems engineering aspects and trade-offs have already been carried out as discussed in Chapter 4. The input for the design then includes the output from those processes. It should be noted that the models used for system design may lack fidelity and so not be suitable for designing or verifying the modules.

It is shown elsewhere and in Section 5.4.1 that the critical design aspects for military autonomous systems concern the critical functions in the targeting process whether performed by a subsystem or module. Therefore, it is essential that the detailed design and specifications identify the relevant function(s), and that verification and Article 36 reviews are included in the aims of the test plans.

Chapter 4 showed the flow-down of customer need to contractual system specifications. Each industry has its own mandatory standards which are included in contractual Terms and Conditions (Ts&Cs). These will flow down into detailed standards and company procedures for the specific design team. These should be included in the work statements for the design, and adherence to them will be checked by the QA organisation as well as at local design reviews. It should be noted that excessive costs can be incurred by specifying unnecessary Ts&Cs at any level in the work breakdown structure, so their management in contract negotiations and later in internal work packages is an important management activity.

---

[11]Authorised power is defined in Panel 3.1 and explained in Section 10.9.

5.5.2   Detailed design

The management organisation of the modules of any subsystem is based on the same basic principles as those described for (sub)system design in Section 5.4. Work breakdown structures will be similar to that in Figure 5.5 which is in turn derived from Figure 5.2. The design department with the greatest expertise required for the subsystem will have the work statement and finance. The department manager will allocate work and funds across his design teams. They can in turn pass on smaller work statements to other teams depending on the company philosophy on inter-team working. One key role is that of the integration authority for the module who may be a member of one of the departments designing one of its parts. This is likely to be a senior engineer who has to ensure all the modules make the contracted subsystem. They may need to place work statements on the other areas of the company supporting the design teams.

There will be an initial local work statement review checking similar factors to those in the system review. Additional items are likely to be:

- Are any aspects of the system's function supported by the module safety-critical?
- Criticality of the use of the module and applicable standards for this level;
- Hardware/software trade-offs;
- Lead times and availability of components in the module;
- Long-term support for future obsolescence of components;
- Long-term support requirements;
- 'Fitted for but not with' decisions (see Section 5.5.3 for details);
- Company policy on reuse of existing smaller items and software;
- Interfaces with other subsystems and modules;
- Interfaces between module and subsystem models;
- Make-or-buy decisions;
- Risk register and policy on costs for contingencies.

Configuration control of all aspects of the design is a standard engineering process. It applies equally in the design of autonomous subsystems. Given the relatively low level of maturity of autonomous systems and the likely reliance on synthetic environments and modelling for design proving, it is essential to have strict

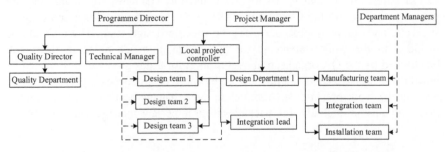

Figure 5.5   A typical work breakdown structure for a subsystem or module
[The solid lines show lines of financial authority. Dotted lines show line management control]

configuration control of both the item under test and the test environment. Without this, it will be difficult if not impossible to recreate test results and make correct comparisons of the performance of the various functions as inputs are varied. See also the discussion on emergent behaviours and wicked problems in Section 4.2.2.

5.5.3 Fitted for, but not with

Long timescales for procurement and platform lifetime cause difficulties for both customers and suppliers who want to see advanced technologies both in the delivered products and throughout their operating life. This was discussed in Section 4.5 and the concept of procurement using capability requirements explained in succeeding sections. Capability procurement can work at system and system-of-systems level. However, it is not too useful as a concept at the design level when decisions must be made for tangible hardware and software implementation at the time of design freeze.

Changes to the design of a module or subsystem become progressively more expensive as the design matures. This is simply due to the fact that as a design matures, more and more decisions are made, hardware is built and software written. Changes to the design may come from a change in original requirement, in which case there will be contractual changes to reflect it. The more difficult reason is a decision to use a different or modified technology in one area, such as replacing a processor with a more advanced successor. These changes may not be covered by a contract change, but some of the current design and production will be redundant and have to be reworked. Some changes can be foreseen, so a precautionary approach is to adopt a 'design for but not with' philosophy in the design.

The concept is to identify likely changes in technology or requirement over at least two timescales:

• The design phase, which may be months, but can be up to a decade for an infrastructure or large military project;
• The in-service lifetime of the system. This is more difficult as the timespan may be decades.

The design concept and work breakdown structure should be examined to see which subsystems or modules will be changed by any of the likely changes. The next question is whether there can be relatively small changes to the design which will make it straightforward to introduce the new technology or requirement change. Then the cost of implementing this design change can be compared with the changes necessary to introduce the new technology or requirement should the current design be implemented.

The concept may be as simple as installing higher capacity communication cables or links in anticipation of increased data flows. More sophisticated approaches are to make the work breakdown structure such that the division of functions between modules is made so that a single module or set of modules can be replaced with an upgrade, but still keeping the same interface specifications. The steady improvements in GPS technology from a federated system of antennas and processors to single integrated units is an example of this, although some of the mechanical interfaces will have changed. Similar approaches can be adopted for software suites. In the latter case, particular care must be taken with precision and rounding errors.

All of the technologies associated with autonomous systems are evolving rapidly, as are customer expectations. It is here that it is particularly important that the system architecture has the autonomous functions separated from those that are not. Both should have their own distinct and identifiable subsystems and/or modules. The likely changes for the technologies in the functions can then be identified and the design implemented taking them into account.

'Fitted for but not with' would appear to be an obvious approach, but it is actually more complex than it might appear. There are several reasons:

- In general, the extra design effort will take time. This is likely to delay project milestones even if costs are not increased. If the change will occur within the delivery timescale of the project, the project manager may accept the delay and extra cost, trading them against later savings;
- The new technology will, of necessity, be at a lower TRL than those in the current design and so there will be a risk that the technology, when at an acceptable TRL, will not be implemented as foreseen at the time of the 'fitted for but not with' decision. The result will then be increased early costs to implement the decision followed, later in the programme, by similar extra costs to those if the original design had not been changed, i.e. the extra cost will be incurred twice;
- A likely change in a requirement from the customer will require a contract amendment. This may happen if the customer sees that the proposed design changes have made the requirement changes more affordable. He or she may then be in the position of being able to afford an upgrade programme that would otherwise be unaffordable, later in the system of systems' lifetime;
- The supplier may see the upgrade to requirements as a future sales opportunity. Its implementation of 'fitted for but not with' is a potential future profitable contract with the current customer and a marketing opportunity in other markets. This view may then feed back into the design decisions based on the predicted dates for the new contracts and the type of enhancement which are attractive in the market place.

Decisions of this nature should be based on hard-headed technological and commercial factors. There are a variety of 'crystal balls' which can be used as a basis for debate and decision-making. All companies will have a view on their long-term place in the world markets and the investments they can make to achieve these aims. External sources of information come from technology roadmaps and commercial forecasts. There is an overview of many relevant ones by Innovate UK [11]. Examples are the DOD Unmanned Systems Integrated Roadmap [12], the EU Robolaw guidelines [13], the FAA Aerospace Forecast 2016–2036 [14] and by NASA for space [15].

Appendix A5   Technology, integration and system readiness levels from different sources

Key: IRL = Integration Readiness Level, ISD = In Service Date, SRD – System Requirements Document, SRL = System Readiness Level, TRL = Technology Readiness Level, URD = User Requirements Document

TRL, IRL or SRL	NASA and DOD (ESA TRLs 5, 6 and 7 differ)	MOD	European Union (EU)	IRL Sauser et al. (2008)	SRL Ross et al.	UK MOD Systems engineering SRL (From larger matrix)
9	Actual system 'flight proven' through successful mission operations.	Actual Technology System qualified through reliability and maintainability demonstration in service.	Actual system proven in operational environment (competitive manufacturing in the case of key enabling technologies; or in space).	Integration is mission proven through successful mission operations.	Production system is used, demonstrated and maintained in an operational environment.	Application of the production system in its final form and under mission conditions.
8	Actual system completed and 'flight qualified' through test and demonstration (ground or space).	Actual technology system completed and qualified through test and demonstration.	System complete and qualified.	Actual integration completed and Mission Qualified through test and demonstration in the system environment.	The production representative system has been demonstrated in an operational environment.	Final system prototype demonstrated in a representative target platform.
7	System prototype demonstration in a space environment.	Technology system prototype demonstration in an operational environment.	System prototype demonstration in operational environment.	The integration of technologies has been Verified and Validated with sufficient detail to be actionable.	An integrated system prototype has been demonstrated and fabricated in an operational/manufacturing environment.	Representative prototype system demonstrated (all major subsystems integrated and operating) in high-fidelity simulated environment such as a vehicle integration test rig.
6	System/subsystem model or prototype demonstration in a relevant environment (ground or space).	Technology system/ subsystem model or prototype demonstration in a relevant environment.	Technology demonstrated in relevant environment (industrially relevant environment in the case of key enabling technologies).	The integrating technologies can accept, translate and structure information for its intended application.	A system prototype has been demonstrated and fabricated in a relevant environment. Interface control has been demonstrated traceable to a deployed environment.	Key subsystems integrated with realistic supporting elements so that subsystems can be tested in a simulated operational (Lab) environment.

(Continues)

(Continued)

TRL, IRL or SRL	NASA and DOD (ESA TRLs 5, 6 and 7 differ)	MOD	European Union (EU)	IRL Sauser et al. (2008)	SRL Ross et al.	UK MOD Systems engineering SRL (From larger matrix)
5	Component and/or breadboard validation in relevant environment.	Technology component and/or basic sub-system validation in relevant environment.	Technology validated in relevant environment (industrially relevant environment in the case of key enabling technologies).	There is sufficient control between technologies necessary to establish, manage and terminate the integration.	All system components have been built and tested in a relevant or emulated production and deployment environment. Components with simulated interfaces have been tested.	Subsystems are demonstrated to perform as required when subjected to simulated system (Lab) conditions.
4	Component and/or breadboard validation in laboratory environment.	Technology component and/or basic technology sub-system validation in laboratory environment.	Technology validated in lab.	There is sufficient detail in the quality and assurance of the integration between technologies.	All system components have been built and tested in a laboratory environment separately. Numerical studies show component compatibility.	Mature subsystem design documents with defined integration and test plans.
3	Analytical and experimental critical function and/or characteristic proof of concept.	Analytical and experimental critical function and/or characteristic proof of concept.	Experimental proof of concept.	There is Compatibility (i.e. common language) between technologies to orderly and efficiently integrate and interact.	Experimental evidence has been obtained that the system is possible in principle to develop and manufacture.	Mature architecture design document(s) (ADD) with defined integration test and acceptance strategy.

TRL, IRL or SRL	NASA and DOD (ESA TRLs 5, 6 and 7 differ)	MOD	European Union (EU)	IRL Sauser et al. (2008)	SRL Ross et al.	UK MOD Systems engineering SRL (From larger matrix)
2	Technology concept and/or application formulated.	Technology concept and/or application formulated.	Technology concept formulated.	There is some level of specificity to characterise the Interaction (i.e. the ability to influence) between technologies through their interface.	Subsystem technology path identified to include a specific technology solution. Technology, manufacturing, and interface drivers understood.	Mature and verifiable SRD under configuration control with clear linkage to URD.
1	Basic principles observed and reported.	Basic principles observed and reported.	Basic principles observed.	An interface between technologies has been identified with sufficient detail to allow characterisation of the relationship.	The system concept has been identified to include the subsystems. Overall system functional requirements are qualitatively understood.	Mature and verifiable URD under configuration control with defined ISD.

References

[1]   International Standards Organisation ISO 31000:2009. 'Risk management — Principles and guidelines'.

[2]   Mankins J.C. 'Technology readiness assessments: A retrospective', *Acta Astronautica*. 2009, vol. 65(9–10): pp. 1216–1223.

[3]   Blanchette S Jr., Cecilia Albert C., and Garcia-Miller S. '*Beyond Technology Readiness Levels for Software: U.S. Army Workshop Report*', Technical Report: CMU/SEI-2010-TR-044, ESC-TR-2010-109, 2010.

[4]   Sauser B., Ranirez-Marquez J.E., Magnaye R., and Tau, W. 2008, 'A systems approach to expanding the technology readiness level within defence acquisition', *International Journal of Defense Acquisition Management. 2008*, vol. 1, pp. 39–58.

[5]   Ross S, 2016, Application of system and integration readiness levels to Department of Defense research and development, Defense ARJ, Vol. 23 No. 3, pp. 248–273.

[6]   Austin M.F., and York D.M. 'System readiness assessment (SRA), an illustrative example'. *Procedia Computer Science*. 2015, vol. 44: pp. 486–496.

[7]   The UK MOD SRL tool is available at https://www.defencegateway.mod.uk/home/ in the ASG app, but access requires the reader to set up an account with them. [Accessed on 13 June 2018].

[8]   International Committee of the Red Cross (ICRC). 'Autonomous Weapon Systems: Technical, Military, Legal and Humanitarian Aspects', *Report on Expert Meeting; 26–28 March 2014*, Geneva, Switzerland, November 2014; Part A, Section 3.1, pp. 61–63.

[9]   Gillespie T. 'Systems engineering as a solution to legal problems for highly-automated military systems', *Presented at RAeS conference: Autonomy – are we there yet?*; London, June 2016.

[10]   Gillespie A.R., and Hailes S. In preparation 2018.

[11]   UK Transport Knowledge transfer Network (KTN) 2015 webpage at https://connect.innovateuk.org/web/technology-roadmap [Accessed on 1 April 2018].

[12]   Department of Defense 2011, *Unmanned systems integrated roadmap 2011-2036*. Washington 2011.

[13]   The Robolaw project produced its 2015 report: *Regulating Emerging Robotic Technologies in Europe: Robotics facing Law and Ethics*, available at www.robolaw.eu [Accessed on 1 April 2018].

[14]   This is available at https://www.faa.gov/data_research/aviation/aerospace_forecasts/media/FY2016-36_FAA_Aerospace_Forecast.pdf

[15]   NASA Technology Roadmaps, TA4: *Robotics and Autonomous Systems*, 2015, https://www.nasa.gov/sites/default/files/atoms/files/2015_nasa_technology_roadmaps_ta_4_robotics_and_autonomous_systems_final.pdf [Accessed on 1 April 2018].

Chapter 6

Ethics, civil law and engineering

6.1 Introduction

New technologies which provide innovative products and service usually raise questions about the ethics of their use in many fields. There are many papers and books about the ethics of autonomous systems, especially those using Artificial Intelligence (AI) [1–3]. The fast pace of technical developments in AI and automation is matched by an extensive public and legislative debate with proposals for legislation. See, for example, the 2014 report by the International Committee of the Red Cross (ICRC) [4] and the Robolaw report from the European Union [5]. This book takes the pragmatic view that although ethical standards will eventually be implemented as laws and regulations, engineers currently engaged in autonomous systems projects will have to make design decisions now.

It is almost self-evident that society expects technology to be used safely, and be safe when in normal use. This leads directly to laws which apply to all products, whether for military or civilian use. There are then regulations and standards which are specific to particular industries, product type and operating environment. This chapter describes their common aspects and then looks at areas where specific laws, regulations and standards will affect the design and operation of autonomous systems. It is not a definitive guide to the law, but is intended to give engineers an indication of areas where they should seek advice about their responsibilities.

It should not be assumed that the standards and regulations discussed in this chapter do not apply to military personnel and systems. Laws will vary between nations but, in general, civil laws will apply except in operational areas. As an example, UK military personnel have exemption from civil law in very limited circumstances [6], which are summarised here to illustrate how limited they are.

- Operations where members of the armed forces come under attack or the threat of attack or violent resistance;
- Activities preparing for, or in direct support of, the above operations;
- Hazardous training considered necessary to improve or maintain the effectiveness of the armed forces in support of the above operations;
- Special forces are treated separately.

It is clear that the civil legislative framework will apply to military autonomous systems when they are used outside operational areas and military training ranges. Military personnel will be trained in both their use and logistical support, but there will be civilian staff employed both by government and contractors where the immunity applicable to military personnel does not apply. Therefore, engineers designing military systems should be conversant with the law as it applies in civilian fields as well as the relevant parts of IHL at a level commensurate with their roles. This means that all engineers should periodically check with their management chain about any developments in the regulatory environment for their work.

6.2   Ethical background

Ethics is a large field of study for philosophers and public debate. It could be argued that ethics are not of direct concern to engineers. This is false on two grounds: first, ethical discussions have a large impact on a nation's laws and consequently on the use of any product; second, there is a growing awareness that the design process must take ethical issues such as long-term responsibilities into account. One example is climate change which may be considered an ethical issue; it is an important engineering issue with numerous international undertakings by most nations, and extensive national laws. Developments in IT such as Big Data and privacy are of concern to the public and specialists. Ethical implications are discussed in the book by Sarah Spiekermann [7] which is based on an approach known as value-based design, which is discussed in Section 13.3.2. These ethical concerns are in addition to the normal ethical guidelines for any profession such as engineering [8].

It can be argued that there are some underpinning psychological values which are recognised by the whole of humanity. Different cultures then express them in different virtues, narratives and institutions. These are known as 'moral foundations', proposed by Haidt *et al.* [9,10] and are as follows:

1. Care/harm
2. Fairness/cheating
3. Loyalty/betrayal
4. Authority/subversion
5. Sanctity/degradation
6. Liberty/oppression.

The ethics of autonomous systems are an important public issue. One widely discussed ethical problem is how an autonomous system can be expected to make an acceptable choice when presented with an ethical dilemma. This appears in many guises, but is essentially the 'trolley' problem illustrated in Panel 2.1. There is no general agreement about either the solution or the approach. Philosophers can give different answers depending on the specific philosophical approach applied. These are not discussed here, instead it is assumed that the designer of each specific autonomous system will include fail-safe or other definable reactions to an unexpected event, and that they are acceptable to the society for that application. The autonomous system's reactions will probably be based on those of a reasonable[1] person following applicable, existing or planned laws and regulations, or societal standards.

---

[1]The definition of a 'reasonable person' might be the subject of legal debate. Here, it is taken as a person who is not a criminal and who wants to minimise harm to others.

Recognising that ethical principles have an effect on all aspects of engineering, it is important to understand how they may have an impact on autonomous systems at all stages of their life cycle. The engineering profession is beginning to address the wider engineering concerns, with professional bodies and academies publishing numerous reports either independently or on behalf of government for their guidance. Some of their conclusions are finding their way into laws and regulations and will continue to do so.

One of the most important documents encapsulating basic ethical principles as they affect humans is the United Nations Charter of Human Rights [11]. This is considered by some to be based on Western Christian doctrines, so there is also the Cairo Declaration [12] which may need to be considered. Both documents were influenced by national cultures and have direct and indirect influences on national laws.

6.3   Regulations, standards and certification

All procurement, from studies, through research, to product design and support is based on contracts in an agreed legal framework. Product and service specifications rarely include explicit extracts from laws, instead, technical quality is specified by recognised standards and adherence to regulations. These then flow down into system and product specifications through the processes described in Chapters 4 and 5. Quality systems in the supply chain ensure that these standards are adhered to. An important part of design reviews is to confirm adherence to requirements and that testing has verified it thoroughly.

This section explains the difference between regulations and standards in the context of autonomous systems. It should also be noted that regulations, standards and technology are intertwined. Section 6.4 introduces the new standards which are evolving for autonomous systems. These cover both civil and military applications so it can be anticipated that they will affect the evolution of autonomous technologies in all fields.

6.3.1   Regulations

A regulation is a rule or directive made by a government or by an authority acting on their behalf. Breach of them may be an offence according to the law of that country. Even if it is not a criminal offence, a breach of regulations will give a lawyer a solid starting point in the pursuit of financial compensation.

The authority may be a national one, but it may also be an international body. International bodies setting regulations and standards which must be complied with by any system operating in their regime are:

- The International Maritime Organisation (IMO) (www.imo.org)
  This agency covers all aspects of maritime operations. Their main instrument is the United Nations Convention on the Law of the Sea (UNCLOS) [13]. This mainly concerns areas of potential dispute between ships and nations, covering the sea bed up to, and including some aspects of airspace. It does give rules about scientific research in the maritime environment. The IMO also have

within its auspices, the International Convention for the Safety of Life at Sea (SOLAS), 1974 [14]. This specifies minimum standards for the construction, equipment and operation of ships.

- The International Civil Aviation Authority (ICAO) (www.icao.int)
  This agency covers all aspects of aeronautical operations. It was established to oversee the implementation and observance of the treaty usually referred to as The Chicago Convention [15]. Its technical remit includes standards for:
  o Communications systems and air navigation aids
  o Rules of the air and air traffic control practices
  o Licensing of operating and mechanical personnel
  o Airworthiness of aircraft
  o Aeronautical maps and charts
  o 'and such other matters concerned with the safety, regularity, and efficiency of air navigation as may from time to time appear appropriate'.
- The International Telecommunications Union (ITU) (www.itu.int)
  This agency develops the technical standards that ensure networks and related technologies can interact seamlessly. It allocates global radio spectrum and satellite orbits. The private sector is involved through membership of companies and academic institutions in partnership with the governments of the United Nations member states.

These bodies are all established under the United Nations and derive their authority from this. Although their role is dominantly for civilian regulation, there is extensive collaboration and working with the equivalent national military authorities. This is necessary both for operation of military assets in non-military areas (the majority of the world) and also for the regulation of both types of platform around and into regions with conflicts. Interpretation of the regulations can be extremely complex, especially the UNCLOS, so engineers should seek advice about their interpretation in any design process.

There is no equivalent agency or organisation which covers land vehicles. Instead, each country has its own set of regulations for road vehicles plus appropriate health and safety regulations for agricultural and other industrial equipment. These are not relevant in combat, but may be mandatory requirements for military vehicles when used on roads away from operational areas.

The International Organization for Standards (ISO) is an independent, non-governmental organisation with a membership of 164 national standards bodies. It writes standards by bringing together experts from its membership and other knowledgeable bodies to form a technical committee. The standard is then agreed by consensus, approved and published.

The military equivalent of regulations is their manuals and handbooks which provide rules for the conduct of military personnel. These may be considered to be national interpretations of IHL. They do not affect engineering design in the same way as regulations because they govern the conduct of personnel using equipment, not the operation of the equipment itself.

Autonomous systems will have to be designed to meet the regulations set by the relevant agencies. It may be possible to conduct experiments in controlled areas with strict boundaries on the areas where the test system can operate. Even in these cases, the design will need to have built-in limitations, rapid human intervention and fail-safe modes. Test plans and their risk assessments and safety cases should explicitly include these aspects.

There are many areas of activity which are highly regulated but, like road vehicles, many of the regulations are aimed at controlling human behaviour, not the technology. The associated technology is not often autonomous according to the definitions in Panel 1.1 even though it can be very sophisticated. Examples include medical robotics, home care and shipping. The relationship between technology and regulation is closer to the 'drone' one in Section 6.3.3 than to the 'large Unmanned Air Vehicle (UAV)' one. These will be areas where engineers will have to work closely with regulators at national and probably international levels.

6.3.2   Standards

A standard is an agreed method of working which is applied consistently. They have been in existence for over a century. An early standard is the Whitworth thread standard for nuts and bolts which was proposed in 1841, in common use by 1858, but did not become an official standard until 1880 [16].

The ISO defines standards in the following way [17]:

A standard is a document that provides requirements, specifications, guidelines or characteristics that can be used consistently to ensure that materials, products, processes and services are fit for their purpose.

They can be agreed at local, national or international level. Local standards may be set by a company for their own processes. These are not considered further in this book, but it is worth noting that ISO are introducing a set of system and software standards for very small entities (VSEs), ISO/IECTR 29110 [18].

National standards are set by bodies established by their national governments but not usually funded by them. When one nation has a dominant position in a market through its size, either as a customer or supplier, or both, its standards can become *de facto*, the international standards. This is the case with US Military Specifications (MIL SPECs) although other nations may have their own, such as the UK's Defence Standards (DEF STANs). Compatibility of standards and proving adherence to different sets may be a complex problem for a supplier selling one product sold into several markets.

Globalisation requires international standards. This was recognised in the 1940s with ISO being established in 1947 as a non-government international organisation. An ISO standard is produced through their internationally agreed process and then implemented in countries by their national standards organisation. Often there is no change, but sometimes there will be significant differences, so care must be taken to ensure that the design adheres to the correct and up-to-date standard.

Standards do not refer to regulations as the latter are controlled through the regulatory authorities and have the force of law. Regulations can refer to standards for their technical implementation. Both regulations and standards will evolve for autonomous systems and AI over the next years, so engineers are well-advised to understand both the current environment and trends in the standards relevant to their domain.

6.3.3   Drones – an illustration of the extremes of regulatory problems

The word 'drone' came to prominence in the early part of the twenty-first century to describe the medium-sized and large UAVs used in military operations. There were ethical debates about their use, but they are now a generally accepted part of the military inventory[2]. UAVs of this size are anticipated to be used for air transport purposes among others. The word is now more generally taken to mean the small UAV which can be purchased at low cost (tens to a few thousand pounds or dollars) from many suppliers and are in widespread use around the world. The two types of 'drone' have very different regulatory regimes.

There are numerous benefits in having no pilot on board. A UAV can fly for longer times and so fly further if needed. It can carry more payload as there is no pilot weight or life-support systems. Widespread civilian uses have been proposed and it could be argued that as most airliners have sophisticated autopilots that can carry out most parts of a flight, including take-off and landing, then it will be straightforward to make equivalent large UAVs. The problem is replacing the pilot's autonomy and oversight with remote human supervision, recognising that airspace is heavily regulated as well as all aspects of aircraft design and operation.

The first regulatory extreme is that for large UAVs with existing regulations being based on assumptions that do not apply to them and, arguably correctly, inhibit their introduction. The second extreme is that of small drones with initially no regulation, giving an almost anarchistic introduction into use, with the frantic (for regulators) attempts to prevent serious accidents and criminal activities before they become prevalent.

Air regulations assume that there is a human in the cockpit to oversee all operations. Examples include: programming and, if necessary, overriding the autopilot; monitoring the health of critical aircraft systems; constant monitoring of local airspace for possible collisions[3]; final decision-making in the case of imminent collision; removal of ambiguities in the Air Traffic Control (ATC) system by direct voice communication with the air traffic controller. All UAVs that operate in controlled airspace will have to be able to carry out all the pilot's functions autonomously. The most intractable of these functions is that of collision avoidance, the

---

[2]There are still several organisations which oppose the use of drones and Autonomous Weapon System (AWS). The debate about banning AWS is carried on through the United Nations meetings of a Group of Governmental Experts (GGE) under the umbrella of the 1980 Convention on Prohibitions or Restrictions on the Use of Certain Conventional Weapons Which May Be Deemed to Be Excessively Injurious or to Have Indiscriminate Effects (CCW).

[3]It is mandatory for most aircraft to be fitted with a Traffic alert and Collision Avoidance System (TCAS), also known as an Airborne Collision Avoidance System (ACAS), which gives pilots warnings of possible collisions with other aircraft if they are also fitted with TCAS.

problem being that of reliably sensing, and then avoiding other aircraft and obstacles, the UAV 'sense and avoid' problem.

Airspace below 500 ft is generally uncontrolled with far fewer regulations than controlled airspace. There are excluded areas such as those around airfields. Model aircraft have used this airspace for decades with minimal regulations, such as the aircraft must weigh less than 25 kg and always be in sight of its operator. The last ten years have seen the development of small (<~10 kg) drones using technological developments in automated navigation systems, lightweight materials, motors and batteries. These have proliferated and their numbers must now be in the hundreds of thousands around the world. Users include emergency services, hobbyists, press and the media, and utility companies among many others. Initially unregulated, their widespread use has led to three main types of concern: safety issues as operators are generally unlicensed, with numerous near-misses with passenger aircraft[4]; criminal use for transporting drugs and other illegal materials; and invasion of personal privacy. The result has been that national regulators, mainly in the UK and the US, have had to introduce regulations quickly to address these concerns. The first steps have been to introduce pilot licences and registration of the aircraft, but there are numerous practical problems in their application as well as the questions of regulating already-existing pilots and drones.

6.3.4   Certification

All products must meet the appropriate national regulations and standards. The formal process demonstrating that a product meets all relevant national standards is known as certification. Certification is defined by the ISO as [19]:

> The provision by an independent body of written assurance (a certificate) that the product, service or system in question meets specific requirements.

The earlier parts of this section show that the regulatory position for autonomous systems in general is not yet clear. This makes certification problematic with solutions requiring inputs from regulators, standards bodies, and research and development organisations. The state of the art is moving rapidly, so any company or design team considering a product that need certification must, *as a minimum*, consult their national regulatory authority and conduct an extensive literature review before committing significant funds.

The first pragmatic step in finding a route to certification is to understand how current regulations apply to the proposed product. Then it is possible to identify the necessary changes to them. The ASTRAEA programme in Panel 6.1 is an example of a successful programme examining certification for autonomous systems in the air domain.

---

[4]The UK AIRPROX Board reports no near-misses in 2012 and 2013, but 6 in 2014, 29 in 2015, 71 in 2016 and 93 in 2017. Information from https://www.airproxboard.org.uk/Topical-issues-and-themes/Drones/ and one collision in 2016. (Accessed on 6 October 2018).

Panel 6.1:   The ASTRAEA Programme

The Autonomous Systems Technology Related Airborne Evaluation and Assessment (ASTRAEA) research programme emerged from the UK Government's Aerospace Innovation and Growth Strategy programme in 2001.

The strategy had identified that autonomy was a significant emerging technology that would impact the civil aerospace sector. Consequently, the major UK aerospace companies came together to understand what would be required to introduce large UAVs into all classes of shared airspace before embarking on high risk and high cost development programmes without the certainty that certification could be achieved. The resultant ASTRAEA programme ran from 2006 to 2015 and was jointly funded by the UK Government and industry.

The programme was undertaken in phases, beginning with capturing the technical requirements that arise as a result of moving the pilot from the vehicle to the ground and the consequent needs of the air vehicle, the remote pilot station and the infrastructure as well as the route to certification. This was followed by two major work strands, Autonomy and Decision Making and Separation Assurance and Control, developing the necessary technologies, such as detect and avoid and the communications and control strategies, to enable the seamless introduction of large UAVs into shared airspace. It made extensive use of simulation, ground and airborne laboratory testing. The presumption was that the air vehicle would fly itself but remain under the supervision of a ground based 'pilot'.

A variable autonomy concept was investigated where the UAV would seek instruction from the supervising pilot in the event of an anomaly, such as a potentially impending collision or systems unserviceability, but with the ability for the UAV to take over full decision-making in the event of a loss of the communication link to the ground. A fundamental requirement was that it should be able to operate in any airspace or from any airfield without impacting other air users or Air Traffic Management (ATM).

Throughout the programme, the industry members worked closely with the UK's Civil Aviation Authority (CAA) and the European Aerospace Standards Organisation (EUROCAE) in developing the standards and the route to certification for the subsystems required. This was then taken to the full system level by undertaking a complete virtual certification of an existing small passenger aircraft (Part 23 of FAA certification regulations) and converting it into a UAV. This demonstrated that the existing certification processes could be developed for large UAVs and it captured, in some detail, all the outstanding certification requirements.

6.4   Current standards for autonomous systems

Panel 3.1 in Chapter 3 gives definitions of several key words used in this book as it is important to have consistency in their use. It is even more important that a set of standards have consistent use across them. This is achieved by using a standard vocabulary and an ontology. An ontology is much more than a dictionary. The

Oxford dictionary defines it as *a set of concepts and categories in a subject area or domain that shows their properties and the relations between them.* A well-constructed ontology can be interpreted by a computer and used as the basis for human–computer interaction.

Robotics and autonomy are relatively new engineering fields, but there are many application areas. Consequently standard writers have had to restrict themselves to specific areas with pressing problems. At the time of writing, there are three published standards for robotics. These are defined in Sections 6.4.1–6.4.3.

6.4.1 ISO 8373-2012 robots and robotic devices – vocabulary

This is a set of definitions of 169 terms for robots, where a robot is defined to be an *actuated mechanism programmable in two or more axes with a degree of autonomy, moving within its environment, to perform intended tasks.*

The standard's aim is to cover robots in both industrial and non-industrial applications, but the assumptions underlying its definitions makes many of its definitions irrelevant for military and aerospace applications. However, some of the definitions can provide part of an unambiguous vocabulary for use in the design of autonomous systems in all applications. (See the discussion in Chapter 3 on definitions.)

6.4.2 IEEE 1872–2015, IEEE standard ontologies for robotics and automation

This standard, like ISO 8373-2012, assumes that a robot is a physical entity accomplishing one or more predetermined tasks. The group writing this standard tried to make it as general as possible, but concentrated on industrial and service robotics as the most pressing areas that need an ontology. This means that, as written, it is valuable but should be used with care in defining autonomous systems.

The aim of the standard is to provide a methodology and a core ontology for robotics and automation rather than just provide a vocabulary. There are 34 terms defined in the standard expressed in a form suitable for use in a specific ontology (Suggested Upper Merged Ontology – SUMO).

The standard was used as a basis for defining autonomy in Panels 1.1 and 3.1 with some of its terms and others from BS8611:2016, given in section A3.1 of Appendix A3 to Chapter 3.

6.4.3 BS8611:2016, robots and robotic devices. Guide to the ethical design and application of robots and robotic systems

The increasing public perception that robots and autonomous systems may have a large impact on all aspects of human activity has led to ethical debates. BS8611:2016 is the first edition of a new standard addressing ethics. It does not attempt to anticipate developments. Consequently it is given as a guide and recommendations not as a formal specification, so claims of compliance cannot be made to it.

The standard builds on ISO 8373-2012. It takes all the definitions in that document and adds ten extra ones. This does mean that there is an implicit assumption about physical instantiation, i.e. it applies to a robot, so care must be taken in using the terms for specific applications.

The standard includes the following widely applicable paragraphs in Section 5.1, *Societal ethical guidelines and measures*:

5.1.7 Legal issues
The roles, responsibilities and legal liabilities should be clearly identified for all stages of the robot's life cycle. It should always be possible to easily discover the person(s) legally responsible for the robot and its behaviour during all stages of the life cycle.

5.1.12 Informed command
On receipt of a command, the robot should be able to construe it into a properly-constructed command, and check it for coherence and compatibility with its internal constraints (including any ethical constraints). If the command does not appear to be properly constructed, the robot should be programmed to pause and query the command.

Note: For examples of properly-constructed commands, see NATO's C-BML. For further information, see Reference [20]. (This is reference [10] in the standard.)

These two paragraphs give requirements for clarity of responsibilities and control procedures. If the design of an autonomous system complies with Paragraph 5.1.7, the liabilities discussed in Section 6.6 below can be assigned to defined organisations and even individuals. The concept of authorised power [21], defined in Panels 1.1 and 3.1, discussed in detail in Chapter 10, Sections within 10.8 and Section 10.9, gives an approach that can ensure a design meets this requirement.

Despite the robotic limitations, military applications are included. In section *5.2 Application of ethical recommendations and measures*, the standard states:

5.2.4 Military use
The use of robots in military applications should not remove responsibility and accountability from a human.
The deployment of robots should be in accordance with international humanitarian laws and laws governing armed conflict.

This is an explicit statement of the applicability of IHL to the deployment and use of weapons. It does not give any detail so it does not put any constraints on the designer in meeting the standard.

6.5   Future regulations and standards for autonomous systems

Development of a new standard has to be a careful process with widespread consultation. Consequently it takes several years for the initial version, followed by a series of new issues as the standard is applied widely and unexpected problems or ambiguities are found. Usually regulation follows behind the technology, but there are exceptions. One example is the development of both the technology and the legal framework for autonomous vehicles. Here, there is a potentially huge market,

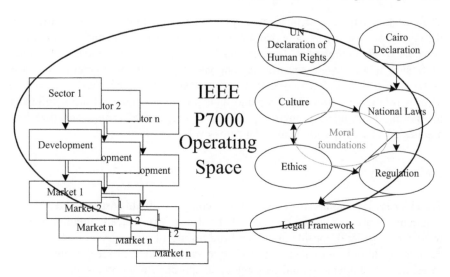

Figure 6.1   Ethics, culture, engineering and standards

but one where safety is clearly paramount. This led to governments and industry working together in the countries that have an industry that wishes to exploit the technology but needs clear regulatory guidance.

Probably the first engineering standard that addresses ethics as a general issue is IEEE P7000 *Model process for addressing ethical concerns during system design* [22]. This is part of a wider IEEE initiative, the IEEE Global Initiative on Ethics of Autonomous and Intelligent Systems [23], and is one of a family of nine draft standards on various aspects of autonomous systems and AI. The related IEEE document Ethically Aided Design [24] gives statements of ethical issues for thirteen topics that engineers and others should consider for all work in autonomous systems.

Figure 6.1 shows several relationships. The right-hand side shows how national laws result from human rights documents, moral foundations and the nation's culture. Laws are then the framework for regulations which usually have the force of law in their implementation. Ethics and culture are closely related and both evolve over a period of years and decades. Regulations for autonomous systems are strongly influenced by both laws and ethics, especially for their use and implicitly in their design, support and disposal. The ellipse shows the likely coverage of the IEEE P7000 series of standards.

Section 6.3.1 gives some of the regulatory authorities for transport. Regulations in these and other areas cover virtually all disciplines in their sectors, including the design, testing and use of autonomous systems. It can be confidently predicted that regulations will evolve symbiotically with the technologies. There may be problems and even some potentially serious problems during this process so engineers must be alert to any potential risks and bring them to the attention of the relevant authorities.

Autonomous systems are starting to have applications in many other fields such as medicine, caring for the elderly and disabled people and security guards.

These sectors may have sophisticated systems and are strictly regulated. However, the regulations usually assume that all decisions are made by qualified individuals and none by their equipment. The personnel follow professional guidelines for their profession as well as any specific regulations. Regulations will emerge for these wider sectors, but anyone associated with introducing autonomous systems into them will have to ensure that they and their project management are continually monitoring the regulatory environment and be prepared to change specifications and design as the regulations and standards evolve.

6.6   Safety

The business and engineering processes in a company should ensure that legislative requirements are met in the course of normal business. Instead, attention here is restricted to general considerations and the testing of autonomous systems. This section should not be treated as a definitive as it is intended for guidance. In all cases, definitive advice should be sought and appropriate action taken.

Safety is subject to considerable legislation and enforcement in every country. It is the responsibility of dedicated organisations and management procedures and will not be covered here in its wide sense. All engineers should understand, as a minimum, that systems should be assessed for the severity of the consequences of a system failure. When the results will be catastrophic, generally those that involve loss of life or serious injury, then the system is safety-critical and special mandated processes must be followed. One essential system standard is ISO 15026 [25] which has a bibliography of related standards. Software has its own set of safety standards and the most suitable one should be selected for a project [26]. Increased autonomy in control systems does have safety implications; these cannot be ignored and the system integrity and security levels should be reviewed.

When assessing safety, all aspects must be considered. Failure to do so can lead to failures arising from multiple occurrences of low-probability failures. This is known as the Swiss-cheese argument and is illustrated in Figure 6.2. Take many slices of Swiss cheese which all have random holes in them and a light shining on them from one end. Each slice is a safety mechanism or process and the holes are

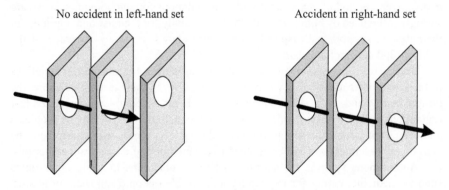

No accident in left-hand set   Accident in right-hand set

Figure 6.2   The Swiss-cheese model of safety

breaches in it. These breaches will change over time. Light cannot pass through them all if the holes do not align. This is the initial case. If there are enough holes and their size and position change over time, then there is eventually a high risk of them all aligning and light passing through them all. This illustrates the evolution of risk and how it can lead to system failures. In autonomous systems there may be continual upgrades and new information being used for decisions, giving more holes in the cheese. This makes continuous safety assessments essential.

A well-documented example of an accident occurring due to multiple risks all happening in rapid succession was the loss of the tanker aircraft XV230 with all 14 crew members over Afghanistan in 2006. Although this was a military aircraft in an operational area, the causes of the accident were not due to its operational use. There was an accumulation of problems, each of which gave a small risk of catastrophic failure – the slices in the Swiss-cheese model. There was a wide-ranging enquiry by a senior lawyer, Charles Haddon-Cave QC. Although there were no autonomous systems involved, the report highlights in a very forceful way how multiple, relatively small events can lead to disaster [27]. It led to a complete revision of the UK safety regime and the establishment of the UK Military Aviation Authority (MAA). The lessons from the report should be applied in all aspects of autonomous systems regardless of their application area.

Tests of any system should only be conducted after considering the risk of harm to people and other objects. This should be conducted formally and the result written up as a risk assessment. This lists the potential sources of harm, the chance of them happening as high, medium or low, and mitigation action. This should be part of every company's processes, the detail and complexity depending on the particular activity.

A system that is mobile and under test should have a test or trials plan written for the test programme. This gives the objective of the trials programme, with all necessary details for conducting them. Safety is demonstrated by a written safety case which must provide evidence. A literature review for assurance cases is in Reference [28] which has been prepared to assist in assessing how to structure safety cases. One definition of a safety case is in the UK Defence Standard 00-56 [29]:

> a structured argument, supported by a body of evidence, that provides a compelling, comprehensible and valid case that a system is safe for a given application in a given environment.

The safety case must show how every risk in the risk assessment has been mitigated and reduced to an acceptable level. A risk of an event happening can still be high, but the safety case should show how the severity of its consequences have been reduced to an acceptable level. Risk cannot be reduced to zero; instead there is the concept of As Low As Reasonably Practicable (ALARP) which is discussed in Section 6.8.

The problem with autonomous systems or functions is the lack of certainty in its actions. The safety case must show the range of possible actions, how they are limited, and evidence to show that the limits are valid. Military trials of unmanned systems have been conducted for many years and some of the lessons are applicable to civilian tests and given below.

A general guide to safety in military systems can be found in a UK MOD booklet, *An introduction to system safety management in MOD* [30]. Some of the materials in this section are based on the booklet. Weapons have detailed procedures for their design and use, such as explosive-handling ones, which are also not relevant to this book.

It could easily, but incorrectly, be surmised that civil safety rules do not apply to military systems as their applications will be either in or, in support of, combat operations where military and IHL laws will apply. This assumption is false because all military systems, including weapon systems are used for many purposes outside military operations. Military training areas and test ranges are restricted in size even if they are in less-densely populated areas. The risk to people and objects in adjacent areas must be minimised. The risks can be due to objects leaving the trials area and causing damage, but it can also be due to radio interference, noise or any other disturbance. Risks due to unmanned air, marine or land vehicles moving outside the trials area, or harming personnel in the area, must be thoroughly assessed with fail-safe procedures, and specialised equipment built into the unmanned assets. All radio and optical signals used during the trial and preparations must follow ITU rules, noting that there are usually extra regulations that apply near airfields.

There have been debates in the UK about how far employer liability and Health and Safety (H&S) legislation applies to troops undergoing training [31] and operating from large military bases in countries where there are combat operations. This has led to close collaboration between MOD and the H&S Executive (the UK body responsible for the enforcement of H&S laws and regulations) with clarification of which law applies in given circumstances. This outcome can be generalised to postulate that analogous understandings will be reached in most nations.

6.7 Liabilities

The company that provides any product that does not behave according to its specification, or fails due to a design fault or omission is liable to face litigation by those who suffer loss. For most products, the litigation is for financial compensation, but there are also cases when it can be a criminal prosecution of the company and their board of directors. Breaches of IHL can lead to prosecution of the commander at the International Criminal Court. With increased autonomy in systems and products, it is not yet clear how the liability is to be identified as belonging to the user, supplier or elsewhere,

It is possible to identify a core of evidence that will be sought by lawyers acting for either party during litigation. It should be generated as a normal part of the design and test phases, but if it is not be recorded or archived for later use, the production organisation will appear deficient. This section takes a military procedure (JSP454) [32] as it illustrates the type of information that any government or organisation will require in order to operate a system safely. It provides a good check list for civilian applications.

JSP 454, in Paragraph 36, lays down duties for staff to discharge at each stage of the CADMID cycle[5]. This is to ensure that MOD can meet its responsibilities and identify any potential liabilities. They are typical of the duties that any organisation delivering a sophisticated system should carry out and are expressed here as responsibilities placed on the suppliers and support contractors:

1. All suppliers or designers who are involved in the project are competent in their field, and that they fully understand the application of the system, recognising that security rules may limit the knowledge of operations that can be imparted to the supply chain;
2. Provide evidence and demonstrate that any risks associated with the safety of the system have been eliminated or reduced to a level that is ALARP and potential risks to the environment are managed as far as reasonably practicable;
3. Provide information and data so that adequate instructions, training and maintenance arrangements can be put in place to control those risks within defined limits;
4. Provide a level of support so that MOD can ensure that the system remains in a good and acceptably safe condition throughout its service life;
5. Provide data and information so that adequate means can be put in place to monitor the safety and environmental performance of the system to ensure continual improvement;
6. Provide data and information so that the system is managed in accordance with the Safety and Environmental Management System and the recommendations and controls stipulated in the Safety and Environmental Case(s);
7. Provide data and information so that instructions and training can be provided for use of the equipment in as safe a manner as is reasonably practicable. Attention should be drawn to those authorities responsible for upkeep and/or operation of any deficiencies or shortfalls in safety performance;
8. Provide data and information so that MOD, when disposing of the equipment, can fulfil its duty of care to those agents and authorities acquiring the system and that the environmental implications of waste are captured.

The requirement for competence should be self-evident, but is essential. Given the rapidly evolving nature of autonomous systems, there is an implication that Continuous Professional Development (CPD) for all staff should include keeping abreast of developments in their fields.

It should be noted that data alone may not be sufficient as it will almost certainly need some interpretation. The customer requires the information needed to set up their processes to ensure safe operation in normal use. The volume and quality of information provided to the customer and support agencies is a contractual issue, but must be sufficient to meet the above criteria. There are usually several contracts involved in the development, supply and in-service operation of

---

[5]The CADMID cycle, discussed in Section 5.4.2, is the whole product life cycle from Concept to Disposal.

complex equipment so the requirements and cost may be negotiated during the supply process.

An autonomous system will be dominated by software. This may be bespoke or Commercial Off The Shelf (COTS). The two will need to be clearly separated. Assessment of their reliability and safety will be either by process or product-based techniques. The chosen technique and supporting evidence from the software supplier must be presented so that field trials can address any potential deficiencies in performance due to limitations in the software design approach.

A supplier must provide the type of information given above for any complex system or system of systems. Failure to do so, resulting in the users being unable to discharge their duties, may lead to legal liability and penalties, especially in the event of a mishap.

6.8   Risks, benefits and ALARP

Increased autonomy is introduced to bring benefits to the system or process. The benefit will have been quantified in some way and the cost of introducing it needs to be estimated. The financial constraint on cost is usually the available budget. If the likely cost is well below the available funds, the budget holder is likely to reallocate the surplus to other needs, leading to a cap on the project cost and the same pressures as if the available funds were only just sufficient.

There are suggestions that the benefits of a product should be replaced by values which include not only the market value but also the more intangible values to society as a whole. They are discussed in Reference [7] and are part of the draft IEEE P7000 standard.

Business cases are based on cost/benefit analyses of the proposed system or service. Budget holders and procurement executives have well-established techniques for assessing and ranking proposals. Risk includes: changes in available finance; failure to achieve quality requirements; and slips in delivery dates. These are not considered here as they are straightforward programme risks and should be taken care of in the processes described in Chapters 4 and 5.

This section looks at another fundamental aspect of the risks for a new system which are particularly relevant to an autonomous one, the risk that it will cause harm. First we must understand what is meant by risk in this case. We will base our discussion on the definition of ethical harm in BS8622:

- Risk is the probability of harm occurring from the frequency and severity of exposure to a hazard;
- Harm is injury or damage to health of humans.

Harm can be extended to include harm to other systems or equipment, but again this is part of standard engineering design processes. Equipment or component failure rates should be an output of design, calculation and test. These figures can form part of the safety case and risk assessment process discussed in Sections 6.6 and 6.7.

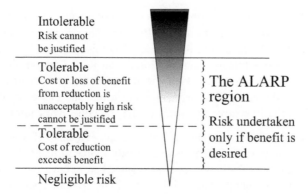

Figure 6.3   Tolerable and acceptable risk

The basic question is: what level of risk can be tolerated? Zero risk is impossible, but we want to make it as low as possible. However, this is also impossible unless we have a semi-infinite budget. The discussion then becomes one of seeing how low it can be made within the budget, timescale and staff skill levels for the system, both in design and test prior to delivery and in-service. The system must still be able to deliver the benefits to operations that it promised as well as being safe.

The next question is: can this level of risk be tolerated? i.e. is it acceptable? Figure 6.3 [33] illustrates the factors to be considered in answering this question. An intolerable risk should never be accepted. Negligible risks should not be ignored, but should be capable of management by use of existing processes.

Success from answering the two questions above, which may be a long process, is that the risk has been reduced to a value which gives a probability of harm that is ALARP. The ALARP region is where most engineering products lie. Considerable judgement is used to decide what the risks are, their classification and mitigation. The final decisions for a given product will be made by the company's Design Authority.

The Design Authority role, or equivalent, such as Technical Director in a small company, is a formal one. It will have defined responsibilities in the company' procedures and the holder has legal responsibility for them. Normally the Design Authority will 'sign off' the final deliverable after ensuring that all design reviews have been completed or any non-compliances agreed and accepted by the customer. Smaller companies may not have a Design Authority who can authorise certain types of system or components. This possibility should be checked at the start of a bidding process. If it occurs, partnering should be agreed with a company with that authority and it must be accepted by the customer.

A key factor in any post-delivery enquiry is the level of evidence that a company can produce. The data gathered during design, test and verification will be needed as well as justification that this was sufficient to show that the risk had been reduced to a demonstrably acceptable level. Evidence will also need to include assessment of factors outside those considered in the engineering design process. The Design Authority needs to ensure that a wide review of risks has been completed and the risk register or equivalent document includes these factors.

The ALARP concept is a complex one and not always understood. One mis-understanding is that a risk that is ALARP cannot be accepted in normal use and cause the system to be taken out of service. This misunderstanding occurred in a high-profile Coroner's Inquest[6] into the loss of an RAF Nimrod aircraft with 14 crews mentioned in Section 6.6. This and the need for breadth of ALARP con-siderations were shown by a subsequent enquiry into the causes of the accident. Paragraph 23 of the Executive Summary of the Haddon-Cave enquiry [26] stated:

> The Coroner's Inquest produced little factual evidence of value to the Review. The Coroner's finding as to the likely source of fuel did not accord with the realistic probabilities, or the evidence before him, and his Rule 43 recommendation (that the Nimrod fleet should be grounded pending certain repairs) was based on his misunderstanding of the meaning of As Low as Reasonably Practicable (ALARP). The Coroner's widely-publicised remark that the MOD had a 'cavalier approach to safety' was unjustified. The fun-damental problems are ones of structure, culture, and procedure, not indif-ference. (Chapter 16)

The enquiry's recommendations for a complete restructuring of the UK mili-tary air processes were implemented fully. The evidence at the enquiry showed that everyone involved with systems should be pro-active about safety concerns and not let them pass without being acted on. This applies to everyone reading this book!

6.9   Safety cases for autonomous systems

How do Sections 6.5 to 6.8 relate to an autonomous system? This question is best answered by considering how to draw up the safety case or risk assessment for an autonomous system. This section will only give some initial guidance. Specifics will depend on the system and reference to the relevant standards and regulations will also give a guide to the best way to identify what it should contain.

A good starting point is the top-level questions, taken from Section 7.2 of Reference [34]. These are given in Table 6.1 with the answers from that reference. They are equally applicable for the case of an autonomous system. The autonomy in the system leads to less certainty in the detailed answers as there may not necessarily be complete certainty about the response of the system to changes in its environment.

The safety case will need to identify failure modes both in it and in its wider system-of-systems and estimate the probability of it happening. The case must provide evidence about the limits on the autonomous system's behaviour for each of these failure modes, the resultant harm and the probability of it happening. The consequences of all responses to all questions and the derived ones from Table 6.1

---

[6]In the English legal system, the initial legal enquiry into any death which is not clearly from natural causes is conducted by a coroner as an inquest. They have wide-ranging powers. The result of the inquest can trigger formal legal procedures and criminal investigations. This coroner's inquest was into the deaths of the fourteen service personnel who died in the accident and pre-dated the Haddon-Cave enquiry.

Table 6.1 Questions to be answered by a safety case

Question	Potential answer
What are we looking at?	System description
What could go wrong?	Hazard identification and analysis
How bad could it be and what are the major threats?	Risk estimation
What has been or can be done about it?	Risk and ALARP evaluation, risk reduction and acceptance
What if it happens?	Emergency and contingency arrangements

must be thought through. There will be a need to develop mitigation plans in the normal way.

If the autonomous system is a military one with kinetic effects, there may be no practical mitigation plan for some failures. In these cases, the risk and consequences will have to be identified and the operational consequences evaluated. The results will almost certainly be an input into the setting-to-work programme and the procurement authority may need to seek legal advice if there are potential breaches of IHL.

References

[1]   Bostrom N. *Superintelligence*. Oxford, UK: Oxford University Press; 2014.

[2]   Lin P., Abney K., and Bekey G.A. *Robot ethics*. Intelligent robotics and autonomous agents series. Cambridge Massachusetts; MIT Press; 2012.

[3]   Wallach W., and Allen C. *Moral machines*. New York: Oxford University Press; 2009.

[4]   International Committee of the Red Cross. Part 3 o*f Report on expert meeting on autonomous weapon systems, technical, military, legal and humanitarian aspects, 26–28 March 2014.*

[5]   European Union, Project Robolaw, Report D6.2 Guidelines on Regulating Robotics, 2014.

[6]   UK Joint Service Procedure (JSP) 454 Paragraph 33.

[7]   Spiekermann S. *Ethical IT innovation: A value-based system design approach*. Boca Raton: CRC Press; 2016.

[8]   The RAEng and IET ethical guidelines can be found at: www.theiet.org.uk and www.raeng.org.uk [Accessed on 6th October 2018].

[9]   Graham J., Haidt J., Koleva S., *et al*. 'Moral foundations theory: The pragmatic validity of moral pluralism'. *Advances in Experimental Social Psychology*. 2013, vol. 47: pp. 55–130.

[10] General details are available at the Moral Foundations website www.moralfoundations.org [Accessed on 22 January 2018].

[11] UN Charter of Human Rights, 1948. Available at http://www.un.org/en/universal-declaration-human-rights/ [Accessed on 15 January 2019].

[12]   Cairo declaration on human rights in Islam, Adopted and issued at the nineteenth Islamic conference of foreign ministers at Cairo on 5 August 1990. Text online at https://www.oic-iphrc.org/en/data/docs/legal_instruments/OIC_HRRIT/571230.pdf

[13]   United Nations Convention on the Law of the Sea, available at http://www.un.org/depts/los/convention_agreements/convention_overview_convention.htm [Accessed on 26 February 2017].

[14]   Convention for the Safety of Life at Sea. A description is given at http://www.imo.org/en/About/Conventions/ListOfConventions/Pages/International-Convention-for-the-Safety-of-Life-at-Sea-(SOLAS),-1974.aspx [Accessed on 26 February 2017]. The consolidated set of documents is available from IMO publications.

[15]   Convention on International Civil Aviation, 2006, also known as The Chicago Convention, Published by the International Civil Aviation Organization. This is updated from time to time.

[16]   Information taken from http://www.whitworthsociety.org/history.php?page=2 [Accessed on 21 February 2017].

[17]   http://www.iso.org/iso/home/standards.htm [Accessed 15 January 2019].

[18]   Available from the ISO website [17] and entering VSE into the search engine.

[19]   International Organisation for Standards website www.iso.org/certification.html [Accessed on 14 October 2018].

[20]   Hieb M.R., and Schade U. *Applying A Formal Language of Command and Control For Interoperability Between Systems*. George Mason Univ Fairfax Va Center for Excellence in Command Control Communications Computers-Intelligence [Accessed on 21 May 2008].

[21]   Gillespie T, 'Systems engineering as a solution to legal problems for highly-automated military systems'. *Presented at RAeS conference: Autonomy - are we there yet?* [Accessed on 17 June 2016].

[22]   IEEE P7000 Model process for addressing ethical concerns during system design. This is due for release in 2019 but information is available at: https://standards.ieee.org/develop/project/7000.html

[23]   The IEEE Global Initiative on Ethics of Autonomous and Intelligent Systems. Details at: https://standards.ieee.org/develop/indconn/ec/autonomous_systems.html [Accessed 15 January 2019].

[24]   Ethically aided design, V2.0, IEEE, available at: https://ethicsinaction.ieee.org/ [Accessed 15 January 2019].

[25]   ISO 15026 Systems and software engineering – Systems and software assurance: *Part 1, Concepts and vocabulary; Part 2, Assurance case; Part 3, System integrity levels; Part 4, Assurance in the life cycle*.

[26]   Wong W.E., Gidvani T., Lopez, A., Gao, R., and Horn M. *Evaluating software safety standards: A systematic review and comparison*; 2014 IEEE Eighth International Conference on Software Security and Reliability-Companion; 2014. pp. 78– 87.

[27]   Haddon-Cave, Charles. *The Nimrod Review*. HM Government report ordered by the House of Commons to be printed 28 October 2009, Available from

https://www.gov.uk/government/uploads/system/uploads/attachment_data/file/229037/1025.pdf

[28]   Nair S., de la Vara J.L., Sabetzadeh M., and Briand L. 'Classification, structuring, and assessment of evidence for safety – A systematic literature review'. *2013 IEEE Sixth International Conference on Software Testing, Verification and Validation, Luxembourg*; 2013, pp. 94–103.

[29]   UK Ministry of Defence, *Def-Stan 00-56 Safety management requirements for defence systems, Part 1 Requirements*, Issue 4, paragraph 9.1, June 2007.

[30]   UK Ministry of Defence booklet. *An introduction to system safety management in MOD.* Issue 3 January 2011.

[31]   House of Parliament, Commons Select Committee on Defence https://www.publications.parliament.uk/pa/cm201516/cmselect/cmdfence/598/59809.htm [Accessed on 20 February 2017].

[32]   Joint Service Procedure (JSP) 454, *Land systems safety and environmental protection, Part 1: Directive.* Version 7, UK MOD Defence Authority for Health, Safety and Environmental Protection. July 2016.

[33]   This figure is based on Figure 9.1 in the Haddon Cave Report [27]. It is referenced there as Lord Cullen's ALARP diagram with references: Cullen, fig 17.1. The diagram was introduced to the Piper Alpha Inquiry by Dr. M. S. Hogh, Manager of Projects and External Affairs, Group Safety Centre, BP International.

[34]   UK MOD Booklet: *An Introduction to System Safety Management in the MOD.* Issue 3; Defence Ordnance Safety Group – Safety Management Office (DOSG SMO); January 2011.

Chapter 7

Introduction to military legal context and its relevance to engineering

7.1 Introduction

Design standards and regulations are an essential part of every engineer's work. Demonstrating adherence to them is an accepted part of every aspect of a product's life cycle. In a similar way, war is governed by its own set of rules, known as International Humanitarian Law (IIHL), often called the Laws of Armed Conflict (LOAC). IHL and other legal terms are explained in Panel 7.1. Most of IHL relates to the conduct of warfare and its immediate aftermath, but IHL's evolution has been shaped by the technologies used in war. It accepts that war will happen but aims to minimise suffering so the results include both limitations on weapon design and requirements for improved performance. These affect many aspects of weapon design either directly or indirectly.

This chapter describes the legal framework that applies to armed conflict. It shows how specific parts of international law set engineering requirements which must be met by a design. This and other chapters are not intended to be definitive; engineers should always refer to legal advisors for specific guidance. Chapter 8 discusses current targeting processes, their compliance with IHL and the specific issues which arise when increased automation is proposed.

Most IHL protocols have the same structure, with Parts, Sections and Chapters made up of Articles numbered sequentially through the document. References in this book are made to the Article number which is unique within a protocol or treaty. Definitive copies of the latest version of the treaties, protocols and other IHL documents are available from the United Nations (UN) and International Committee of the Red Cross (ICRC) websites.

There is no international treaty on autonomous weapons, but there are concerns about increased autonomy in weapon systems which could lead to Lethal Autonomous Weapon Systems (LAWS). As with many terms, LAWS is not defined precisely, but is generally taken to mean weapons with a high degree of independence, probably with Artificial Intelligence (AI), so that they can find and attack targets with no human supervision after release. These concerns have led to the establishment of UN-sponsored meetings which are explained in Section 9.10.3. There are numerous papers presented and published by governments, Non-Government Organisations (NGOs) and lobby groups about autonomous systems.

They show a wide range of opinions and evolving ideas, so the interested reader should conduct their own literature review for current views. It is advisable to make a critical review of the technical developments and engineering problems for the advances proposed in some of them.

Panel 7.1:   A note on legal terminology

A fuller discussion of these terms, the differences between them and a historical introduction is in Solis [1] Chapter 1. Chapter 13 of that reference discusses rules-of-engagement.

There are several legal terms which, unfortunately, are sometimes used loosely in the defence industry. It is important that engineers understand their correct meaning as engineering data is used for different purposes by legal and military authorities at different times.

Armed conflict is not the same as war. The latter is more restrictive; the differences matter to national leaders in justifying their military actions but probably make no difference to weapon design.

The acronym IHL is used throughout this book and is explained in Section 7.2. It can be considered to be synonymous with the term LOAC for engineering purposes. There are some small legal differences but these do not affect engineering requirements.

Rules-of-engagement are written by a state for its armed forces in a specific scenario. They dictate which weapons will be deployed and how they will be used. They are discussed in Section 7.11.

Article 36 Reviews are reviews of new or novel means or methods of warfare. They are explained in Chapter 11. Article 36 is in Additional Protocol 1 and is explained in Section 7.11.2.

It is worth noting that international law is different to domestic law. The former is agreed between states by mutual consent on their wording and enforcement; states have their own interpretation of these which is usually made public. Domestic law is set by the acts of parliament or other national processes, and adherence to it within the relevant domestic system is mandatory.

7.2   The need for laws of war

Warfare is as old as humanity and there may always have been customary ways to limit the harm that it causes. The book by Solis [1] quotes an agreement between Egypt and Sumeria in 1400 BCE concerning the treatment of prisoners. The first international war-crime prosecution was probably that of Peter von Hagenbach in Breisach, Austria, in 1474. An *ad hoc* tribunal of judges found him guilty of murder, rape and other offences despite a defence that he was only following orders. He was hanged.

One of the most important books written about war is by the Prussian officer, Carl von Clausewitz. This was published posthumously by his widow in 1832, but there are several modern translations into English, for example, Reference [2]. He described war as: *merely the continuation of policy by other means.* Accepting this premise, we can see that the death and destruction of war are not aims in themselves, but are to be considered as one of the tools available to a government in pursuing its chosen policies. This view allows a state to declare war on another merely because the second one will not follow the wishes of the first. The twentieth century saw a reaction against this view, and an increasing desire to outlaw war, initially as a consequence of the First World War. Following World War II, the United Nations Charter requires that States abstain from the threat or use of force against another State, except as a response to an armed attack – which is a term defined under customary international law. However, the UN Security Council can decide to resort to the collective use of force in response to aggression, or threat to, or breach of peace.

There is a natural human desire to avoid causing harm to others. Adding this to the general view stated in the last paragraph that the use of force should be minimised leads to a logical reason for laws governing the conduct of war. Basically, death and destruction should be limited, and halted when one state prevails over another, its opponent.

Specific technical developments may be seen as abhorrent and outlawed before coming into widespread use. The 1868 St Petersburg Declaration renounced the use of explosive bullets; the 1899 Hague Conventions banned the use of asphyxiating gases and expanding bullets. These have been reasonably successful although the use of gas in World War I is a notable exception. However, the early-twentieth century attempts to impose legal controls on the new domain of aerial warfare were less successful. These started in 1899[1] and 1907[2] with attempts to restrict the use of balloons and aircraft as launch platforms for ordnance. The 1923 Hague Rules of Aerial Warfare had so few signatories that they are often referred to as draft rules. There have been so many breaches of the rules that they are not now considered to be relevant except where certain parts are explicitly included in subsequent treaties. Despite this failure, it is wrong to deduce that aerial warfare is not covered by IHL; it still applies, but using its more general principles and the post-World War II conventions.

Greater success in banning specific weapon types has been achieved through the UN, especially a 1980 treaty known as CCW and summarised in Section 7.10.7. New or existing weapons can be brought into its remit by the use of specific protocols such as Protocol IV in 1995 banning blinding laser weapons. LAWS are under consideration to be a new Protocol in CCW and discussed more extensively in Section 9.10.3 [3].

Military leaders always seek to gain an advantage over potential or actual adversaries. One of the limits to their strategies and tactics is the number of forces available to them and the effectiveness of their weapons. Advances in technology are

---

[1]The 1899 Hague Declaration prohibited the launching of projectiles and explosives from balloons. It came into force in 1900 but had a five-year time limit and expired in 1905.

[2]The 1907 Hague Declaration was similar to the 1899 Hague Declaration. It can be interpreted as applying to aircraft, but many nations did not sign it and practice has overtaken it.

utilised to give new weapons or increase the capabilities of existing ones. One note-worthy advance is the linking of different systems to achieve an overall aim that cannot be achieved by the individual components. The resultant system can achieve the same effect as increasing the number of forces and/or the effectiveness of their weapons. Panel 7.2 gives an example of how a technology, in this case radar, can give different military advantages when integrated with other capabilities in different ways.

A modern Integrated Air Defence System (IADS) would appear to be a good example of a complex system which would not raise any moral problems when used. In general this is true, but not necessarily so. Advances in signal processing for Non-Cooperative Target Identification (NCTI) could soon lead to the position where individual aircraft types can be classified into hostile and friendly with no human input. (The current state of the art for NCTI can be found in Reference [4].) It would be a small technical step to link these NCTI outputs to an automated Surface to Air Missile (SAM) control system to give an automated IADS, but there would need to be very high confidence in the NCTI results before this could happen. Even with this confidence, the IADS would need to be controlled by humans with a comprehensive knowledge of its operation and the probability of it incorrectly classifying a friendly or neutral aircraft as hostile. This is one example of an area where legal authorities need to liaise with engineers who can communicate engineering concepts such as false-alarm probabilities in a way which can inform law-makers and legal practitioners when they decide what is reasonable and acceptable in law.

Panel 7.2:   An example of enhanced effectiveness from system integration

The use of radar in the first part of World War II shows how the same technology can be used to give different advantages. Germany developed radar as an aid to Luftwaffe aircraft supporting ground troops in the overall strategy of Blitzkrieg, but the radars were not suitable for air combat or area defence against other aircraft.

British airborne radar was primitive by comparison with German radar. However, their investment was in a ground-based surveillance system called Home Chain. A good non-technical description of the development of Home Chain is given in Reference [5]. The critical advantage was that it was integrated nationally with visual observers and used assets such as the telephone system for communications, and the phase of the national electricity supply as a common time reference. The air picture was then presented to the RAF Fighter Command on a very simple large map table. They were then connected directly to the air-bases who could respond rapidly with accurate knowledge of the enemy positions and intent. The extent of the integration can be seen in Reference [6].

This was an early example of an IADS which gives the defending air commanders the strategic advantage of locating and tracking hostile aircraft in sufficient time to be able to direct their limited resources to the points where they can achieve maximum effect.

7.3   Legality and legitimacy

There is an important distinction between these two concepts. Something that is legal may be considered illegitimate by the public. This is of relevance to engineers because public perceptions change and a legal weapon that is considered to be legitimate may come to be considered as illegitimate and eventually made illegal.

There are at least two types of weapon where this has happened, cluster munitions and landmines that cannot be cleared easily. Cluster munitions, i.e. a munition that contains submunitions which are dispersed near or at the target point, were developed to give a variety of capabilities such as area denial during the late twentieth century. The problem was the large number of submunitions which did not explode and caused death or injury after the end of the conflict, especially to uninvolved civilians.

There are several exceptions for submunitions which are discussed in Section 7.10.8. Similar concerns about post-conflict death and injuries from uncleared minefields gave rise to a well-publicised campaign for a ban on landmines. This was not fully successful, but did produce Protocol V, Explosive Remnants of War to the 1980 CCW Protocol, also discussed in Section 7.10.7.

7.4   International Humanitarian Law

IHL is the legal framework covering war. It is also called LOAC or simply as the laws of war. As shown in Section 7.2, IHL has evolved over many centuries. The major components of IHL as now understood were defined in the twentieth century. The descriptions in this chapter draw on ICRC publications [7, 8] and legal textbooks which are referenced in the text.

The sources of IHL are treaties; conventions; customary principles and rules; judicial decisions; and 'writings of the most highly qualified legal specialists'. States have manuals of military law which are their application of IHL to their own forces. IHL applies once war has started, regardless of the legitimacy or otherwise of the conflict. It applies to wars between states, hostilities between a state and armed groups within it, and hostilities between armed groups.

When a treaty is signed by a state it does not have the force of domestic law within that state, so its citizens cannot necessarily be prosecuted by the state's criminal courts. Ratification is the process where a state enacts the treaty into its domestic law. The Geneva Conventions are universally ratified, but it is worth noting that not all states have signed all the relevant treaties and a smaller number have ratified those which they have signed.

When a state signs an IHL treaty, it also publishes any reservations it has about specific provisions in the treaty, and may also notify other state parties of its interpretation of particular terms or provisions in the treaty. A compilation of treaties up to the year 2000, with an explanation of their background, details of the states signing them and reservations expressed by particular states, is in the book by Roberts and Guelff [9].

IHL has evolved and will continue to evolve. There are several causes of change: new treaties; court rulings giving interpretations of the law in new circumstances; and the gradual adoption and acceptance of practices which eventually become the 'customary practice' within the scope of IHL. In general, the parts of IHL which affect the design of military systems are contained in treaties and conventions agreed since World War II. These are discussed in Section 7.10. Section 9.11 discusses how lawyers and ethicists are looking ahead in attempts to change IHL in symbiosis with technical developments.

There is an ICRC summary of IHL giving seven rules[3] applicable to armed conflict which is reprinted with notes as Document 26 in Reference [9]. The first five show that IHL is mainly about respecting civilians, wounded and prisoners of war. These aspects are probably the most well-known to the general public but do not have any direct effect on weapon design. However, there are implications for defence engineering arising from Rules 6 and 7:

Rule 6. Parties to a conflict and members of their armed forces do not have an unlimited choice of methods and means of warfare. It is prohibited to employ weapons or methods of warfare of a nature to cause unnecessary losses or excessive suffering.

Rule 7. Parties to a conflict shall at all times distinguish between the civilian population and combatants in order to spare civilian populations and property. Neither the civilian population as such nor civilian persons shall be the object of attack. Attack shall be directed solely against military objectives.

Rule 6 is directly relevant to highly automated weapon systems. It requires that weapon systems be limited in their effect and that their commanders will be aware of these limitations. It may be difficult to limit their autonomous destructive power when in situations that were not foreseen during procurement; or the commander may not fully understand the range of decisions and actions that such a system can take.

A further consideration is the implication in both paragraphs that weapons must be under control: fundamental to this in modern warfare is the integrity of the Command and Control (C2) chain with contingencies in the event of failures, which is an engineering problem.

Rule 7 gives a clear requirement for sensing systems, in their widest sense, to provide the information needed by a commander to distinguish military and civilian materiel and infrastructure. Target recognition algorithms and software are widely used and continually evolving, bringing improved performance. However, the rise in automation levels in decision aids makes an argument for the development of equivalent algorithms for the recognition of non-targets, i.e. objects such as ambulances, petrol tankers and other civilian objects or people in the vicinity of a target.

An extreme hypothetical example of an unacceptable method of warfare would be to attack a small group of armed hostile combatants in the middle of a large crowd of civilians with a bomb from an UAV. A human UAV operator would not

---

[3]Note that although the ICRC use the word 'rule', this does not imply that their form of words have the weight of law as only the exact words of the treaties form the law. Despite this, their documents are very useful and informative. See also the 'basic rule' in Article 35 of API in Section 7.10.2.

release the weapon, but a UAV would if it were allowed to release its weapon based solely on on-board target recognition software with no calculation of the balance of the relative advantage of targeting hostile combatants against expected civilian casualties. This illustrates how a straightforward technical step of integrating two functions could produce an illegal capability and the need for Article 36 reviews for system upgrades as well as new weapons.

Although not of direct relevance to engineering design, it should be noted that IHL is not the same as International Human Rights Law (IHRL). The former has the presumption that war will cause suffering and damage and seeks to limit these to the minimum necessary to achieve the political and military aim. The latter has the presumption that human life must be respected under all circumstances. There is a lack of agreement between states about the extent to which IHRL applies in time of war [10].

7.5 The Geneva Conventions, their protocols and subsequent agreements

The aftermath of World War II led to four Geneva Conventions:

- 1949 Geneva Convention I. For the amelioration of the condition of the wounded and sick in armed forces in the field;
- 1949 Geneva Convention II. For the amelioration of the condition of wounded, sick and shipwrecked members of armed forces at sea;
- 1949 Geneva Convention III. Relative to the treatment of prisoners of war;
- 1949 Geneva Convention IV. Relative to the protection of civilian persons in time of war.

Most post-World War II conflicts had a different character to the two world wars. The Geneva Conventions above are for the protection of people who are not participating in the conflict, not for the regulation of the conduct of hostilities. Recognition of this led, in 1977, to two additional protocols to the 1949 Geneva Conventions, generally called API and APII. APIII was added in 2005 to define the various symbols (emblems) used to mark areas or objects which must not be attacked. Their exact titles are:

- 1977 Protocol additional to the Geneva Conventions of 12 August 1949, and relating to the protection of victims of international armed conflict (Protocol 1);
- 1977 Protocol additional to the Geneva Conventions of 12 August 1949, and relating to the protection of victims of non-international armed conflict (Protocol II);
- 2005 Protocol additional to the Geneva Conventions of 12 August 1949, and relating to the Adoption of an Additional Distinctive Emblem (Protocol III).

These are generally known as API, APII and APIII, respectively. Several articles from API which are relevant to weapon design are reproduced in Appendix A7.

Article 36 of API requires that every new means or method of warfare must be reviewed for adherence to IHL. This process is often called an Article 36 review and should comprise reviews at each major design or upgrade review. Guidance on holding these reviews is available from the ICRC [11] which is reproduced in whole in a separate annex in this book. An earlier, practical approach to these reviews is in Reference [12]. It is not known if all states conduct Article 36 Reviews, but the Stockholm International Peace Research Institute (SIPRI) have published a compendium of the way nine states conduct Article 36 Reviews based on information from those states [13].

When weapons were simple kinetic devices reviews were relatively straightforward, but this is no longer the case as modern weapon systems are complex with highly automated sub-systems. Article 36 reviews are discussed in detail in Chapter 11. The problems associated with them include both technical and legal ones and it is essential that lawyers and engineers work together on them. This collaboration and other problems for new complex weapons are discussed in Reference [14].

The law does take into account the cost of procuring a weapon, which does put a spotlight on the engineering and programme judgements made in balancing the cost-benefit analysis of each procurement phase. This is implicit at the end of Section 2.3 of Reference [11], where acknowledgement is made that unforeseen changes during procurement can make a weapon more costly or even illegal.

Another important part of IHL is explicit bans on particular types of weapon. These are:

- 1976 Convention on the prohibition of military or any other hostile use of environmental modification techniques;
- 1980 UN convention on prohibitions or restrictions on the use of certain conventional weapons which may be deemed to be excessively injurious or to have indiscriminate effects;
- 2006 Convention on cluster munitions.

Although not legally binding, the 1994 San Remo Manual on International Law Applicable to Armed Conflict at Sea is a restatement of the law up to that date with some developments. (This can be found in reference [9].) There is a related Explanation which discusses each paragraph and its origin. The Manual was the first restatement of the law since 1913 and according to the President of the International Institute of Humanitarian Law (Quoted in Reference [9]) it took into account:

> state practice, technological developments and the effect of related areas of the law, in particular, the United Nations Charter, the 1982 Law of the Sea Convention, air law and environmental law.

Engineering implications are identified in Section 7.11.

7.6   Customary principles and rules

There are four fundamental principles which underpin IHL:

1. **Military necessity.** A state is permitted to use force that is not otherwise banned to achieve its military aims in the shortest possible time and with minimum loss of life and damage. This is a long-standing principle and is in several treaties since the nineteenth century;

2. **Humanity.** This prohibits the infliction of suffering, injury or destruction except for the achievement of the military objective. It probably originates from the rules of chivalry and appears in treaties from the early twentieth century;
3. **Proportionality.** The losses, suffering and damage caused by a military action should not be excessive when compared with the military advantage arising from the action. This first appears in API of 1977 in articles 51 and 57, forbidding indiscriminate attacks and requiring protection of the civilian population and infrastructure;
4. **Distinction.** This requires an armed force to distinguish between civilians and combatants, and between military objectives and civilian objects. It is stated in API in a section on the general protection of the civilian population against the effects of hostilities.

These four principles, and the way they are expressed in IHL, especially API, are at the core of how weapons are used in conflict. The relevant parts of API are given as Appendix A7. Automation of functions in the intelligence, detection and attack or non-attack of potential targets can enhance adherence to these principles. Unreliable or defective automation can lead to their breach or extra effort in reconfirming a situation as well as wasted resources in attacking false targets. Unjustified attacks must be avoided and it is important that engineers and procurement authorities fully understand the limitations of the automated functions and that the military user also understands them.

The four principles are discussed in detail in Section 10.3 where they are used to derive the generic engineering requirements for autonomous weapon systems.

The ICRC published a report in 2005 [15] giving their view of the customary rules of HL. One of its authors published an article [16] giving the rationale behind the report which was in response to a 1995 mandate to:

> prepare, with the assistance of experts in international humanitarian law representing various geographical regions and different legal systems, and in consultation with experts from governments and international organisations, a report on customary rules of international humanitarian law applicable in international and non-international armed conflicts, and to circulate the report to States and competent international bodies.

The report gives 161 rules which are can be interpreted as customary rules. The report has a section, *Proportionality In Attack*, which includes the following:

> Rule 14. Launching an attack which may be expected to cause incidental loss of civilian life, injury to civilians, damage to civilian objects, or a combination thereof, which would be excessive in relation to the concrete and direct military advantage anticipated, is prohibited.
>
> Summary: State practice establishes this rule as a norm of customary international law applicable in both international and non-international armed conflicts.

This shows that IHL recognises that there may be civilian suffering and damage to civilian objects in an armed conflict. This is known as collateral damage and is not illegal although not specifically allowed. In practice, collateral damage is acceptable in law provided it is not excessive compared with the military advantage anticipated from the successful mission. This is the legal concept of proportionality. Reference [11] defines (codifies) its terms for proportionality in attack by reference to Article 51(5)(b) and Article 57 of API. (These Articles are reproduced with others in Appendix A7.)

Decisions about the acceptability of collateral damage is a matter for lawyers and beyond the scope of the engineering profession. However, the military user, their legal advisers and courts must have access to the quantified performance of every weapon system used in an attack in order to understand the damage it will, or has, cause(d) when used.

Proportionality for smart munitions which travel a route to a defined target is more problematic. They are discussed in the UK Joint Service Procedure JSP383 [17]:

Applying the principle of proportionality

2.7 Modern, smart weaponry has increased the options available to the military planner. He needs not only to assess what feasible precautions can be taken to minimize incidental loss but also to make a comparison between different methods of conducting operations, so as to be able to choose the least damaging method compatible with military success.

2.7.1 The application of the proportionality principle is not always straightforward. Sometimes a method of attack that would minimize the risk to civilians may involve increased risk to the attacking forces. The law is not clear as to the degree of risk that the attacker must accept. The proportionality principle does not itself require the attacker to accept increased risk. Rather, it requires him to refrain from attacks that may be expected to cause excessive collateral damage. It will be a question of fact whether alternative, practically possible methods of attack would reduce the collateral risks. If they would, the attacker may have to accept the increased risk as being the only way of pursuing an attack in a proportionate way.

This shows the complexity of the problems faced by military planners and commanders when they use what may be considered as a low-autonomy-level weapon. A smart weapon such as the cruise missiles available when JSP383 was published (2004) had no autonomous capability as now understood. They were fired at previously chosen well-defined targets and flew there along pre-defined routes. The planners ensured that any changes to the target area between weapon firing and impact would be unlikely to change the proportionality justification. An intelligent weapon with the ability to choose its target, means of approach and possibly timing of its attack presents a much more complex problem. The procurement chain from specification through design to trials in realistic scenarios must provide the information that the military users, their legal advisers and a court passing judgement after a possibly controversial incident will need to make informed judgements.

7.7 Judicial decisions

The Nuremberg and Tokyo International Military Tribunals following World War II produced the first extensive body of case law. The United Nations Security Council has established two international tribunals since then:

• The International Criminal Tribunal for the Former Yugoslavia;
• The International Criminal Tribunal for Rwanda.

The International Criminal Court was set up in 1998 to try grave breaches and serious violations of IHL and gives rulings on the specific cases tried before it. Clearly, any interpretation of rulings from the above bodies should not be considered by anyone involved in defence design without detailed consultation with appropriate lawyers.

International-court decisions do become precedents for their later cases, but do not necessarily have the power of precedent for a state's courts. However, their rulings can be used in a national court as evidence of customary rules of international law [18]. Lawyers also assume that IHL includes practices that are so widely adopted that they can be regarded as 'customary'. Rulings from international courts can also influence state behaviour so that it becomes customary practice.

7.8 Expert opinions

Treaties and conventions are intended to be clear, but there may be differences of opinion about their interpretation, especially some years after the original agreement. There may also be questions about the relationship of particular treaties between states and more general conventions. Published opinions by experts, especially where they agree, may be taken as evidence of where the law stands at a particular time. They may also clarify how the law applies to new situations and technical developments. Published specialist opinion may be given as one of the sources of IHL in references such as References [9] and [19], but they are not to be taken as having the weight of law. Their use is outside the scope of any engineering interpretation of IHL and they should be discussed with legal specialists if there is any reason to consult them.

7.9 Military manuals

Military manuals can give a clear exposition of a state's interpretation of IHL. The UK's 2004 Manual of the Law of Armed Conflict [16] and the US 2016 Law of War Manual [20] are good examples. (The US manual includes a section on cyber war whist the UK does not.) Particularly important for engineers are their sections on weapons. Their sections on principles (necessity, humanity, proportionality and distinction; with the US adding honour) are directly relevant to systems engineers as they can be used to derive engineering requirements for military systems of all kinds, both offensive and defensive. This is the subject of Chapter 10.

Military manuals are recognised by the international community as a way of implementing IHL. This is seen most clearly in the attempts to widen the scope of the law to cover short-term environmental effects such as those from damaged oil wells in the 1991 Gulf War. A variety of reasons led to a decision not to revise the 1976 UN Convention on the Prohibition of Military or Any Other Hostile Use of Environmental Modification Techniques. Instead, the ICRC and United Nations General Assembly produced the 1994 ICRC/UNGA Guidelines for Military Manuals and Instructions on the Protection of the Environment in Times of Armed Conflict. Both these documents are in Reference [9] and are explained in more detail in Section 7.10.6.

7.10   Engineering requirements from selected conventions in Section 7.5

7.10.1   General

All of the treaties in Section 7.5 apply to the use of weapons and weapon systems in armed conflicts but not all of them place specific requirements on engineering design. The ones that do are identified here and attention drawn to those which are relevant to highly automated systems.

One important consideration is the use of the outputs from automated decision-aids by a commander when making a lethal decision. It is almost inconceivable that commanders will not use some form of computer support in analysing and assessing data before making critical decisions, especially when in a complex scenario. The commander needs to be aware of the limitations of the particular decision-aid and criteria for independent corroboration of any output from it. Mandating corroboration for particular types of data is likely to be one of the rules-of-engagement for a campaign.

API and APII are discussed in Section 7.11. They form the legal framework which weapon designs must meet; adherence to them is audited through reviews conducted under Article 36 of API.

7.10.2   *1949 Geneva Convention 1. For the amelioration of the condition of the wounded and sick in armed forces in the field*

Articles 19 and 20 give protection[4] from attack to 'Fixed establishments and mobile medical units of the Medical Service' and 'Hospital ships entitled to the protection of the Geneva Convention'. Articles 24 to 26 extend this protection to specific classes of individuals. Articles 35 and 36 cover medical transports and medical aircraft. Geneva Convention II gives similar protection to medical transports at sea. These should all be distinguished by displaying a 'Distinctive Emblem'. This is one of the well-known symbols: Red Cross; Red Crescent; or Red Shield of David.

---

[4]'Protection' means legal protection, i.e. not attacked; not physical protection or defended by military or technical means.

There are also signs added in later protocols distinguishing: prisoner of war camps; internment camps; hospital and safety zones and localities; cultural property; civil defence; works and installations containing dangerous forces; and areas containing mines. The *1954 Hague Convention for the protection of Cultural Property in the Event of Armed Conflict* added another emblem ('A shield, pointed below, per saltire blue and white') to mark immovable cultural property and places where movable cultural property is stored. This was supplemented by a *Second Hague Protocol for the Protection of Cultural Property in the Event of Armed Conflict* in 1999 which includes Article 7 *Precautions in Attack.* This brings a requirement for identification of protected cultural property similar to those for civilians during an attack.

There are also protective light and radio signals for medical transports defined by an amendment to API discussed in Section 7.11.

Clearly there is a requirement for the ability to recognise these distinctive emblems and electromagnetic signals. Human observers considering the use of lethal force would expect to be able to recognise these visually or from previous declarations of protected area and, for example, marked on maps. Automated target recognition systems which do not pass an image to their human commander should therefore have the ability to recognise these emblems or signals, or the commander should know that they do not and so be able to make an informed judgement whether they can use the automated system.

7.10.3   1949 Geneva Convention II. For the amelioration of the condition of wounded, sick and shipwrecked members of armed forces at sea

This mainly covers direct protection for those stated in its title. However, the convention also gives protection to hospital ships, medical supply ships, aircraft used for transporting sick, wounded personnel and medical supplies. The protection also extends to lifeboats and lifeboat stations. The convention gives the means of identification for these protected categories such as prior registration, white colour, their distinctive emblem and radio transmissions not being in code.

7.10.4   1949 Geneva Convention III. Relative to the treatment of prisoners of war

The only part of this convention that is relevant to targeting is that prisoner of war camps shall be marked, 'whenever military considerations permit', in day-time by the letters PW or PG placed so they are visible from the air. This gives another emblem that must be recognised by an automated sensor identification algorithm.

7.10.5   1949 Geneva Convention IV. Relative to the protection of civilian persons in time of war

This provides protection to hospitals, categories of civilians and civilian areas by agreement between the warring parties. There are detailed differences from the first and second conventions, but the nature of the protection and markings for hospitals are the same. It assumes a conventional war where one side occupies territory which the other has taken by force and the territory has a civilian population. Its scope has been widened considerably by API and APII in 1977 which place much more stringent conditions on armed forces whilst conducting hostilities.

7.10.6   1976 Convention and 1994 Guidance on environmental modification techniques

This 1976 Convention on the Prohibition of Military or any other Hostile Use of Environmental Modification Techniques has a general scope. It should be clear to engineers if they are being asked to design a banned weapon as it defines 'environmental modification technique' to be:

> Any technique for changing – through the deliberate manipulation of natural processes – the dynamics, composition or structure of the Earth, including its biota, lithosphere, hydrosphere and atmosphere, or of outer space.

This convention was criticised because its enforcement may be weak for three main reasons: it uses broad, vague terms; it does not ban development of such weapons, only their use; and the potential use of a veto in the UN Security Council.

The 1991 Gulf Wars had environmental impacts with the destruction of oil wells and the release of oil into marshlands and the sea. Rather than a new treaty, the result of diplomatic negotiations was the publication in 1994 of the ICRC/UNGA (United Nations General Assembly) Guidelines for Military Manuals and Instructions on the Protection of the Environment in Times of Armed Conflict. This document gives guidelines based on the existing legal obligations and state practice that should give protection to the environment in a more precise way than the 1976 convention. It extends the general principles of distinction and proportionality to provide environmental protection and states that:

> In particular, only military objectives may be attacked and no methods or means of warfare which cause excessive damage shall be employed. Precautions shall be taken in military operations as required by international law.

Specific rules give categories of object where attack must be avoided such as:

> works or installations containing dangerous forces, namely dams, dykes and nuclear electrical generating stations, even where they are military objectives, if such attack may cause the release of dangerous forces and consequent severe losses among the civilian population and as long as such works are entitled to special protection under Protocol I additional to the Geneva Conventions.

It also adds historic monuments, works of art or places of worship. These objects should be known before any campaign but may not be in every database, so the databases which generate information for a weapon operator or commander must have the ability to be updated with these locations and flag them as non-targets.

The 1994 Guidelines, in Section IV Implementation and Dissemination Paragraph 18, gives a slightly modified version of API Article 36 to include the environment in its scope:

> In the study, development, acquisition or adoption of a new weapon, means or method of warfare, States are under an obligation to determine whether its employment would, in some or all circumstances, be prohibited by applicable rules of international law, including those providing protection to the environment in times of armed conflict.

As a result, engineers must consider whether any new system, automated or not, will have environmental effects, and make predictions about them for review by their legal authorities.

7.10.7   1980 UN convention on prohibitions or restrictions on the use of certain conventional weapons which may be deemed to be excessively injurious or to have indiscriminate effects

This convention, known as CCW, is much more precise than API and is likely to have an effect on engineering design even if its requirements are not specifically called up in a requirements specification. Article 1 was amended in 2003, but should not have any effect on the design aspects of weapons.

A new protocol banning or restricting the use of autonomous weapons is often suggested for CCW. It was the subject of an ICRC meeting in Geneva on 11-15 April 2016, the CCW Meeting of the Group of Government Experts (GGE) on LAWS held [21]. More recently there have been GGE meetings under UN auspices in April and August 2018[5]. More details of these meetings and their possible implications for autonomous weapon systems are given in Section 9.10.3.

All military autonomous system designers should ensure that they have a way of being informed of the implications of developments in IHL and that some element of future-proofing should be considered for their designs.

The weapon types that are banned or have restricted use under CCW protocols are, the time of writing:

1. Protocol on non-detectable fragments (1980 Protocol I)
   This simply states that:

   > It is prohibited to use any weapon the primary effect of which is to injure by fragments which in the human body escape detection by X-rays.

   It does not seem directly relevant to autonomous systems although it does place limitations on the type of material used in weapons.

---

[5]Reports and papers from these meetings were not available at the time of writing, June 2018. The UNOG website has a page: *2018 Group of Governmental Experts on Lethal Autonomous Weapons Systems (LAWS)* which will have the latest information.

2. Protocol on prohibitions or restrictions on the use of mines, booby-traps and other devices (1980 Protocol II).

This protocol has a technical annex which covers detectability, markings, deactivation means and booby traps. It was amended later (Protocol II to the 1980 Convention as amended on 3 May 1996) and the comments below are based on this version.

It is unlikely that an autonomous system will be defined as a mine, booby-trap or other device as defined in the protocol; these are all devices placed on, under, or near the ground which are triggered by the proximity of a person or vehicle. However, the level of automation in mines will probably increase in the future. Engineers engaged in any activities with mines must ensure that their actions and any design that they make comply with IHL.

It is feasible that an autonomous system could be used in preference to a human to place or disperse mines in an area. In this case it is highly likely that they will be considered to be remotely delivered mines and fall under Article 6 of the protocol. Deployment by the autonomous system will almost certainly have to meet other articles restricting the use of different types of mine as well as recording their location and creating any required marking of the mined areas.

Any person in a procurement chain for an autonomous system delivering mines, or increasing the automation level of a mine type, should be certain that the requirements have been approved by competent legal authorities and that Article 36 Reviews are in place. Close liaison between the designers and the legal authorities would seem to be essential for any combination of autonomy and mines.

3. Protocol on prohibitions or restrictions on the use of incendiary weapons (1980 Protocol III)

This does not have any direct relevance to autonomous systems beyond those from API that would not be expected to be covered by the rules-of-engagement for the incendiary weapon.

4. Protocol on blinding laser weapons (1995 Protocol IV)

This bans weapons designed to cause permanent damage to unaided vision, but it may have relevance for autonomous systems using lasers, for example, in laser guidance systems as Article 2 states:

In the employment of laser systems, the High Contracting Parties shall take all feasible precautions to avoid the incidence of permanent blindness to unenhanced vision. Such precautions shall include training of their armed forces and other practical measures.

There is a danger of collateral damage to humans in the target area if the guidance laser is powerful enough to cause blindness and its direction is not controlled adequately. A human operator would be able to minimise this risk as they, almost by definition, can see where the laser is pointing. It can be expected that the design and Article 36 review teams for an autonomous system which includes a laser system will need to decide on the engineering specifications for the control of the pointing and power levels for it.

5. Protocol on explosive remnants of war (1980 V)

This does not have any direct relevance to autonomous systems.

7.10.8   2008 Convention on cluster munitions

This was drawn up because of the large number of fatalities and injuries arising from unexploded submunitions that were not cleared during conflict. The definition of a cluster munition is very precise and excludes some types of submunition:

> 'Cluster munition' means a conventional munition that is designed to disperse or release explosive submunitions each weighing less than 20 kilograms, and includes those explosive submunitions. It does not mean the following:

(a) A munition or submunition designed to dispense flares, smoke, pyrotechnics or chaff; or a munition designed exclusively for an air defence role;

(b) A munition or submunition designed to produce electrical or electronic effects;

(c) A munition that, in order to avoid indiscriminate area effects and the risks posed by unexploded submunitions, has all of the following characteristics:

  (i)   Each munition contains fewer than ten explosive submunitions;
  (ii)   Each explosive submunition weighs more than four kilograms;
  (iii) Each explosive submunition is designed to detect and engage a single target object;
  (iv)   Each explosive submunition is equipped with an electronic self-destruction mechanism;
  (v)   Each explosive submunition is equipped with an electronic self-deactivating feature;

This does give scope for highly automated submunitions which have sufficient intelligence to identify specific target types and attack them either kinetically or electronically. These would need to be treated in the same way as the other highly automated weapon systems in this book. Additionally, the designer would need to agree with their national legal authority how Paragraph (c) and its subparagraphs are to be interpreted.

7.11   The additional protocols to the 1949 Geneva conventions

7.11.1   Introduction

The Additional Protocols were written decades before highly automated weapons were a practical proposition, although they are still generally accepted to be a new or novel method of warfare. Section 7.9.5 showed that there is some international opinion that autonomous weapons should be the subject of a new protocol under CCW. International discussions and negotiations will continue and engineers must respond to advice from their legal advisers. Certainly, engineers must work on the basis that the additional protocols (APs), and the four principles of IHL given in Section 7.6 apply. However, it is intended that the principles given in this book will help in the discussions that will occur.

> 1977 Protocol additional to the Geneva Conventions of 12 August 1949, and relating to the protection of victims of international armed conflict (Protocol 1)

This is known as API and is one of the most important IHL documents for engineers in all parts of the weapon-supply chain. It covers all means of warfare so highly automated and autonomous weapons are included. However, there are formal international legal debates about whether there is a need for additional treaties or protocols to cover them. These are under the auspices of the United Nations and are explained in Section 9.10.3.

1977 Protocol additional to the Geneva Conventions of 12 August 1949, and relating to the protection of victims of non-international armed conflict (Protocol 2)

This is known as APII and is much shorter than API. Basically it extends the protection of civilians in time of internal conflict to be very similar to their protection in times of inter-state war.

An amendment to API was published in 1993 [22] which revises an annex to API. Engineers should note that this has two chapters Articles 6 to 12 which define lights and radio signals which identify medical transports (air, sea and land). These must be recognised by all combatants. Automatic target recognition algorithms must be able to identify these, or the commander must know that they do not.

7.11.2   General requirements from API and APII

API starts with a Preamble. Part 1 – General Provisions, which has seven Articles defining API's applicability, and the diplomatic channels for ensuring its implementation during a conflict.

Articles 8 to 34 form Part II – *Wounded, Sick and Shipwrecked*. They define: the terminology in the protocol; its applicability; details about the protection it offers to specific groups such as medical transportation; and processes for dead and missing personnel.

Part III – *Methods and Means of Warfare, Combatant and Prisoner of War Status* is important to engineers. The first two Articles (Articles 35 and 36) are the core of the engineering requirements placed by IHL on the whole procurement process. It has implications from the R&D phases right through to the engineering and trials work in setting the weapon to work in service. Their interpretation and applicability to a new weapon are decided by a state's legal authorities, but they will need assistance and advice from technical authorities and advisers. The Articles are given here as well as in Appendix A7:

Article 35 – Basic rules

1. In any armed conflict, the right of the Parties to the conflict to choose methods or means of warfare is not unlimited.
2. It is prohibited to employ weapons, projectiles and material and methods of warfare of a nature to cause superfluous injury or unnecessary suffering.
3. It is prohibited to employ methods or means of warfare which are intended, or may be expected, to cause widespread, long-term and severe damage to the natural environment.

Article 36 – New weapons

In the study, development, acquisition or adoption of a new weapon, means or method of warfare, a High Contracting Party is under an obligation to determine whether its employment would in some or all circumstances, be prohibited by this Protocol or by any other rule of international law applicable to the High Contracting Party.

As noted in Section 7.10.5, this is very similar to Section IV, Paragraph 168 of the 1994 *ICRC/UNGA Guidelines for Military Manuals and Instructions on the Protection of the Environment in Times of Armed Conflict*.

The term 'new means or methods of warfare' is much wider in scope than simply weapons. In engineering terms, it extends the scope to new concepts of all types, whether the development of a new technology, or increasing military capability by either an upgrade or by integrating existing materiel. The integration of subsystems that have previously been reviewed separately under Article 36 will need a further Article 36 Review as the integrated system or system-of-systems is almost certain to give an increased or new capability.

Demonstrating compliance under Article 36 is not a trivial task. Reference [13] gives the methodology for nine nations. New technologies are increasing the complexity of the reviews with a detailed discussion of cyber and autonomous weapons in a SIPRI document [23] and a summary of expert opinions in another [24]. (These are not legally binding expert opinions under IHL.) The necessary collaboration between lawyers, military staff and engineers is explained in reference [25]. Article 36 Reviews for autonomous systems are discussed in Section 8.6 with examples of their application to a specific example in Section 11.5.

The remainder of API Part III concerns the conduct of military personnel during the conflict.

7.11.3   More specific requirements from API and APII

There are two parts of API which bring out requirements for weapon systems which can be turned into engineering requirements. The first is Part IV – *Civilian Population*, Section I – *General Protection Against Effects of Hostilities*. Articles 48 to 58 are reproduced in Appendix A7 and discussed here.

The Basic Rule (Article 48) requires a distinction to be made between:

• Civilians and combatants and
• Civilian objects and military objectives.

The distinctions are to be made at all times, so it is not acceptable to simply survey the target area, identify the military and civilian objects, and then continue only to monitor the military objects if it is likely that the area will change and that one or both types of object will move with consequent increased risk to the civilians or civilian property.

Dealing with target areas that change over short timescales is, in principle, not a problem when the timescale between the most recent observation and the effect of the weapon is short. The simplest example is a soldier firing at enemy forces in a town which may have civilians present. He or she, or a close comrade, will have seen the enemy immediately before they fire so they can be expected to have seen and probably recognised any civilian who has the misfortune to enter the combat area. Contrast that with self-guided projectiles discussed in Section 8.2.4 and the restrictions on their use.

One important challenge for engineers arising from highly automated systems is the information flows in the C2 system. The data from sensors and other sources must be processed and presented to the human commander with sufficient clarity and time to authorise or allow the automated system to release its weapon and still meet IHL requirements. If, as may happen in the future, there are algorithmic decision-makers in the chain, they must be able to assess the information presented to them and pass control back to their commander. This information must be timely, and in a format that the human can interpret if the decision is outside the algorithms' authority level.

Two useful exercises for the reader which should help understand the implications of API are:

1. Identify what design and trials information would need to be presented to an Article 36 Review for the Phalanx system when fitted to a ship of their own nation. (Phalanx is described in Section 3.4.);
2. Design a military C2 system and information network which will allow a UAV to release a missile based on automatic target recognition algorithms on a nearby ground target and still meet IHL. (This is part of the scenario used in Chapters 11 and 12.) What would you set as rules-of-engagement?

Part V – *Execution of the Conventions and of this Protocol*, Section 1 – *General Provisions*, has one article that can give rise to demands for engineering data which must be responded to.

> Article 82 — Legal advisers in armed forces. The High Contracting Parties at all times, and the Parties to the conflict in time of armed conflict, shall ensure that legal advisers are available, when necessary, to advise military commanders at the appropriate level on the application of the Conventions and this Protocol and on the appropriate instruction to be given to the armed forces on this subject.

Legal advisers can only give advice about the use of a weapon when their knowledge includes their technical performance. With conventional weapons, this could be straightforward data such as the weight and explosive power of a warhead and circular error probable (CEP) of the delivery system. It may be complex to calculate expected damage to nearby buildings or vehicles, but human judgement can be used.

The position may be different if the weapon system has some level of autonomy in its decision-making. The engineering information provided to the legal adviser must be clear about the predictability of the system's decisions. This information may be used in drawing up rules-of-engagement, so it must be based on actual measured performance. Gathering the evidence can be expensive and time consuming if carried out shortly before or during a campaign. It is much better to plan it into the design, final testing, and introduction-into-service phases of the new weapon or capability.

7.12   The law at sea

All aspects of the law at sea except armed conflict are covered by the: *United Nations Convention on the Law of the Sea of 10 December 1982* (UNCLOS).

Armed conflicts at sea were covered by eight Hague Conventions in 1907 and the 1913 Oxford Manual. The 1949 Geneva Conventions and API apply to wars on land although they do contain Articles giving protection to the wounded and shipwrecked.

Legal experts considered it necessary to have an updated version of the Oxford Manual. The result was the: *1994 San Remo Manual on International Law Applicable to Armed Conflicts at Sea.* This is given in Reference [9] as Document 30. It is not a legally binding document but is a restatement of existing laws and some developments. From an engineering perspective, there are two important parts of this treaty which are discussed below. The manual has sections giving protection to people and areas which are analogous to those in API and also give protection to the maritime environment and sensitive areas.

Part 3 has nine paragraphs covering basic rules and target discrimination, which are very similar to the equivalent Articles in API. It has two sections on basic rules and precautions in attack. These ensure that the principles of necessity and proportionality apply to armed conflict at sea.

Part 4 on methods and means of warfare at sea has specific paragraphs on torpedoes and mines. The relevant ones are reproduced here as any level of auto-mation will probably include requirements derived from these paragraphs:

- Torpedoes
  - 79 It is prohibited to use torpedoes which do not sink or otherwise become harmless when they have completed their use.

- Mines
  - 81 Without prejudice to the rules set out in paragraph 82, the parties to the conflict shall not lay mines unless effective neutralisation occurs when they have become detached or control over them is otherwise lost.
  - 82 It is forbidden to use free-floating mines unless:
    - They are directed against a military target; and
    - They become harmless within an hour after loss of control over them.

  - 84 Belligerents shall record the locations where they have laid mines.
  - 90 and 91 impose an obligation on parties to the conflict to collaborate with each other and international bodies where appropriate on clearing mines, including exchanging technical information.

7.13   Rules-of-engagement

Rules-of-engagement are instructions to combatants for their conduct during times of armed conflict. They are issued by a state's armed forces and must be obeyed. There are standing rules-of-engagement which are generic for a particular campaign and may be published. The 'Yellow Card' given to the UK soldiers in Northern Ireland during The Troubles in the late twentieth century is one example. This was for the guidance of soldiers about whether to open fire with small arms. It was originally classified but after much debate it was published before the end of military operations.

Rules-of-engagement that are more illustrative for designers of automated systems are those on the card issued by the US for their forces in the 1991 operation against Iraq. This is shown in Panel 7.3. This shows the emphasis on distinction between legitimate military targets and others. The use of words such as 'unless' and 'except in self-defence' shows that combatants must exercise considerable judgement. Any weapon system comprising a human combatant and one or more automated weapons must still be in a position to use human judgement to ensure that any action meets the rules-of-engagement. This places requirements on the weapon design and clarity in the military understanding of the weapon so that the combatant will comply with rules-of-engagement.

Particular operations may have specific rules-of-engagement (supplemental rules-of-engagement). They can be very detailed with the probability of different rules-of-engagement applying at different places and phases of a mission. Normally they are classified, as an enemy can turn them to their advantage, for example, by exploiting situations when a commander cannot fire.

Engineers are not normally a party to the drawing up of rules-of-engagement as campaigns are planned using weapons that are part of the current inventory with well-understood performance. New weapons may be introduced during a campaign. If they are, they are usually introduced initially as an Urgent Operational Requirement or Need (UOR or UON). Their engineering aspects are discussed in Section 9.5.

Although not directly involved in writing rules-of-engagement, engineers will provide design and performance information which will be used to draw them up. This information will be provided through the procurement authority and be integrated with knowledge from military experience. With increased weapon complexity and autonomy there is a heightened risk of the weapon system exhibiting behaviours which may not have been seen in previous use. Mitigation of this type of risk must be taken into account in the specification, design and proving stages of procurement as a minimum.

There must be a way of ensuring that non-technical military personnel understand the likely behaviour of their weapon in representative circumstances. The technical advice will depend on the specific system, but some idea of the areas where advice is likely to be needed can be seen from Section 7.11.3.

Panel 7.3: Desert Storm Rules of Engagement

ALL ENEMY MILITARY PERSONNEL AND VEHICLES TRANSPORTING THE ENEMY OR THEIR SUPPLIES MAY BE ENGAGED SUBJECT TO THE FOLLOWING RESTRICTIONS:

(A)   Do not engage anyone who has surrendered, is out of battle due to sickness or wounds, is shipwrecked, or is an aircrew member descending by parachute from a disabled aircraft.

(B)   Avoid harming civilians unless necessary to save the US lives. Do not fire into civilian populated areas or buildings which are not defended or being used for military purposes.

(C)   Hospitals, churches, shrines, schools, museums, national monuments, and any other historical or cultural sites will not be engaged except in self-defence.

(D)   Hospitals will be given special protection. Do not engage hospitals unless the enemy uses the hospital to commit acts harmful to the US forces, and then only after giving a warning and allowing a reasonable time to expire before engaging, if the tactical situation permits.

(E)   Booby traps may be used to protect friendly positions or to impede the progress of enemy forces. They may not be used on civilian personal property. They will be recovered or destroyed when the military necessity for their use no longer exists.

(F)   Looting and the taking of war trophies are prohibited.

(G)   Avoid harming civilian property unless necessary to save the US lives. Do not attack traditional civilian objects such as houses, unless they are being used by the enemy for military purposes and neutralization assists in mission accomplishment.

(H)   Treat all civilians and their property with respect and dignity. Before using privately owned property, check to see if publicly owned property can substitute. No requisitioning of civilian property, including vehicles, without permission of a company level commander and without giving a receipt. If an ordering officer can contract the property, then do not requisition it.

(I)   Treat all prisoners humanely and with respect and dignity.

(J)   Rules-of-engagement Annex to the OPLAN provides more detail. Conflicts between this card and the OPLAN should be resolved in favour of the OPLAN.

REMEMBER

1.   FIGHT ONLY COMBATANTS.
2.   ATTACK ONLY MILITARY TARGETS
3.   SPARE CIVILIAN PERSONS AND OBJECTS
4.   RESTRICT DESTRUCTION TO WHAT YOUR MISSION REQUIRES

7.14 An example – design changes to move Phalanx from sea to land

The Phalanx Close In Weapon System (CIWS) is described in Section 3.4 as an example to illustrate the difficulties in defining a weapon system as either automatic or autonomous. It also makes a useful example of the design changes necessary when a weapon system is used for a different application but legal considerations necessitate significant design changes.

Large bases were established for operations in Afghanistan and Iraq and soon came under attack from rockets and artillery fire. The launchers could move rapidly into and out of position, and the launch site could be in civilian areas, so return fire had a high risk of causing unacceptable collateral damage. This gave a requirement for a CIWS that could destroy the rocket or ordnance in the air.

The rapid-fire high-accuracy gun used for Phalanx appears to make it the ideal solution. However, the problems are:

- Its projectiles have a range of several kilometres. Any civilian person or structure would have a high chance of being hit anywhere in the guns arc of fire out to this range;
- The bases have a large number of movements of rotary and fixed-wing aircraft. These aircraft try to avoid using fixed approach routes to reduce their vulnerability to hostile shoulder-launched SAMs. A CWIS relying purely on radar cues based on a simple interpretation of speed and approach to the defended area would automatically fire at them unless overridden by a human supervisor.

The original Phalanx was completely automatic with a wide arc of fire. This version would have a high chance of firing at civilian or friendly aircraft and its spent bullets landing in populated areas but still with enough energy to kill people. However, the Block 1B discussed in Section 3.4 does have an infra-red sensor, allowing the human operator to check if the target is an aircraft before the gun fires. This potentially solves one problem, but the time between the launch and impact of a mortar bomb is short compared with the timescales for projectiles and aircraft approaching a ship.

The result of analysing the requirements for a CIWS for a base with a dynamic air environment and nearby civilian population and infrastructure is the Counter-Rocket, Artillery and Mortar (C-RAM) system which has been deployed for more than 10 years with no reported civilian casualties [26]. The final system is a complex system-of-systems, with less automation than the original Phalanx system as it relies on operator and commander confirmation that the threat must be destroyed.

The final C-RAM system comprises:

- A suite of air surveillance radars which monitor all air traffic movements and update the C-RAM command centre on all non-threatening aircraft every 23 ms;
- The radars identify a potential incoming threat and track it and warn the C-RAM commander;

- One or more 20 mm M61A1 Gatling guns using a different ammunition to the naval CIWS;
- The air traffic information is used to set up-to-date no-fire arcs;
- The commander decides whether the threat must be destroyed. If so, a command is issued to the operator to allow the system to engage the threat, but not to fire;
- The C-RAM computers decide the optimum interception location;
- The commander carries out a final check and authorises the operator to allow the gun to fire;
- The operator has also been monitoring the target area and can stop the engagement if he or she considers that the situation has become unsafe;
- The gun fires at the optimum location for minimum number of rounds consistent with destroying the target

Legal reviews were made an integral part of the whole procurement process. Considerations of minimising collateral effects led to increased system complexity, two humans able to stop the engagement at any time, and different ammunition. The ammunition self-destructs in the air after travelling 2 km. This prevents damage on the ground from rounds which did not hit the target. The compromise is that the rounds are not as effective as the naval ones, but C-RAM gives protection from threats launched from civilian areas which would not be possible with naval ammunition.

The conclusion is that IHL considerations must be included throughout the design process and will affect design decisions in the same way that technical ones do. Both technical and legal requirements dictate the concept, system-of-systems design and the innumerable trade-offs that are made during any procurement process.

.

Appendix A7   Extracts from Additional Protocol I (API)

Chapter 7 and some other places in this book make reference to the 1977 additional protocols to the 1949 Geneva Conventions. This appendix gives the referenced articles unless reproduced at that point in the text of book.

PART III – METHODS AND MEANS OF WARFARE, COMBATANT AND PRISONER-OF-WAR STATUS

SECTION I – METHODS AND MEANS OF WARFARE

Article 35 – Basic rules

1. In any armed conflict, the right of the Parties to the conflict to choose methods or means of warfare is not unlimited.
2. It is prohibited to employ weapons, projectiles and material and methods of warfare of a nature to cause superfluous injury or unnecessary suffering.
3. It is prohibited to employ methods or means of warfare which are intended, or may be expected, to cause widespread, long-term and severe damage to the natural environment.

Article 36 – New weapons

In the study, development, acquisition or adoption of a new weapon, means or method of warfare, a High Contracting Party is under an obligation to determine whether its employment would in some or all circumstances be prohibited by this Protocol or by any other rule of international law applicable to the High Contracting Party.

PART IV – CIVILIAN POPULATION

SECTION I – GENERAL PROTECTION AGAINST EFFECTS OF HOSTILITIES

CHAPTER I - BASIC RULE AND FIELD OF APPLICATION

Article 48 – Basic rule

In order to ensure respect for and protection of the civilian population and civilian objects, the Parties to the conflict shall at all times distinguish between the civilian population and combatants and between civilian objects and military objectives and accordingly shall direct their operations only against military objectives.

Article 49 – Definition of attacks and scope of application

1. 'Attacks' means acts of violence against the adversary, whether in offence or in defence.
2. The provisions of this Protocol with respect to attacks apply to all attacks in whatever territory conducted, including the national territory belonging to a Party to the conflict but under the control of an adverse Party.
3. The provisions of this Section apply to any land, air or sea warfare which may affect the civilian population, individual civilians or civilian objects on land. They further apply to all attacks from the sea or from the air against objectives on land but do not otherwise affect the rules of international law applicable in armed conflict at sea or in the air.
4. The provisions of this Section are additional to the rules concerning humanitarian protection contained in the Fourth Convention, particularly in Part II thereof, and in other international agreements binding upon the High Contracting Parties, as well as to other rules of international law relating to the protection of civilians and civilian objects on land, at sea or in the air against the effects of hostilities.

CHAPTER II - CIVILIANS AND CIVILIAN POPULATION

Article 50 — Definition of civilians and civilian population

1. A civilian is any person who does not belong to one of the categories of persons referred to in Article 4 A (1), (2), (3) and (6) of the Third Convention and in Article 43 of this Protocol. In case of doubt whether a person is a civilian, that person shall be considered to be a civilian.
2. The civilian population comprises all persons who are civilians.

3. The presence within the civilian population of individuals who do not come within the definition of civilians does not deprive the population of its civilian character.

Article 51 – Protection of the civilian population

1. The civilian population and individual civilians shall enjoy general protection against dangers arising from military operations. To give effect to this protection, the following rules, which are additional to other applicable rules of international law, shall be observed in all circumstances.
2. The civilian population as such, as well as individual civilians, shall not be the object of attack. Acts or threats of violence the primary purpose of which is to spread terror among the civilian population are prohibited.
3. Civilians shall enjoy the protection afforded by this Section, unless and for such time as they take a direct part in hostilities.
4. Indiscriminate attacks are prohibited. Indiscriminate attacks are:
    (a) those which are not directed at a specific military objective;
    (b) those which employ a method or means of combat which cannot be directed at a specific military objective; or
    (c) those which employ a method or means of combat the effects of which cannot be limited as required by this Protocol;

    and consequently, in each such case, are of a nature to strike military objectives and civilians or civilian objects without distinction.
5. Among others, the following types of attacks are to be considered as indiscriminate:
    (a) an attack by bombardment by any methods or means which treats as a single military objective a number of clearly separated and distinct military objectives located in a city, town, village or other area containing a similar concentration of civilians or civilian objects; and
    (b) an attack which may be expected to cause incidental loss of civilian life, injury to civilians, damage to civilian objects, or a combination thereof, which would be excessive in relation to the concrete and direct military advantage anticipated.

6. Attacks against the civilian population or civilians by way of reprisals are prohibited.
7. The presence or movements of the civilian population or individual civilians shall not be used to render certain points or areas immune from military operations, in particular in attempts to shield military objectives from attacks or to shield, favour or impede military operations. The Parties to the conflict shall not direct the movement of the civilian population or individual civilians in order to attempt to shield military objectives from attacks or to shield military operations.
8. Any violations of these prohibitions shall not release the Parties to the conflict from their legal obligations with respect to the civilian population and civilians, including the obligation to take the precautionary measures provided for in Article 57.

CHAPTER III – CIVILIAN OBJECTS

Article 52 — General protection of civilian objects

1. Civilian objects shall not be the object of attack or of reprisals. Civilian objects are all objects which are not military objectives as defined in paragraph 2.
2. Attacks shall be limited strictly to military objectives. In so far as objects are concerned, military objectives are limited to those objects which by their nature, location, purpose or use make an effective contribution to military action and whose total or partial destruction, capture or neutralization, in the circumstances ruling at the time, offers a definite military advantage.
3. In case of doubt whether an object which is normally dedicated to civilian purposes, such as a place of worship, a house or other dwelling or a school, is being used to make an effective contribution to military action, it shall be presumed not to be so used.

Article 53 – Protection of cultural objects and of places of worship

Without prejudice to the provisions of the Hague Convention for the Protection of Cultural Property in the Event of Armed Conflict of 14 May 1954, and of other relevant international instruments, it is prohibited: (a) to commit any acts of hostility directed against the historic monuments, works of art or places of worship which constitute the cultural or spiritual heritage of peoples; (b) to use such objects in support of the military effort; and (c) to make such objects the object of reprisals.

Article 54 – Protection of objects indispensable to the survival of the civilian population

1. Starvation of civilians as a method of warfare is prohibited.
2. It is prohibited to attack, destroy, remove or render useless objects indispensable to the survival of the civilian population, such as foodstuffs, agricultural areas for the production of foodstuffs, crops, livestock, drinking water installations and supplies and irrigation works, for the specific purpose of denying them for their sustenance value to the civilian population or to the adverse Party, whatever the motive, whether in order to starve out civilians, to cause them to move away, or for any other motive.
3. The prohibitions in paragraph 2 shall not apply to such of the objects covered by it as are used by an adverse Party:
    (a) as sustenance solely for the members of its armed forces; or
    (b) if not as sustenance, then in direct support of military action, provided, however, that in no event shall actions against these objects be taken which may be expected to leave the civilian population with such inadequate food or water as to cause its starvation or force its movement.
4. These objects shall not be made the object of reprisals.
5. In recognition of the vital requirements of any Party to the conflict in the defence of its national territory against invasion, derogation from the prohibitions contained in paragraph 2 may be made by a Party to the conflict within such territory under its own control where required by imperative military necessity.

Article 55 – Protection of the natural environment

1. Care shall be taken in warfare to protect the natural environment against widespread, long-term and severe damage. This protection includes a prohibition of the use of methods or means of warfare which are intended or may be expected to cause such damage to the natural environment and thereby to prejudice the health or survival of the population.
2. Attacks against the natural environment by way of reprisals are prohibited.

Article 56 – Protection of works and installations containing dangerous forces

1. Works or installations containing dangerous forces, namely dams, dykes and nuclear electrical generating stations, shall not be made the object of attack, even where these objects are military objectives, if such attack may cause the release of dangerous forces and consequent severe losses among the civilian population. Other military objectives located at or in the vicinity of these works or installations shall not be made the object of attack if such attack may cause the release of dangerous forces from the works or installations and consequent severe losses among the civilian population.
2. The special protection against attack provided by paragraph 1 shall cease:
    (a) for a dam or a dyke only if it is used for other than its normal function and in regular, significant and direct support of military operations and if such an attack is the only feasible way to terminate such support;
    (b) for a nuclear electrical generating station only if it provides electric power in regular, significant and direct support of military operations and if such attack is the only feasible way to terminate such support;
    (c) for other military objectives located at or in the vicinity of these works or installations only if they are used in regular, significant and direct support of military operations and if such attack is the only feasible way to terminate such support.
3. In all cases, the civilian population and individual civilians shall remain entitled to all the protection accorded them by international law, including the protection of the precautionary measures provided for in Article 57. If the protection ceases and any of the works, installations or military objectives mentioned in paragraph 1 is attacked, all practical precautions shall be taken to avoid the release of the dangerous forces.
4. It is prohibited to make any of the works, installations or military objectives mentioned in paragraph 1 the object of reprisals.
5. The Parties to the conflict shall endeavour to avoid locating any military objectives in the vicinity of the works or installations mentioned in paragraph 1. Nevertheless, installations erected for the sole purpose of defending the protected works or installations from attack are permissible and shall not themselves be made the object of attack, provided that they are not used in hostilities except for defensive actions necessary to respond to attacks against the protected works or installations and that their armament is limited to

weapons capable only of repelling hostile action against the protected works or installations.

6. The High Contracting Parties and the Parties to the conflict are urged to conclude further agreements among themselves to provide additional protection for objects containing dangerous forces.

7. In order to facilitate the identification of the objects protected by this Article, the Parties to the conflict may mark them with a special sign consisting of a group of three bright orange circles placed on the same axis, as specified in Article 16 of Annex 1 to this Protocol. The absence of such marking in no way relieves any Party to the conflict of its obligations under this Article.

CHAPTER IV – PRECAUTIONARY MEASURES

Article 57 – Precautions in attack

1. In the conduct of military operations, constant care shall be taken to spare the civilian population, civilians and civilian objects.

2. With respect to attacks, the following precautions shall be taken:

    (a) those who plan or decide upon an attack shall:

        (i)   do everything feasible to verify that the objectives to be attacked are neither civilians nor civilian objects and are not subject to special protection but are military objectives within the meaning of paragraph 2 of Article 52 and that it is not prohibited by the provisions of this Protocol to attack them;

        (ii) take all feasible precautions in the choice of means and methods of attack with a view to avoiding, and in any event to minimizing, incidental loss of civilian life ,injury to civilians and damage to civilian objects;

        (iii) refrain from deciding to launch any attack which may be expected to cause incidental loss of civilian life, injury to civilians, damage to civilian objects, or a combination thereof, which would be excessive in relation to the concrete and direct military advantage anticipated;

    (b) an attack shall be cancelled or suspended if it becomes apparent that the objective is not a military one or is subject to special protection or that the attack may be expected to cause incidental loss of civilian life, injury to civilians, damage to civilian objects, or a combination thereof, which would be excessive in relation to the concrete and direct military advantage anticipated;

    (c) effective advance warning shall be given of attacks which may affect the civilian population, unless circumstances do not permit.

3. When a choice is possible between several military objectives for obtaining a similar military advantage, the objective to be selected shall be that the attack on which may be expected to cause the least danger to civilian lives and to civilian objects.

4. In the conduct of military operations at sea or in the air, each Party to the conflict shall, in conformity with its rights and duties under the rules of international law applicable in armed conflict, take all reasonable precautions to avoid losses of civilian lives and damage to civilian objects.
5. No provision of this Article may be construed as authorizing any attacks against the civilian population, civilians or civilian objects.

Article 58 – Precautions against the effects of attacks

The Parties to the conflict shall, to the maximum extent feasible:

(a) without prejudice to Article 49 of the Fourth Convention, endeavour to remove the civilian population, individual civilians and civilian objects under their control from the vicinity of military objectives;
(b) avoid locating military objectives within or near densely populated areas;
(c) take the other necessary precautions to protect the civilian population, individual civilians and civilian objects under their control against the dangers resulting from military operations.

References

[1] Solis, G.D. *The law of armed conflict*. Cambridge: Cambridge University Press; 2010.
[2] Heuser, B., and Clausewitz, C.V. On war. Oxford World Classics, Oxford: Oxford University Press; 2008.
[3] United Nations Office for Disarmament Affairs, *Background on lethal autonomous weapons*. https://www.un.org/disarmament/geneva/ccw/background-on-lethal-autonomous-weapons-systems/ [Accessed on 10 May 2017].
[4] Blacknell D., and Griffiths H. *Radar Automatic Target Recognition (ATR) and Non-Cooperative Target Recognition (NCTR)*. Institution of Engineering and Technology, Stevenage, UK. 2013.
[5] Holland J. *The Battle of Britain – Five months that changed history, May-September 1940*. Chapter 26 *Getting Ready*. Bantam Press; London, 2010.
[6] Ibid. Chapter 28, *Bringing it all together*.
[7] ICRC Advisory Service on International Humanitarian Law, Legal factsheet: *What Is International Humanitarian Law?* 2014, Available from: https://www.icrc.org/en/document/what-international-humanitarian-law [Accessed on 14 May 2017].
[8] ICRC Advisory Service on International Humanitarian Law, *The Basics of International Humanitarian Law - ref. 0850-ebook*, Available from: https://shop.icrc.org/l-039-essentiel-du-droit-international-humanitaire-2438.html International Humanitarian Law, Answers to your questions.
[9] Roberts A., and Guelff R. *Documents on the laws of war*, 3rd edition. Oxford, UK: Oxford University Press, 2008.

[10]   Solis G.D. *The law of armed conflict*. New York, USA: Cambridge University Press, 1st edition 2010. Section 1.4.2, pp. 22–26.

[11]   Lawand K. *A Guide to the Legal Review of New Weapons, Means and Methods of Warfare - Measures to Implement Article 36 of Additional Protocol I of 1977*. ICRC report, Revised November 2006.

[12]   McClelland J. 'The review of weapons in accordance with Article 36 of Additional Protocol I', *International Review of the Red Cross*'. 2003, vol. 85 (850): pp. 397–415.

[13]   Boulanin V., and Verbruggen M. *SIPRI compendium on Article 36 reviews*. SIPRI Background Paper, December 2017.

[14]   Backstrom A. and Henderson I. 'New capabilities in warfare: an overview of contemporary technological developments and the associated legal and engineering issues in Article 36 weapons reviews'. *International Review of the Red Cross*. 2012, vol. 94(886): pp. 483–514.

[15]   Henckaerts J-M., and Doswald-Beck L. *Customary International Humanitarian Law*, Volume 1, Rules, Cambridge, UK: Cambridge University Press, 2005.

[16]   Henckaerts J-M. 'Study on customary international humanitarian law: A contribution to the understanding and respect for the rule of law in armed conflict'. *International Review of The Red Cross*. 2005, vol. 87(857): pp. 175–212.

[17]   UK Ministry of Defence. *Joint Services Publication 383 (JSP383), The joint service manual of the law of armed conflict*, 2004, with amendments of 21 May 2014.

[18]   Henckaerts J-M. 'Study on customary international humanitarian law: A contribution to the understanding and respect for the rule of law in armed conflict'. *International Review of The Red Cross*. 2005, vol. 87(857): pp. 175–212, p. 179.

[19]   Boothby W.H. *The law of targeting*. Oxford, UK: Oxford University Press, 2012.

[20]   US Department of Defense. *Law of war manual*. December 2016. Available from https://www.defense.gov/Portals/1/Documents/pubs/DoD%20Law%20of%20War%20Manual%20-%20June%202015%20Updated%20Dec%202016.pdf?ver=2016-12-13-172036-190 [Accessed on 14th May 2017].

[21]   International Committee of the Red Cross. *Expert meeting: Autonomous weapon systems implications of increasing autonomy in the critical functions of weapons. Versoix, Switzerland, 15–16 March 2016*. ICRC Geneva, 2016.

[22]   API Annex 1, Regulations concerning identification (as amended on 30 November 1993).

[23]   Boulanin V., and Verbruggen M. *Article 36 Reviews, dealing with the challenges posed by emerging technologies*. Report by SIPRI summarising a 2017 conference. Available at https://www.sipri.org/publications/2017/other-publications/article-36-reviews-dealing-challenges-posed-emerging-technologies [Accessed on 18 June 2018].

[24]   Boulanin V. *Implementing Article 36 Reviews in the light of increasing autonomy in weapon systems*. SIPRI Insights on Peace and Security No. 2015/1, November 2015.

[25]   Gillespie T. 'New Technologies and Design for the Laws of Armed Conflict', *The RUSI Journal*. 2015, vol. 160(6): pp. 50–56.

[26]   Presentation made by US delegation to the April 2018 meeting of the United Nations CCW LAWS Group of Government Experts (GGE). Available at https://geneva.usmission.gov/2018/04/13/u-s-slide-presentation-at-ccw-gge-counter-rocket-artillery-and-mortar-system-c-ram/ [Accessed on 24 April 2018].

Chapter 8

Targeting

8.1 Introduction to targeting

It is important that the terminology is clear, so we will use the definitions from the NATO Standard (STANAG) AAP-6 [1] which is the reference used for most military terminology in this book. These are given in Panel 8.1 and their meanings are very similar to their normal English ones. We will also use a generic term, non-target, for objects which are similar to legitimate targets but are difficult to distinguish from them, and objects which are not to be attacked due to the nature of their use, treaty obligations or being of limited or no military value.

Panel 8.1:   Definitions from STANAG AAP-06

Target

The object of a particular action, for example, a geographic area, a complex, an installation, a force, equipment, an individual, a group or a system, planned for capture, exploitation, neutralization or destruction by military forces.

Targeting

It is the process of selecting and prioritizing targets and matching the appropriate response to them, taking into account operational requirements and capabilities.

Target intelligence

It is the intelligence which portrays and locates the components of a target or target complex and indicates its vulnerability and relative importance.

Target list

It is a tabulation of confirmed or suspected targets maintained by any echelon for information and fire support planning purposes.

It can be seen from Panel 8.1 that targeting is a formal process which needs to operate at a speed appropriate to the circumstances. An attack on a major installation such as a harbour or airfield is likely to take several days or weeks to prepare, so large amounts of data can be assimilated and even special surveillance operations undertaken to provide missing information. In contrast, an attack on a fleeting target of opportunity has to be completed in seconds but still comply with International Humanitarian Law (IHL). The targeting process is now generally considered to include assessing damage to the target, battle damage assessment and post operation reviews.

Combat engagement, which includes actions against an adversary and self-defence responses, is not part of the targeting process, but the use of weapons must still meet rules-of-engagement. The weapons used in these circumstances will be from the same inventory and order-of-battle for the campaign. Therefore, from an engineering perspective, it can be assumed that all weapon designs must meet API Article 35 fully and be reviewed under API Article 36.

Every nation will have its own targeting process which matches its resources and its abilities to gather target intelligence. The nation will also set rules-of-engagement that follow its own interpretation of IHL. Many campaigns are fought by coalitions where several nations participate and coordinate their activities. This does not give problems with rules-of-engagement if the nations operate in separate areas or carry out different types of operation. However, it is likely that on some occasions they will attack the same target in collaboration, so it is essential that coalition members agree common rules and processes. Without this, there could be confusion between coalition partners at critical moments in an operation.

NATO has a joint targeting process which has been drawn up to enable joint targeting and attacks, and to allow operations where there may also be stabilisation and reconstruction activities. It is described in AJP-3.9 [2] and is a practical solution to the problem of coalition targeting operations. It is argued by Roorda [3] that this process ensures adherence to IHL, but mainly considers the criteria for initial selection of targets and keeps human control of the lethal weapon-release decision. The discussions in this chapter are concerned with making functions in the identification and attack processes more automated.

An example of one nation's targeting process is given in Reference [4]. Note that some of the definitions in this document and AJP-3.9 are not the same as those in AAP-06, Panel 8.1, which emphasises the need for engineers to take legal or military advice when necessary.

Military decisions about the type of target chosen for attack are beyond the scope of this book. The starting point here is that an approved target list has been drawn up and agreed with unit commanders. There will also be a list of targets with restrictions on attacking them, called restricted targets, and a no-strike list. The local commander's problem is attacking the allowed targets whilst acting within IHL and meeting the extant rules-of-engagement. Tracking a target when detected is an important part of the process with significant engineering implications if automated. Restricted targets may include ones which need higher-level approval before weapon-release even when correctly identified and collateral risk is considered acceptable by the commander.

There are two main types of targeting:

- *Deliberate targeting* is where a future attack is planned against a known target. There is considerable knowledge of the target and its environment so the force used, and its timing, can be matched to achieve success;

- *Dynamic targeting* is carried out against targets which need to be attacked but there is, or may be, insufficient time to go through the deliberate targeting process. This type can be further split into: unplanned; unanticipated; emerging; and time sensitive targets. Example targets include: known targets which are being tracked but are in areas with a high risk of collateral damage; ones which are probably in an area, but details will be incomplete; and unexpected targets or threats which appear and need a rapid response. If the threat is a weapon fired at the blue[1] asset, then the response is one of self-defence. The force used is likely to be whatever can be assigned to the attack at short notice.

There are also high-value targets and high pay-off targets whose neutralisation or destruction contributes significantly to the success of the mission or campaign respectively.

Although the time required to make a decision is very different for the two types of targeting, the common feature is that there is a list of approved targets which have been compiled by humans. Attacks on these will meet IHL provided that the weapon used meets defined criteria for proportionality and humanity. Collateral damage is acceptable against clear criteria set out in the rules-of-engagement. Again this will be decided at the mission planning stage by humans. Section 2.8.2 has already illustrated the scope of targeting directives. These are produced during the planning process, identify legitimate targets to commanders and may place restrictions on their authority to act.

Accurate targeting is a very important military requirement and has been one of the principal technology drivers for centuries. The principal needs are: target identification; location; tracking; and speed of response. Speed of response may be needed if the target is about to launch an attack at the attacking platform, but it is also necessary in order to disrupt the enemy's decision process and so gain a more strategic advantage. Automation is an obvious way of improving the speed of response and is used as far as possible.

When a target and its surroundings are visible to a human, either directly or remotely by sensors and data link, there may be no further need for automation and the weapon release decision will be made by a human with all necessary information available to them. This can be considered as an ideal scenario, but almost always only occurs after a chain of events including searching for the target, positive identification of it among clutter, assessing likely collateral damage at attack time, and tracking both the target and surrounding objects. The chain may take weeks if intelligence has to be built up prior to the attack. It should be obvious that algorithms and automation will be used at most if not all stages of the process. Even if the target is directly visible, the identification passed to the weapon commander may be based on algorithms to separate the target from non-targets in the vicinity. These algorithmic assessments are decision aids which must be considered as part of the weapon system. They are discussed in this chapter in the context of the various military targeting processes, such as the Observe, Orient, Decide and Act (OODA) process.

---

[1]This book uses the convention that own and allied forces are called blue forces and hostile ones are red.

8.2   Types of weapon used in attack

8.2.1   Classes of attack

Historic reasons mean that attacks using projectiles are often separated into direct and indirect fire attacks. The former being where the target can be seen by the aimer and the latter where it cannot. The traditional example being artillery fire which may be direct, usually at fairly short range at a visible target, or indirect fire over a hill, possibly with a spotter relaying information to the battery commander.

The advent of remote surveillance systems relaying images to aimers located a long way from both the weapon and the target makes this separation less useful. Additionally, many weapons now have a guidance system which guides the, for example, missile onto the target, both increasing its range and accuracy and allowing the weapon to home in on a moving target. Automation of most guidance functions has already taken place and will continue, so the distinction taken here is between those which have ballistic trajectories, i.e. no guidance after launch, and those which have post-launch guidance systems.

Tracking potential targets over periods of minutes or hours before weapon launch brings a different, but related set of problems, that of correct initial identification followed by maintaining lock on that object for a period of time. The target may pass close to similar objects or disappear for short periods, for example, it may be a road vehicle in traffic which disappears behind a building and then reappears. Tracking a potential target is the type of activity that may be best carried out by an automated system. However, the performance must be verified to a high standard and also tested for scenarios where a hostile opponent may use deception techniques which would not fool a human observer but might fool an automated system.

8.2.2   Ballistic projectiles

The classic examples of these are bullets and shells fired from guns. Their trajectory is dictated by the gun's aiming, and local atmospheric and meteorological conditions. Accuracy is quoted in CEP[2]. Bombs dropped from aircraft will also follow a ballistic trajectory if they have no further guidance system, the so-called dumb bombs. High-level bombing operations usually use guided bombs rather than dumb ones, but these may have a short period of ballistic flight before the guidance system locks on to the target position. Loss of lock may also be a problem which must be considered by the pilot before releasing the bomb.

The autonomous capability most likely to be associated with short-range ballistic projectiles is locating and tracking the target. The Phalanx system discussed in Section 3.6 is a good example of this type of system. The ethical problem facing the designer is ensuring that the system correctly identifies the target and that the

---

[2]The CEP associated with an aim point (x,y) is the radius of the smallest circle with centre at (x,y) which has 50% probability of containing the impact points. It is not itself a probability and may not have a Gaussian distribution.

lethal decision can only come from a correctly authorised source, i.e. the decision meets API Article 35 and the weapon system can pass an Article 36 Review.

The tracking problem mentioned in Section 8.1 will apply if the release of a ballistic projectile is preceded by target tracking.

8.2.3   *Externally guided projectiles*

Externally guided projectiles are taken here to mean ones that are guided to a target which is visible to the commander operating the system. (Visible includes detection and tracking by radar, sonar or other sensors as well as by optical means.) The commander fires or releases the weapon and guides it to the target by methods such as:

*   Tube-launched Optically-tracked Wire-guided (TOW) missiles. Guidance is by a thin wire connecting the missile to the operator which spools out after launch. Guided range, clearly, is limited by the length of the wire. Beyond that, the missile trajectory is ballistic;
*   Beam-riding missile systems where the missile detects its location relative to a matrix of overlapping laser beams which are locked on to the target;
*   Laser-guided bombs and missiles, where a laser is shone on the target, either from the weapon-carrying aircraft, another aircraft, or a forward air controller. The bomb has a laser sensor and homes in on the laser spot. It must be released so that its ballistic trajectory takes it into the capture basket, i.e. the volume of space where its sensor will detect the reflected laser beam, and the guidance system will be able to respond and guide the bomb onto the target.

8.2.4   *Self-guided projectiles*

Dumb bombs have relatively low accuracy and are being superseded by GPS-guided ones. The bomb is released from a volume of (air) space which places the aim-point and/or target in an area of the ground which can be hit by the bomb. This volume is considerably larger than the single release point for a ballistic trajectory, simplifying the pilot's problems on his or her attack run. The bomb is given the GPS coordinates of the aim-point and it is steered to it by control of its guidance fins.

The original 'smart missiles' used in Gulf War I had an autopilot and on-board GPS-based routing system which flew the missile to its target. The flight path could be complex but was pre-planned and loaded onto the missile prior to launch. Once fired, the target and route could not be changed, so the time for making the rules-of-engagement decisions is when the missile is launched. More modern cruise missiles have data links to them so the target can be changed during the flight. This complicates the commander's problems under IHL: an assessment will be made at launch time, but the target area may need to be monitored up to the last time at which aim-point changes can be sent to the missile and correctly executed.

Modern guided missiles usually have a seeker head for the last phase of their flight. The missile is guided towards the target by some other means. When close, it uses either Infra-Red (IR) imaging or radar to track its target. The warhead is finally detonated by a fuse.

Homing torpedoes work in an analogous way to guided missiles, but they have the additional advantage that they may be designed as passive systems; the homing head detects and tracks the acoustic signature emitted by the target ship or submarine. A torpedo can have an active or semi-active guidance system which uses a reflected acoustic signal from the target, either generated by itself or from another source.

8.3   Targeting law

It should be clear from Chapter 7 that IHL demands that all lethal decisions must be based on human judgement of the often contradictory criteria to be met. It can also be seen from Section 8.2 that the modern targeting process up to weapon release is heavily dependent on technology. Processor-based automation is used at most, if not all stages and its role can only increase. Developments in Artificial Intelligence (AI) cannot be ignored and it must be assumed that AI will appear in some phases of targeting. At present, there are no clear rules about how AI-based systems can be reviewed under IHL, so the designer would have to work with the procurement and legal authorities for their country or user's country.

Possible ways that laws may change are discussed in Chapter 9, with on-going discussions in international fora. Most of this book is based on current IHL and this section describes the parts relevant to targeting. There is a legal book [5] which discusses targeting law in great detail which has been used as a reference in several parts of this book. The UK Joint Service Procedure JSP383 [6] and the US 2016 Law of War Manual [7] interpret IHL for their armed forces and give more explicit guidance for military personnel.

There are four principles underpinning the law as it affects lethal attacks and targeting. These are covered in more detail in Section 7.6, but are summarised here:

- *Military necessity* which permits the use of force only if it is necessary to achieve its military purpose in the conflict with the minimum possible loss of life and damage to property;
- *Distinction* requires all parties to distinguish between combatants and civilian populations as well as military objectives and civilian objects;
- *Humanity* restricts death, injury and damage to property to the minimum necessary to achieve the required military effect;
- *Proportionality* requires that the losses resulting from a military action should not be excessive in relation to the expected military advantage.

Chapter 10 shows how these give rise to detailed design requirements in each military domain. Here, it is only necessary to understand that there are requirements for the commander to:

- Be part of a command chain;
- Have the capability to identify targets;
- Understand the context of the target area;
- Use the weapon causing the least collateral damage if they have a choice;
- Call off the attack if collateral damage will be excessive for the military benefit;
- Meet explicit criteria set by their rules-of-engagement.

Supplementary to these is API Article 57 'Precautions in Attack'. This mandates that commanders will apply the above criteria, but also adds that the attack must be suspended if it becomes apparent that the target is not a legitimate one. The implications of this are that the commander must be able to stop the attack up to the last time that they have control over the weapon. After that point, assuming they have followed their rules-of-engagement, their decisions will be assessed based on information at the time that the decision was made. However, if they can control some aspects of the weapon's operation whilst *en route* to the target area, they must take appropriate action if more information is presented to them about the target.

It is illegal to deliberately prevent information reaching the commander, but latency in the communication network will be an issue. The commander releasing the weapon is likely to be at the lowest levels of the command chain and may not have immediate access to information available at higher levels. Ideally, current information about the target area should be relayed directly to the commander who has the capability to make post-release changes to the weapon's aimpoint even if he or she has the visibility of it. The new information could be, for example, from intelligence sources. This is a system of systems issue where collaboration between designers, military staff and legal advisers is essential in order to write IHL-compliant requirement specifications and test plans.

The development of IHL for many decades assumed a simple form of targeting, with a sole weapon commander acting on prior information and visual or radar imaging of the target area. Collateral damage was assessed prior to the mission allowing for the time interval between commencing the mission and the weapon's impact. Information during the mission would probably be relayed to the weapon commander by a voice link from their immediate commander. Weapon accuracy meant the CEP, and the probability of the weapon malfunctioning, but this is no longer sufficient.

Modern military operations may involve identifying targets in complex, evolving environments and tracking them until conditions are suitable for a strike. This adds uncertainties in the results of imaging and tracking algorithms to considerations of overall accuracy. There are two aspects to this problem: the probability of correct classification of a target and the probability of falsely identifying a non-target as a target. The two will be related but not necessarily in a simple manner. The designer will express these probabilities as percentages under quantified target scene conditions. Tests and trials will be conducted to assess actual performance in these terms which are very different to the IHL terminology given as the four principles at the start of this section. This is an area where collaboration is needed between the military user, their legal advisers and the designer. The trials data from the actual system should be available in a suitable form for the setting of rules-of-engagement for every campaign. Considerable judgement will be involved in setting the various probability thresholds in the implementations of the classification algorithms for the anticipated military operations.

Data may come from the immediate command chain, but it will also come from the wider coalition network and on-board databases which may be regularly updated. There will be time delays for integration necessitating configuration control of information flow around the network. It may not be possible to reduce these delays, but they should be quantified in some suitable way for the envisaged operations so that informed plans can be made.

The procurement authority must be able to produce verified and validated performance figures for all the subsystems and the wider system-of-systems as well as the traditional CEP, blast range and reliability figures for an Article 36 review.

8.4   Targeting processes and cycles

8.4.1   A range of targeting processes

The most well-known targeting cycle is the OODA loop which was introduced in Section 2.7 as well as the more recent Dynamic Observe, Orient, Decide and Act (DOODA) loop. As explained there, OODA is a process rather than a loop. It does not set the attack in its wider context as it will have immediate and longer-term effects which can give intelligence analysts data leading to another legitimate target. Pre- and post-targeting processing activities need to be taken into account and considered to be part of a targeting cycle. This led to suggestions in the late 1990s [8] that the future approach to targeting for the twenty-first century should be based on a more complete process: Find, Fix, Track, Target, Engage and Assess (F2T2EA). The intention being that post-strike assessment will lead to information about future potential targets and their locations. They can then be attacked, completing the cycle.

F2T2EA was conceived for air strikes against both pre-planned targets and time-sensitive targets where there is likely to be a known type of target which has to be destroyed quickly after its detection. A similar process was developed by the US army: Decide, Detect, Deliver and Assess (D3A). They have proved successful for deliberate targeting, but do need dynamic targeting constructs for rapid responses. As a result, variations of the parts and order of F2T2EA have been developed as well as adding in-depth post-assessment phases. Other proposed processes are: Find, Fix, Finish, Exploit and Assess (F3EA) [9,10] and Find, Fix, Finish, Exploit, Analyse and Disseminate (F3EAD) [11]. A useful discussion of the various processes is given by Ferry in Reference [10].

Figure 8.1 shows these targeting processes alongside each other. One common time is when the weapon hits its target or nearby 'Impact' and can be used as a reference. This is shown in Figure 8.1 with the time when the targeting directive is issued as a second, earlier, reference time. The starting times of the phases in each process do not necessarily coincide even if they have the same name. One important common feature is that there is a list of potential targets which have been selected using human judgement. Attacks on them must follow rules-of-engagement and the targeting directive in order to be legitimate in IHL.

Deliberate targeting usually does not need the rapid assessments required for dynamic targeting. The target and its surroundings should be well-understood so the human decisions concern the status of the area at the time of attack and differences that would, for example, cause unacceptable collateral damage, i.e. the

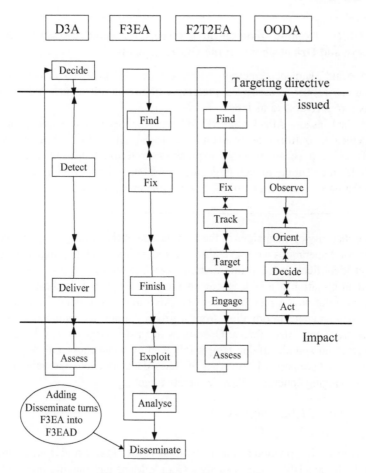

Figure 8.1   Targeting processes

OO, *Observe and Orient*, phases of OODA are not complex so most emphasis is on the D, *Decide*, phase.

It was stated in Section 2.7 that, in the author's opinion, OODA is the best framework for engineers to use for discussions with military staff on targeting. This is true, but for more detailed considerations of engineering aspects of targeting, the F2T2EA process splits the OO phases of OODA into functions that can be related more directly to engineering design with current and future technologies.

In order to give some context to the discussion below, some example scenarios for the attack of time-sensitive targets are given in Panel 8.2. These examples have an emphasis on the quick reactions required to strike the target. The F, *Find*, phase may be comparatively long. It may rely on a rapid response to deviations from a picture of normality built up over many weeks or months of observations. Therefore it is convenient to examine decision aids using the F2T2EA functions as a framework. This examination is in Section 8.4.2.

Panel 8.2:   Example time-sensitive targets for rapid assessment in the Observe and Orient phases of the OODA process

1. A Scud[3] launcher emerging from a hide to launch its missile(s). The launcher must be destroyed, preferably before missile launch, but certainly before it returns to its hide;
2. A Fast Inshore Attack Craft (FIAC) approaching a ship where it must be detected, confirmed as hostile and destroyed before it launches its attack[4];
3. Tracking a consignment of explosives or terrorist operatives leaving a hideout and going to their target locations, one or both locations being in an urban area with a high probability of collateral damage.

The majority of campaigns in the first decade of the twenty-first century have entailed joint operations with several forces and nations operating together, leading to target identification using information from a range of sources. There may be detailed information available about an area from updated databases, but more immediate information may be necessary. Air support to ground forces in action will need immediate information from a ground controller with those forces to separate 'red' and 'blue' forces. Operations against insurgents may have Special Forces tracking and identifying targets followed by requesting an air strike, possibly in confused scenarios. These types of attack demand much more rapid response times to changing conditions than deliberate targeting.

8.4.2   The F2T2EA phases

8.4.2.1   Find

There is always information available about any part of the world from a variety of commercially available sources such as Google Maps and satellite information at both optical and radar wavelengths as well as from military surveillance sources. The provenance and age of commercial data is not always clear, producing uncertainty in its suitability for military use.

The 'Find' phase can be surveillance, by one or more assets, of an area followed by a change in the scene. Identification of the one or more genuine targets from the clutter and false alarms is based on a set of signatures and behaviours. There are many tools available to automate change detection, target indication and some for target identification[5], but the parameters that they use and the various thresholds used must be understood. The requirement for low 'false positives' (i.e. declaring a non-target as a target) must be balanced against the contradictory

---

[3]Scud is a NATO name designating a tactical missile which is usually launched from a mobile launcher.
[4]The FIAC can be an unmanned vessel.
[5]Note that there is a distinction between target indication, i.e. this object looks as if it may be a suitable object for further examination, and identification, i.e. this object is a tank which is almost certainly of type N and hostile. The data required is much larger for the latter than for the former, as is the cost of the tools to automate the processes. See also Panel 8.3.

requirement for a low threshold so that a real target is not missed. There must be a high probability of correct identification for an object to be declared as a target. This is an area where human judgement relies on many subjective factors as well as the objective ones provided by software tools.

The characteristics identifying a target are not necessarily physical ones from optical or radar sensors, but may also be based on an object's behaviour and the functions it is performing. A command centre may be identified by the radio traffic into and out of a building or vehicle with no other distinctive features. A hostile observer need not be obviously armed, but may be identified by their appearance at critical times and use of civilian communication methods such as mobile phones. However, the identification is carried out, the commander releasing the weapon must have a method of unambiguously identifying the target and assessing the potential collateral damage from the weapon so that he or she can make the release decision.

When identified, a decision must be made about the target's importance and the action required. The options include do nothing, track and attack. Referral to a higher node in the Command and Control (C2) chain is also an option, especially if the objective criteria from the tools are close to the thresholds for identification.

It is possible that unexpected threats will appear that need a rapid response. The threat may be from an object or person not on the target list and may be in a location chosen to give a high chance of collateral damage. The targeting process must allow the commander under threat to make a rapid response under difficult circumstances. A pilot whose Missile Attack Warning System (MAWS) gives an alarm may have several options and a low number of seconds to choose. A foot patrol commander will have less time to respond to an alert by a system that detects sniper fire and the error on the location of the sniper may be fairly high. Reliable and repeatable operation of these and similar warning systems is essential with error budgets that are known and understood by the users. This is an engineering challenge as the sensor input may be from one event in high clutter levels giving a high false alarm rate.

8.4.2.2 Fix

When identified, the target should be tagged and geo-location carried out. Fix should also include the identification of other objects in the target area so that possible collateral damage can be assessed. An attack decision may be made at this point if prior authorisation has been given, and rules-of-engagement are met. If moving, its track has to be determined so that the weapon can be laid on correctly.

The most basic step in combining data about an area is to ensure that there is a common reference frame and that this has a clear relationship to that used by other coalition forces and the weapons. Map coordinate systems are not always consistent, particularly non-GPS ones, and even GPS coordinates can be in error due to signal propagation errors and compass calibration faults. This is not usually a problem, but different users have different priorities. A pilot knows the aircraft's position and trajectory in World Geodetic System 1984 (WGS-84) coordinates, almost certainly using GPS for his or her own position. The target coordinates must also refer to the same datum to give the correct bomb trajectory. An artillery commander wants to know the position of his/her target relative to his or her gun and does not worry if they are not in WGS-84 coordinates provided the distance and direction to the target are correct.

Automated image and map fusion engines must ensure that the data to be fused has the same reference frames and units before combining the data. Images produced by different sensor types may be in different coordinate frames. Optical and IR images have perspective, with the same cross-range distance giving smaller angular sizes at larger ranges. A synthetic aperture radar image of an area is in latitude and longitude with no further processing. However, there is usually an assumption that the mapped area is at constant altitude; slopes and hills can give errors in the image. Even without these errors, combining a synthetic-aperture-radar image with an optical image taken at anything other than directly overhead will need a coordinate transformation for accurate geo-location of objects in the image.

A further problem with combining radar, IR or optical data is that the reflections from objects are wavelength dependant. The dominant feature at one wavelength may be almost invisible at another so two images may not be obviously of the same object. It is well-known that IR image intensity depends on the objects temperature. It is not so well known that a dihedral or trihedral reflector gives a strong radar return at almost any angle. This means that wet steps on a building, the angles between a turret and top plate of an armoured vehicle, or vehicle tracks can give a strong radar return but not be obvious at other wavelengths.

8.4.2.3   Track

Tracking a target, or other object, may be achieved by locking a sensor onto it so that its position is followed continuously. The sensor may be part of the weapon system used to attack it, but this is not always the case, with target data passed from the tracking sensor to the weapon system. Track data can come from radar Moving Target Indication (MTI) techniques, but it may also come from measurements of successive optical or infra-red outputs.

Tracking an object is straightforward if it is in open territory. However, an adversary will take precautions such as operating in an environment with buildings, trees and other clutter, and mixing with similar vehicles in a busy scene. This is to take advantage of the performance of both humans and automated trackers which may assume that the first object reacquired after a brief loss will be the same object. This is because the criteria used by an automated tracker to reacquire its 'target' may not include repeating a full identification process.

8.4.2.4   Target

This is the final decision to engage the target before weapon release. It can be considered to be a mini-OODA loop as a final check must be made that rules-of-engagement are satisfied and the risk of collateral damage is still acceptable.

There should be a confirmation that the target is the same as in previous phases and that its identification is sufficiently unambiguous for an attack to meet rules-of-engagement. It must be positively identified if the first cue was on the basis of an indication rather than identification.

There are object-identification tools available to the weapon operator which should have been used prior to this phase of the attack. These must be considered as decision-aids for the operator and not ones that give definitive information and hence authority to proceed to target engagement. There is a human bias to believe an automated software suite rather than one's own experience so designers should consider whether there is a way to show the confidence level, or lack of confidence, to the operator during the 'Target' phase.

8.4.2.5   Engagement

This is locking the weapon(s) onto the target, their release, and hitting or missing the target. The reliability of the 'lock' algorithms is critical and the operator should have reasons for confidence in their operation as well as contingency plans if they do not. The locking-on and following process may use an on-board sensor, or it may use indirect methods such as following reflections from a laser beam illuminating the target. The advantage of the latter method is that there is the possibility of redirecting the weapon to another point if something happens in the target area so that rules-of-engagement are no longer met. An example could be the unexpected appearance of civilians in an area that previously was only occupied by combatants.

The preceding paragraph assumes that there is a short time interval between the release of the weapon and it reaching the target. The principles are the same for a cruise missile or torpedo although the time interval may be up to several hours. More modern systems can have the capability for a mid-course retargeting to attack a different target. This raises the problem of confirming that rules-of-engagement are met not only at weapon release, but also up to the last time when retargeting can be effective. It is unlikely that the weapon operator will have continuous surveillance of the target area, or even any direct surveillance capability after release. The operation planners must address the question of who has authority to confirm or retarget the weapon and how they are given the information for their decision. This is analogous to, but more complex than, the pilot of an aircraft handing over control of a laser-guided bomb to a forward air controller and leaving the target area.

8.4.2.6   Assessment

This is often called battle damage assessment and is the assessment of the success of the strike. Battle damage assessment carried out immediately after the strike leads directly to the decision to break off or re-engage the target. The assessment may be carried out over a longer time, depending on the nature of the target. A strike on a communication centre may cause a temporary cessation of signal transmission for a few minutes but would be considered unsuccessful if transmissions resumed soon afterwards. Destruction of an underground bunker complex with several entrances over a wide area might take some days to assess.

The assessment may need integrated data from several sources as the strike asset may be at high risk of a response from defending forces and so it is only able to provide immediate data. Smoke and debris from the explosion may render its data useless for a decision on the target's residual capabilities, i.e. is the target still a useful asset to the opponent.

Change detection algorithms may be used for this phase, but it is highly likely that the images being compared will be from different directions, and possibly at different wavelengths. The significant changes required to make re-attack decisions may be small in comparison with the large changes due to the weapon strike, so change detection algorithms may produce accurate but misleading results.

8.5   Automating targeting processes

8.5.1   General considerations

This section will cover the issues arising with increasing the automation in the phases of the targeting processes when there is human involvement at one or more stages. Complete removal of the human, giving an autonomously targeting weapon, is discussed in Section 8.6.

With increasingly capable image analysis software and combining temporal information, it will become more difficult to separate the *Find, Fix and Track* parts of the F2T2EA cycle so this section and the next will use the OODA phases. These also align with the three-part models of human decision-making discussed in Chapter 2. Chapters 11 and 12 examine the problems with increasing autonomy in a more specific, but a complex scenario, using the requirements derived from IHL in Chapter 10.

It is important to recognise that whichever description of the targeting process is used, it is sequential. Each step in the process can only be authorised if the preceding one has been completed with a defined level of confidence in its outputs. An object cannot be attacked if there is not a high confidence level in both its identification as a legitimate target and that potential collateral damage is below acceptable limits.

Section 2.5 showed that there are two models of decision-making: the rational one of identifying options, assessing them and choosing the one with the best chance of success judged against specific criteria; and the intuitive one based on experience and analogies, with problem solving by mental simulation. The latter is known as Recognition-Primed Decision-Making (RPDM) which can apply to changing situations. AI with learning systems is more akin to RPDM than rational decision-making. This is the model which is most appropriate to military decision-makers and is shown in Figure 2.6. However, most algorithms used in military operations are based on rational decision-making. This may change as commercial image recognition software is now based on AI, so it is reasonable to assume that machine learning will migrate to the military domain. The timescale for this migration is not clear for an autonomous system. AI-based image recognition used as a decision-aid for a human is straightforward as the human can filter the results based on experience. There will need to be extensive verification of it before it becomes part of an automated targeting chain, a process which will take several years.

Commanders at all levels in a mission will be under considerable stress and have incomplete situational awareness. Situational awareness will be limited by the finite data rate for information reaching them and their own inability to comprehend all factors due to intense workload. They will probably be at Level 7, 8 or 9 on the Bedford Workload Scale shown in Figure 2.5 when they make the decision to release the weapon.

The conclusion from the previous two paragraphs is that the human–machine interface is crucial. It must present information in a clear way that can be immediately assessed by the person, preferably with little mental effort. There is a natural human bias to trust an automated process unless there is a clear reason to doubt it. Should some human input be required, this should be flagged up with the reason, or some guidance given. (See the discussion in Section 8.5.5 and Panel 8.3 on the meanings of ATI, with potential confusion between them.)

8.5.2   Assumptions

The human in the OODA process will be the last human in the command chain for that part of the mission. Military command chains are, of necessity, very simple and very clear. Figure 8.2 shows one with the associated information flows for a C2 chain with several weapon commanders. There will be at least one military network in operation, as shown in Figure 8.2. The most ubiquitous one for NATO operations is called Link 16. This is a secure Time Division Multiple Access (TDMA) network which has relatively low bandwidth, but almost all assets have access to it.

Each block in Figure 8.2 is called a node, whether it is human or machine. Each node in the chain reports to one superior node and has a defined level of responsibility and authority to act. This simplifies discussions when thinking about highly automated functions as they can then be related directly to the nodes in the C2 chain and their role.

It may be necessary, possibly planned, for a higher commander to hand command of one of the weapon commanders to another. This is shown by the dashed lines in Figure 8.2 when line A is replaced by line B. An example of this is when an armed aircraft is to attack a target from high altitude, with a forward air controller giving the target coordinates and using a laser designator to guide a laser-guided bomb onto it.

There will be established rules-of-engagement for decisions by the humans, especially at the lowest node in the C2 chain. There will also be a considerable amount of contextual data and information available about the target area and the target's intent. The opposing forces will also be taking steps to prevent or thwart the attack so that counters to their moves must be considered as part of the mission.

As explained above, targeting is a sequential process requiring completion of one step before the next can be authorised. This means that the authorisation criteria for each step will be different as the steps address different problems. It is possible that parts of the mission rules-of-engagement criteria can be expressed as parameters in the algorithms used in the process, allowing some level of automated authorisation. However, there must always be overall authorisation of the whole process which is currently a human responsibility.

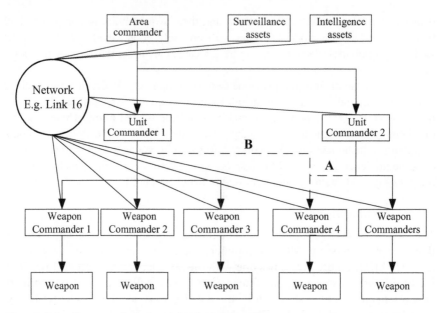

Figure 8.2   Command chain and information flow

[The command normally flows down the thicker lines, including Line A. Under some circumstances, Weapon Commander 4 may come under the command of Unit Commander 1 and Line A is replaced by Line B]

The assumptions made here concern the information available at each relevant node, the rules-of-engagement and integrity of information over time:

1. The weapon commander is the human who makes the decision to release or fire the weapon and has the final IHL responsibility for that action. The commander may be an infantryman with a rifle, or could be the captain of a submarine with a highly automated cruise missile as the weapon;

2. The commander is in a defined command chain such as the one shown with thick lines in Figure 8.2;

3. The commanders in the OODA process have clear and unambiguous rules-of-engagement;

4. The 'blue' communication network is not corrupted by hostile action. Any battle damage to it having been corrected or alternative communication links established, retaining its integrity even if the data rate may have been reduced or latency increased;

5. The off-board data received by the commander has come over reliable data links from accredited sources so it is a legitimate basis for action;

6. There is a recognised picture of the battlespace maintained and authorised by the command chain. Not all information will be available to the weapon commander, but all information about the target area will be passed to them as quickly as is reasonably possible;

7.   There is a local database for each node which is under configuration control and updated as necessary;

8.   The commander has sensors which are under their direct control for targeting purposes. These may be located on one or more platforms including the weapon;

9.   A collateral damage assessment will need to be made by the commander releasing the weapon;

10. The weapon used will be one that has an immediate and limited effect on hitting the target, i.e. it is kinetic and will either explode or will become inert in some way immediately after impact. There may be a time delay of minutes or hours between the last time that the commander can change the instructions to the weapon and impact time.

Operational circumstances may make it necessary to have periods with no communication link between a node and its superior one. Examples are submarines on deep-sea patrol, and a stealthy aircraft entering an area where radio-silence is needed to avoid detection. This does not mean that the node is out of control as it will still operate to agreed instructions and have planned methods of re-establishing its links. There should also be contingency plans if the chain is disrupted by chance or by hostile action.

8.5.3   Observe

A weapon commander who is not in direct line of sight to the target will have to use a sensor suite to give him or her the requisite level of situational awareness. There may be general information available such as satellite or surveillance Unmanned Air Vehicle (UAV) imagery but unless this is available with a direct feed, there will be latency problems. This gives rise to the assumption above that there are one or more sensors under their command.

It is important that the information is presented to the commander in a coherent and understandable way when they are under pressure from the high workloads of being in a hostile environment and likely to come under attack themselves.

Automation of sensor suites depends on their type and raises the following issues:

8.5.3.1   Electro-optical systems

An imager can be centred on a fixed location, or step around several. It can also have an auto-track facility so that it follows a designated object. Zoom facilities allow close examination of an object or area. An automated sensor that removes the need for human control must be capable of operating with positive responses to the following questions, bearing in mind the sensor suite that will be available in operational circumstances:

1.   Is the area under observation one where a strike is allowed?

2.   Is the area observed large enough for predictions of the movements of targets and collateral objects over timescales greater than the time between weapon release and impact?

3.   Has the authorised weapon commander chosen the mode of operation and operating parameters, and are they appropriate for the weapon, target and target area?

A more sophisticated sensor such as a multi-spectral imager or scanner will have a specific set of issues concerning matching its performance to the anticipated target type and surrounding area.

8.5.3.2   Radar and sonar sensors

Both of these sensors have automated systems for optimising their mode of operation to the prevailing circumstances. Recently, however, the level of automation has increased rapidly with cognitive radar systems able to perform many functions simultaneously [12]. Designers of weapon systems incorporating cognitive radars and sonars as part of the surveillance and identification of a legitimate target and non-targets will have to ensure that the sensor design and its identification processes are put through an Article 36 Review, or have clear and recorded reasons why one is not necessary.

8.5.3.3   Other sensor types

There are a variety of intelligence and Electronic Warfare (EW) sensors available to assist the commander in the *Observe* and *Orient* phases of an attack. As with cognitive radar, there will need to be specific issues addressed at an Article 36 Review, or recorded reasons why one was not necessary for an automated system.

8.5.4   Orient

Orient is a data and information fusion process with an output that can be easily understood by a human. Interpretation of the output will almost certainly require specialist training, raising severe questions about the reliability of an automated system producing object identification results and predictions about future changes in the scene.

This step is, in essence, combining all available, probably diverse, data into coherent information about a target area and presenting it to the planners or weapon commander in an intelligible way so that decisions can be made. It does not have a sharply defined boundary with the Observe phase. Sensor fusion may take place in the sensor suite so a multi-spectral image is an input to Orient, or the original images may be separate inputs. Data on the target area, based on pre-existing information, may be combined with the sensor data. In all cases the output must be clearly defined as an information package which can be presented to a human and has sufficient content, reliability and accuracy for targeting decisions to be made.

Figure 8.3 shows one approach to fusing information about a target area and will be used to bring out the type of considerations that a design team must address in engineering terms and ensure that the results are presented in clear military terminology. The actual process for a real system will be very different, but Figure 8.3 should provide the design team with a starting point for interrogating the requirements document.

The starting point is an existing authorised map of the target area from military databases. This will almost certainly comprise a Digital Terrain Elevation Database (DTED) set of heights at specified intervals. This will then have buildings and other features, 'culture', added to give a conventional map. The addition of recent or current images and intelligence information gives an initial world view. The intelligence can come from many sources including Signals Intelligence (SIGINT) and Human

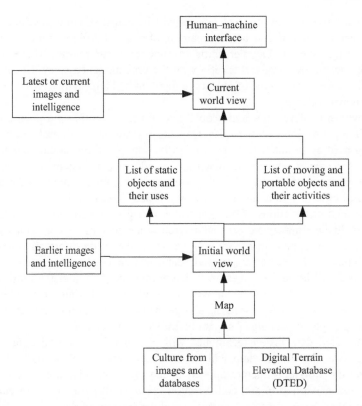

Figure 8.3   An example Orient phase
[This starts with a basic map and builds up a current world view of the target area]

Intelligence (HUMINT). The legal question is the provenance of the data which is not a weapon system design problem. The weapon system designer should ensure that there is a robust configuration control process for the world views and a recorded origin of the initial, authorised one.

The initial world view can be used to generate lists of static objects such as buildings and their uses. This is the point at which statistics becomes important, especially when any automation is introduced into the identification process. Locations and types of building are probably unambiguous, but their use may not be. SIGINT data, in common with other EW geo-locations are based on triangulation or other processes with error bars, giving a probability of false identification. Similarly, vehicle identification techniques are not certain, so regular movements of heavy goods vehicles including ones with cylindrical loads may signify a civilian petrol transport depot rather than a SCUD base. One problem is that a building may be dual-use, i.e. it is used for both civilian and military purposes. Determining that this is the case may require extensive observations over a period such as 24 h; the results may then set criteria for the time of attack. Design criteria for identifying this type of use will be a particular problem for the specifiers and designers.

Moving mobile and transient objects will be detected, either by direct observation over a period of time and comparison of maps at different times, or using a motion sensor such as Doppler shifts in radar or sonar returns. These must be classified according to agreed criteria with the procurement agency. Uncertainties in classification for this type of object are likely to be high, so there will have to be an 'unknown' category.

Movements of vehicles and people can be used to build up a picture of normality in the area, the pattern of life. Deviations from this, with probabilities, may be recognised by an automated system, but the interpretation should be made by a human. A designer who thinks that they can develop a system to perform this interpretation automatically should ensure that the procurement authority has acceptance criteria which are endorsed by the state's legal authority.

Repeated observations of the target area will give updates to the world view. Data should be time-tagged so that its relevance to current conditions can be assessed. The weapon commander will want to gather extra information to reduce uncertainties in the classification and identification of the objects in the area. Object lists will be revised over time. There must be an automatic configuration control process in place as the human will want to concentrate on targeting, but post-operation reviews will need the source of the information used to confirm an object as a target, especially if it is an incorrect identification.

Automated data fusion techniques will play a role at several points in the Orient phase. The design team will have to make choices about their statistical basis and the weighting factors applied to the raw data. This area is one which requires extensive liaison and collaboration between all parts of the procurement chain and their legal advisers. Validation and verification of the technical process, classification thresholds and presentation to the commander will be key aspects. Extensive trials of actual performance in a range of scenarios will be required.

8.5.5   Decide

This process can be reduced to one of matching the targets in the target list to the objects in the target area, identifying which meet pre-set criteria, and making a collateral risk assessment. Meeting rules-of-engagement will be an underpinning requirement that must be considered at the end and probably earlier as well.

The dilemma is that engineers can provide tools to do most of these tasks and provide associated error budgets, but the user needs a binary output: the object is either a legitimate target or it is not, accompanied by a statement of the effect of the weapon on surrounding civilian objects. The ideal system would use its sensors and other information to automatically give a positive target identification and collateral damage estimates to the commander, but it is unlikely that such a system exists and meets the full rigour of an Article 36 Review.

The acronym ATI is sometimes used in discussing and specifying targeting systems. The precise definition must be specified as it has at least four meanings which are shown in Panel 8.3. When specified, that definition must be the only one used in that context. The first three definitions all require that the target scene is examined in more detail and decisions about the validity of an object being on the target list is made by the human weapon commander. The fourth does not. The most appropriate approach at this stage is to ensure in the system design that the final target identification is made by the human weapon commander and the algorithms are decision aids. A match between an object on the target list and an object in the target area should be regarded as a target indication and not a target identification.

Panel 8.3:   The four meanings of ATI

Target indication occurs when the features of an object in a scene are similar to those of known targets in the weapon system's database. The target type may be on the target list for the operation but not necessarily so. The match needs to be confirmed by supplementary information from another source, or repeated observation before being declared as an identification.

Target identification occurs when an object in the scene has a very close match to an object on the target list. The accuracy of the match must be agreed by both design and customer experts and verified by trials of the weapon system in a range of representative scenarios.

Assisted means that the sensor system has the resolution and presentation format to give general indications of a potential target based on fairly loose criteria.

Automatic means that the sensor system has the functionality to mark objects in the area as potential targets using an automated process.

This leads to the following possible definitions of ATI:

1.   Assisted Target Indication
2.   Assisted Target Identification
3.   Automatic Target Indication
4.   Automatic Target Identification.

Rules-of-engagement apply at all times and may prevent weapon release even if a target is identified both as an object on the target list and with a high level of confidence in the result. Rules-of-engagement are normally expressed in words so that humans can implement them. Some parts may be expressible as algorithms and included in the suite of decision aids. In this case, the design, procurement, trials and legal teams will need to work in close collaboration to achieve a practical and legal product. The digitisation of the rules-of-engagement will need to become part of the procedures for the use of that type of weapon system for all future campaigns.

Collateral damage estimates are carried out in accordance with a nation's own processes and their acceptability under proportionality principles is beyond the scope of this book.

8.5.6   Act

This is the action to irrevocably commit a weapon to a strike on a target. It is normally carried out by a human commander, but there are exceptions for defensive systems which need rapid responses which are beyond the capability of humans. Chapters 10 and 13 develop the concept of authorised power which gives a possible approach to automating some of a weapon system's actions.

8.6   Issues for autonomous targeting and Article 36 reviews

8.6.1   The general problem

The assumptions made in Section 8.5.2 included humans at one or more critical points in the targeting process. The question now arises of making most if not all of the targeting process automatic. This is not the same problem as making a completely Autonomous Weapon System (AWS) as there is scope for human intervention at several points in the sequence.

It can be argued that for the general case, IHL dictates that a human must authorise and carry out the *Act* phase. The main question is how this is carried out, the information needed, and if the human's information processing can be automated. Chapter 12 explores these issues and their complexities, with the conclusion that achieving autonomy of a complex capability, even with the limited definition used here, is very difficult.

Sections 7.1 and 9.10.3 show that there is considerable international concern about the legality or otherwise of self-contained weapons and whether new treaties are needed to control them. Although not clearly defined in the international debates, these are weapon systems that are self-contained and do not have human intervention after release. At the simplest level, a cruise missile with the ability to change its target *en route* could be described as an AWS. It can be postulated that current technology can be exploited to produce a self-contained projectile that can select and attack specified types of target after launch. The question that arises is how this can be carried out within the confines of IHL.

8.6.2   A projectile as an Autonomous Weapon System (AWS)

An AWS is defined in Panels 1.1 and 3.1 as a weapon system that can select (i.e. search for or detect, identify, track) and attack (i.e. use force against, neutralise, damage or destroy) targets without human intervention[6]. An intelligent projectile that is launched against a target set after a legal, human-authorised targeting process has been followed will count as an AWS if it relies on its own sensors to select and attack an authorised target.

The problem can be described as one of finding the design constraints that enable an AWS to carry out a complete OODA process using its own facilities but still meeting IHL requirements. The solution is to look on OODA as a set of

---

[6]Note that this definition does not include any interpretation of 'intent'. A human must interpret the particular scenario and decide which type of target can be attacked and interpret rules-of-engagement to remain compliant with IHL.

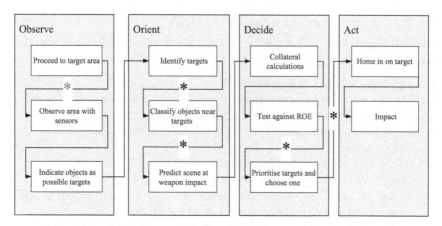

Figure 8.4 OODA tasks for an AWS with limited capabilities
[The asterisks (*) mark times when alternative tasks may be necessary.]

consecutive tasks that the AWS must complete. It cannot start on a new task until it has met success criteria defining successful completion of the previous one. Clearly, the AWS must have pre-planned responses if the success criteria are not met. These could include self-destruction, or communicate with the next higher node in the C2 chain (probably its controller) and request instructions. Design of the latter response must include time for the higher node to respond, and may require that the weapon transmits data with the request.

Figure 10.4 shows an OODA process for a complex military operation with humans making all the key decisions. That operation is complex, so a simpler process is discussed here with fewer decisions. The simpler process is shown in Figure 8.4 and can represent air, maritime or land targeting. Each phase has a restricted set of functions applicable to the specific AWS. There will be intervention points, marked with asterisks in Figure 8.4, when failure to complete a task necessitates that a contingency or alternative task is started.

The AWS has been sent to a specific area which is known to have targets nearby so it starts observations when it reaches that area. Possible targets can be indicated using pre-determined criteria for the specific target set that it is to attack. It then proceeds to classify these as targets or not using stricter criteria, followed by classification of objects near to the identified targets. The categories will be defined before the mission and will almost certainly include 'unknowns'. It is likely that some types of legitimate military object may not be classified as targets due to limitations of the on-board processing and databases. Should there be no identified targets, or a non-strike object nearby, then a contingency task must be started.

Calculations of potential collateral damage have to be carried out and tested against rules-of-engagement before the targets in the area can be prioritised, one chosen and attacked.

8.6.3   Capability restrictions on the AWS

The task sequence in Section 8.6.2 shows that the AWS will have limitations on its capabilities:

1. The target area has to be predefined;
2. The mission must be executed as a series of tasks which have success criteria. The AWS can carry out re-planning and rerouting at any stage;
3. There must be contingency plans for failure to meet success criteria for significant tasks;
4. Only a very limited set of targets can be attacked;
5. Classification of objects near the target will have large uncertainties due to limited on-board databases. The limitations will be due to both processor capacity and security concerns about on-board databases;
6. Collateral calculations involving unknown objects will have to be treated in a way that meets IHL proportionality criteria. This will probably mean that a target near an object classified as 'unknown' cannot be attacked;
7. All necessary rules-of-engagement for the post-weapon-launch part of the mission will have to be expressed in terms that can be interpreted by the AWS's processors. If this is not possible, then the AWS cannot be launched;

There will be tasks such as target indication and identification which will almost certainly be implemented in the future with AI and machine learning systems. Others will continue to use deterministic algorithms. Testing of learning systems with representative scenarios, and measuring their reliability figures will be a significant problem for the foreseeable future. The conclusion is that the success criteria for tasks must be assessed using deterministic algorithms even if AI is used for some tasks. The contingency tasks must also be implemented using deterministic algorithms as they must explicitly comply with all aspects of IHL.

8.6.4   Article 36 reviews of projectile AWSs

The automation in the projectile will have a core of functions and data that will be common to every one produced. When used in a campaign, there will be extra data and possibly extra functions loaded into it. It is likely that values of the parameters used to set decision thresholds and rules-of-engagement criteria will be specific to the campaign and possibly the particular mission. There will be Article 36 reviews of the projectile with its core functions at key points in the development and introduction into service.

The questions, and evidence required to answer them, will be specific to the AWS under review. However, the list of restrictions in Section 8.6.3 can be used as a basis for preparing the questions. The way that the contingency tasks provide an adequate means of human control will be a key issue. Some analogies can be drawn with current systems that have a non-trivial time between launch and impact, but the solutions will need to be more complete.

Article 36 reviews should be carried out at every significant step in the procurement process including the concept and design stages. This can be turned to advantage and cost-saving by providing the key parameters and criteria which can be measured using modelling, simulation and design processes, but also identifying those which require expensive trials data whilst the trials plans are being formulated.

Definitions of the criteria used for lethal decisions and their thresholds will be part of the reviews. The technical evidence presented for these will be needed for setting rules-of-engagement for the projectile in operation. This evidence can be described as mission-specific data. Difficulties may arise if there are campaign-specific functions loaded as well as the data. It is highly likely that they will have been developed to give increased functionality to the weapon and must be reviewed to determine if they represent a 'new or novel means of warfare' and need a specific Article 36 Review. Determining this is best carried out at the concept stage so that reviews can be planned as part of the development programme.

References

[1] NATO Standard. AAP-06 Edition 2015 *NATO Glossary of terms and definitions (English and French)*. North Atlantic Treaty Organisation, NATO Standardisation Office (NSO) 2015.

[2] NATO Standard AJP-3.9 Edition a Version 1. *Allied joint doctrine for joint targeting*. NATO Standardisation Office 2016.

[3] Roorda M. 'NATO's targeting process: ensuring human control over (and lawful use of) "autonomous" weapons' in Williams, Andrew P. and Scharre, Paul D. (Eds.) *Autonomous Systems, Issues for Defence Policymakers*. NATO Capability Engineering and Capability Division, Norfolk Virginia; 2015, pp. 152–68.

[4] Australian Department of Defence. Australian Defence Doctrine Publication (ADDP) 3.14. 2nd Edition 2009, Canberra 2600.

[5] Boothby W.H. *The law of targeting*. Oxford, UK: Oxford University Press; 2012.

[6] UK Ministry of Defence. Joint Services Publication 383 (JSP383), The joint service manual of the law of armed conflict, 2004, with amendments of 21 May 2014.

[7] US Department of Defense. *Law of war manual*. December 2016. Available from https://www.defense.gov/Portals/1/Documents/pubs/DoD%20Law%20of%20War%20Manual%20-%20June%202015%20Updated%20Dec%202016.pdf?ver=2016-12-13-172036-190 [Accessed on 14 May 2017]

[8] Tirpak J.A. 'Find, Fix, Track, Target, Engage and Assessment'. *Air Force Magazine*. July 2000: 24–29.

[9] Flynn M.T., Juergens R., and Cantrell T.L. 'Employing ISR, SOF best practices'. *Joint Forces Quarterly*. 3rd Quarter 2008(50): pp. 56–61.

[10] Ferry M. 'F3EA – A targeting paradigm for contemporary warfare'. *Australian Army Journal*. 2013, vol. x(1): pp. 49–64.

[11] Faint C., and Harris M. 'F3EAD: Ops/Intel fusion "feeds" the SOF targeting process'. *Small Wars Journal*. 2012, vol. 8(1). Available from https://smallwarsjournal.com/jrnl/art/f3ead-opsintel-fusion-"feeds"-the-sof-targeting-process [Accessed 17 January 2019].

[12] Farina F., de Malo A., and Haykin S. (eds). *The impact of cognition on radar technology*. London: IET; 2017.

Chapter 9

Influences on future military autonomous systems

9.1 Introduction

Military technologies, in common with civilian ones, evolve due to the changes in the society arising from economic and social pressures as well as anticipated future operations. Military Research and Development (R&D) policy planners have to make predictions about the future, and where funds should be invested for maximum effect. There are at least three facets to this but only the first is likely to be fully in the public domain:

1. Technology developments that are happening independently of military spending;
2. Developments which can happen with military investment;
3. Likely developments by potential adversaries.

One common technique to aid predictions is to draw up roadmaps for a specific technology or market. These start with a review of the current state of the art and major development programmes, and then predict technology developments over time. The predictions may be plotted graphically, or given as more generic identification of the likely timescales (5, 10 or 15 years) for a particular advance to reach maturity. Roadmaps are a common planning tool and produced for civilian applications [1] as well as military ones [2]. The two references are only examples; an interested reader should carry out an internet search to find the latest ones in their field of interest. All roadmaps are useful aids but must be treated with the same caution and scepticism as any other predictions.

This chapter takes a different approach to that of roadmaps. It discusses how changes in the type of operations conducted in the last twenty years have affected the military's operating environment, as well as looking at longer term trends in politics and society. All these play a role in driving changes in materiel requirements. The actual changes will be driven by the interactions between many factors as well as the perceived military need so only a few speculative suggestions are made. As with roadmaps, these predictions should be treated with caution.

An intelligent weapon of any kind is a compromise between many factors. It has to have propulsion and a guidance system to reach its target area and when there, it must be able to generate its kinetic or other effect. The space and power required for its targeting system are necessarily limited when compared with these other needs. This will limit the on-board processing power and data storage hardware, bringing natural limits to the power of any intelligence in the weapon.

9.2   Recent and current campaigns

When a military campaign or operation is started, it will change the state's R&D programmes dramatically. Military responses by the state's opponents in the campaign show shortcomings in the state's inventory and capability gaps which must be filled rapidly. Techniques to find and remove Improvised Explosive Devices (IEDs) remotely are one example of an urgent need to respond to the asymmetric warfare of the campaigns in Afghanistan and elsewhere.

The campaigns also led to acceleration of the development of new capabilities based on existing R&D work, but well before the planned dates for their introduction into service. Wide area surveillance by Unmanned Air Vehicles (UAVs) was a new, but predicted need. The capability to deliver weapons from high altitude was then a logical, but unforeseen consequence. The priority of surveillance was not predicted by the roadmaps of the late 1990s and early 2000s. They predicted swarms of identical medium to large UAVs, and fast-jet combat UAVs delivering large kinetic effectors.

There will always be a review of the lessons learnt after a campaign as well as a recognition that potential adversaries will also have studied the military operations and learnt lessons. The result is likely to be a change in funding priorities with changes to development programmes for updates to existing equipment and new systems.

Among the 'lessons-learned' is that autonomous systems are highly effective solutions to many, but not all, problems in current campaigns. Initially there was much debate about the ethics and legality of their use, but recognition that a human was always in control of every lethal decision has gradually led to their acceptance and widespread use.

9.3   The military operating environment

9.3.1   The three Ds

When highly automated systems became both a reality and available for routine operations, they were mainly seen as replacements for humans in performing tasks that are one or more of the 3Ds: dull, dirty or dangerous. Now, however, the 3Ds are no longer considered to be an adequate description of the likely uses of autonomous systems, they are but one subset of the tasks which they will have to carry out. Autonomous systems may well be the preferred option for the 3D tasks as there is no risk to personnel but it is better to consider future uses of autonomous systems by analysing likely future military environment using the 5Cs below.

9.3.2   The five Cs

Traditionally, military planners prefer to operate in open areas with few objects present, whether animate or inanimate, giving maximum freedom of action for the armed forces. This may still be the case in some campaigns but not all. Increasingly, operations are mounted in populated areas requiring precise identification of the legitimate targets with equally precise targeting of them, or a decision not to attack due to the presence of civilians, according to the extant rules-of-engagement.

Anticipated tasks can be seen from examination of battlespaces in the likely future operational environments described in terms of: '5Cs' – congested, cluttered, contested, connected and constrained (the 5Cs), with forces expected to achieve success in these conditions [3]. They are also described in *The future operating environment 2035* [4]. They will not all apply to every operation or mission, but one or more will. They are described in the following paragraphs because the need to carry out operations in them will be a driver for new capabilities. Autonomous systems and automated decision-aids will be an important part of many force-mixes making up these capabilities.

9.3.2.1   Congested

Many future target areas, whether land, sea or air, are likely to be congested, i.e. there will be many objects present. Even if the objects, such as buildings are fully understood and do not include civilians, they will still restrict the freedom of operations of the attacking forces. They provide cover for people and vehicles as well as potentially confusing the tracking algorithms used to follow targets moving among them. Databases will need to include more detail than their geometric shape and location so that both a human commander and an automated decision aid will have access to information that may be necessary in interpreting rules-of-engagement. Different building types will give different levels of protection to people in them, influencing collateral damage estimation.

Conversely buildings and other fixed objects may aid automated sensor systems by providing reference points with precisely known coordinates. This will accelerate and improve the quality of the alignment of the sensor image with the existing maps and outputs from other sensors and navigation aids.

The large amount of commercial and other traffic in the air and sea domains make them congested. Although there are regulatory mechanisms to prevent traffic entering an area, planning must be on the basis that there will be civilian ships or aircraft in an operational area. Autonomous systems will need information about civil traffic for context, as well as reliable identification algorithms. There will be two main risks: not detecting an object and misidentification. Normal air traffic control procedures give strict regulation of air space, so misidentification of an aircraft under air traffic control is unlikely. This is not the case for maritime traffic as not all ships have Automatic Identification System (AIS) transponders and there is no maritime equivalent of the compulsory air traffic control regulations and processes.

The electromagnetic spectrum and cyberspace will also be congested, limiting the efficiency of communication channels. This will bring pressure for an unmanned system to carry out as many processes on-board as is possible. The aim is to reduce the data transfer rate by orders of magnitude or to zero. Should the capability of the on-board calculations approach the levels at which a lethal decision can be made, careful consideration will need to be given to the reliability of the data used in making the decision and the quality of human control over the consequent action.

9.3.2.2   Cluttered

A cluttered environment is one where the people, objects or events cannot be readily identified or distinguished from the opponent's military forces. Many of the characteristics are the same as those of contested environments. Understanding a cluttered scene and identifying legitimate targets requires experience, judgement and reasoning. These attributes are often considered to be essentially human and not susceptible to machine assistance. The growth of Artificial Intelligence (AI) may allow better understanding of a cluttered scene and target identification, but the algorithms must be well understood and tested if they are to meet the Article 36 requirements demanded by International Humanitarian Law (IHL).

In the short to medium term (<10 years), it is unlikely that automatic identification algorithms will have sufficient reliability to give anything other than an indication that an object is a target of a known type, such as a tracked vehicle, but it may need a human and context to decide if it is a military vehicle or a construction vehicle such as a crane or heavy lorry. Additional Protocol I (API) demands that military objects are distinguished from civilian ones so an identification process must additionally be able to indicate civilian targets for assessing potential collateral damage. It may also be necessary to identify the type of civilian object so that collateral damage estimates include the level of protection that they give to civilians.

9.3.2.3   Contested

Wars of choice, almost by definition, are only undertaken when the state (Blue) believes that it will win. This does not necessarily mean that their forces have dominance over the opposing (Red) forces but Blue must have plans with a reasonable chance of success. It must be assumed that the physical and cyber battlespace will be contested by Red with equal or superior capabilities in many areas. Red, for example, may compensate for a known weakness by the use of techniques of asymmetric warfare to confuse Blue forces and negate the value of their technical or tactical advantage. It is possible, if not probable that one or both sides' networks may be disrupted, leading to poor quality or false data arriving at automated decision aids used by commanders.

It must also be accepted that both sides will also use automated systems such as small drones or autonomous road vehicles in asymmetric attacks. Counters to them must be available but these may reduce the capability of the defenders' own autonomous systems and will be a factor in the commanders' minds. For example, heavy jamming or GPS-denial may disrupt both sides equally and so bring no net advantage.

9.3.2.4   Connected

It is obvious that all military and civil populations are becoming increasingly connected. Activities of all types now assume good connectivity and rely on ready access to information and Command and Control (C2) systems through networks.

This gives commanders more information prior to making decisions and more tools to analyse it, but it may reduce the ability of front-line commanders to make rapid decisions if they wait for confirmation of proposed actions. Networks are liable to a combination of jamming, cyber-attacks and physical attacks on the bearers or nodes. If successful, these can cause disruption and even loss of control if the C2 system has a vulnerable single point failure location. The human ability to improvise according to local circumstances is allowed in military doctrine when necessary. It is difficult to see how an Autonomous Weapon System (AWS) could do this.

Automated decision aids connected to a network can draw on a range of data for their output to the commander. However, the provenance of the data must be either known or come from an accredited source for any lethal decision to be made and executed. Confirming the validity of information from approved military databases is trivial. Restricting commanders to these information sources may be a critical limitation when one considers the vast amount of data available from commercial databases and search engines. Experienced and critical humans have developed ways of estimating the reliability of information obtained from these sources. It will be a challenge for autonomous systems to generate equivalent reliability estimates.

Autonomous systems controlled over a network will be particularly exposed if they are operating automatically for a time but rely on a communication link to receive their next instructions. Humans can be expected to behave in a rational way if cut off, and follow previous instructions or use their initiative. An autonomous system has no equivalent response unless it is built in and can respond to a wide range of situations.

9.3.2.5 Constrained

Every society places constraints on its armed forces through laws and moral pressures. The rapid distribution of information from an operational area by Blue, Red and/or news organisations is now accepted as normal. The effect of this information on Blue's home population can bring pressure to bear on politicians and commanders to refrain from particular tactics. The information may also be used to bring criminal charges against combatants. Conversely, the knowledge that such information could be made available may prevent illegal and immoral behaviour by combatants.

The constraints on one state's armed forces will not be the same as those on other states. Section 7.13 discusses rules-of-engagement and that these may be different, even between members of the same coalition. The final trials and introduction into service of a new system include advice for those setting rules-of-engagement, so the test plans must be reviewed by military and engineering staff based on that state's interpretation of IHL so that the most appropriate engineering evidence is collected.

Rules-of-engagement will be based on military and political considerations which usually involve human judgement for interpretation. Section 10.4 shows that that some parts of the interpretation of rules-of-engagement may be automated, but the final interpretation must be by a human. The question of who interprets them for a particular lethal decision, and when and where that interpretation is made, will always be a key driver for AWS design and use. It is highly likely that Article 35 and Article 36 review criteria will constrain AWS design more than technology.

9.4   Societal changes

9.4.1   Political changes

There is one significant difference between the development of military technologies and civilian ones. That is the driving force of pressure from national governments. A company or corporation can decide to withdraw from a market or refuse to bid for a particular contract if they do not see a commercial benefit from it and/or it does not fit in their long-term strategy. A state's armed forces do not have this option. They must carry out operations chosen by their government and cannot decline, no matter how much it conflicts with their own plans and strategy.

The armed forces of any nation are configured to meet the needs of their government and the threats that they perceive. There are additional extra tasks such as supporting emergency services and secondment of personnel to allies and the UN. These latter tasks do not usually drive designs for new equipment, but rely on use of existing facilities and materiel.

The Cold War dominated politics and military developments for the last half of the twentieth century. There were two major power blocks (NATO and The Warsaw Pact) with access to funds and the will to spend them on R&D for ever-more sophisticated weapons. A perceived threat from a new development by one block was met with the development of a superior one by the other. This led to increasingly sophisticated platforms and weapons on both sides but also the arms control agreements discussed in Chapter 7.

The examples in Table 4.3 and others show that platform development times lengthened as the platform became more sophisticated. In general, useful lifetimes also became longer. Longer lifetimes led to platforms having a mid-life-update ten to twenty years after their introduction into service. Mid-life-updates have now been superseded by block upgrades. This is where there is an initial capability introduced with a few platforms, usually called Block 1, then improvements are introduced to the platform or major subsystem designs and the next block produced.

The massive post-Cold-War superiority of the NATO alliance over any opponent led to two major developments in warfare: the so-called 'wars of choice'; and asymmetric warfare. However, more recently, new powers are emerging with sophisticated capabilities. China is often quoted in this context, as well as Russia which is rebuilding its military capabilities. Other nations are equipping themselves with sophisticated weapons and are setting up their own defence industries for specialised products. India and Brazil are cases in point.

'Wars of choice' may be in response to a perceived threat to a state or alliance, but generally, legal authority is sought through the UN. Wars of this type, such as those against Iraq in the early 2000s, follow the approach of overwhelming the opponent by the use of massively superior force drawn from the alliance's existing arsenals. This requires extensive, careful preparation, but does not, of itself, generate a demand for major new capabilities as the campaign is initiated, planned and executed in a period of a few months. There will probably be urgent requirements for modifications to existing equipment. These are discussed in Section 9.6 below.

Asymmetric warfare is the term used to describe conflict between armed forces with a large military inventory combined with large numbers of personnel and an opposing force of a completely different nature. These are usually non-state organisations with small, dedicated numbers of personnel, but they conduct operations using modern civilian technology to supplement their limited supply of genuine military materiel. Widespread surveillance using a range of sensors and identifying the normal pattern of life in an area are key features in identifying changes that may be due to hostile intent. Autonomous systems have a significant role to play in the analysis of the information obtained from an area. The provenance of the data and reliability of its interpretation will be a key part of an Article 36 Review of lethal decisions. Another issue which is probably not a technical one is the applicability of International Human Rights Law (IHRL).

A potential new form of asymmetric warfare is one known as 'lawfare'. This is the use by one side of their opponent's legal system or an international court to restrict the latter's freedom of operation in some way. This may be by seeking court rulings that a particular type of operation or use of a weapon type is illegal. It may be similar to, but not the same as creating a public relations campaign against the legitimacy of an operation. The use of selected facts and omission of others can give the impression that AWS are all lethal and out of human control, giving rise to concerns about their legitimacy if not legality. Section 7.3 gives some more information about this problem. As court cases rely on evidence, transparency about the Article 36 Review process could mitigate the opponent's lawfare. Additionally, evidence about and from the AWS' sensors and data management systems needed for decision-making may provide the factual evidence to decide a specific case. Clearly, weapons with increasing levels of automation are likely to be the subject of future court cases which could be initiated by an opponent who sees a possible breach of IHL by the state using the system. This leads to a need to collect and store data for use in future court cases which otherwise might not be collected.

9.4.2   Economic changes

Reduced defence budgets have at least three effects:

- Fewer new capabilities can be developed;
- Existing ones have to be made to last longer;
- Reductions in military and associated civilian personnel numbers.

This has had the effect that despite the Cold War ending in 1991, its military requirements continue to affect capabilities. Many expensive assets from that era have had new equipment added which give capabilities more suited to current activities. However, new, advanced assets still need to be developed as existing platforms become uneconomic to maintain and must be replaced with assets more suited to future world conditions.

The production numbers for military equipment have reduced as a consequence of shrinking budgets. Reduced manufacturing volume may make the product less attractive to companies who see their future in the mass market of consumer products which has and continues to expand rapidly. Military supplies became low-volume expensive commodities and companies withdrew from the market as they considered it uneconomic. The defence industry still has specialised capabilities for making and supporting products where the market is dominated by military users. One example is warheads. Another, more pervasive example is equipment capable of operating in extreme environments. The military standard temperature range is $-55°$ to $125°$, compared with $0°$ to $85°$ for typical commercial devices. Radiation hardening can also be specified as well as resilience against chemical and biological agents. There will also be stretching requirements for replacing failed electronic components in platforms with a life of decades. (This is a growing problem as older components may be produced with obsolete technologies and modern components have short lifetimes.) Effectively many defence products have become a niche market with only a very low number of specialist suppliers and developments proceed only at the pace allowed by military budgets.

In contrast to military R&D expenditure, civil R&D expenditure has increased tremendously over the same timescale. This has the effect of reducing prices and also accelerating technical advances as suppliers use technical excellence as a marketing tool. As well as very capable consumer electronics and Information Technology (IT), reliability for most products has increased well beyond that for comparable products in the twentieth century. The military user now expects his equipment to have comparable capability and reliability. This is underlined when consumer products such as mobile phones and car electronics are now expected to operate outside in adverse weather conditions, which makes their technologies more suitable for military applications than previously. It is likely that governments will accept the use of civilian systems for front-line use as well as 'back-office' facilities. Mixtures of civilian and military products in one system-of-systems will present complex architecture problems. The problem being to gain the capability advantages of civilian development without losing military ones.

9.5   Technology changes

9.5.1   Connectivity

Section 4.3 briefly discussed the contradiction in technology developments for the military. Sophisticated platforms demand long development times and in-service lifetimes, but they must be fitted with advanced capabilities that change on short timescales. Developments of highly integrated system-of-systems using architectural techniques, as discussed in Sections 4.2.3 and 4.9, are one response but are not the only ones.

The rise of the internet is one obvious example where several technologies are integrated to give a capability with unforeseen social consequences. Its economic benefits bring continuous improvements in its capabilities in a virtuous cycle exceeding anything foreseen by its original founders. The military establishment now invests considerable effort into applying the concept for military applications, but with high levels of security. Shared services and storage using cloud computing is now widespread, but is an example of a technology which has many problems for military use. However, it may be possible to use it if suitable security measures are put in place [5]. This will bring considerable changes to operations as it becomes accepted and verified through the widespread use.

Increased reliance on networks, cloud computing and complex computer systems brings vulnerability to cyber-attacks from both opponents and hackers looking for challenges. Many autonomous entities and agents exist in networks to combat these threats but are beyond the scope of this book, except where they impinge on the operation of highly automated weapons and their supporting decision aids. This is certainly an area where there will be considerable activity continuously for the foreseeable future.

Increased wide-bandwidth connectivity offers the possibility of relatively 'dumb' units on the asset, with the 'smart' algorithms and databases in a remote location. Remote surveillance could exploit this approach, with the surveillance asset, UAV or Unmanned Maritime Vessel (UMV) having multiple sensors operating continuously. On-board sensor-processing algorithms would carry out the first stages of processing so that a low bandwidth link could be used to send information to a remote analysis system for further analyses such as target identification and intent. Warfare is never static, so it can be predicted that the hostile forces will continually upgrade their use of Camouflage, Concealment and Deception (CCD) to exploit shortcomings in the on-board sensor processing. This is an argument to keep a fairly high bandwidth.

The use of datalinks does however, increases vulnerability to cyber-attack and gives severe problems if there is a need for no transmissions during phases of a mission. This is highly likely when a platform is in a high-threat environment. Developments in spread spectrum datalinks reduces vulnerability to the opponent interpreting the data, but does not remove the basic problem that the platform is transmitting signals that can be detected.

9.5.2   Artificial Intelligence

When considering AI applied to weapon systems, it is sensible to divide the possible applications into two types:

1. Decision-making which produces action by the system. This is the type shown as 'Intelligent control' in Figure 3.1 and must meet regulatory and legal requirements;
2. Provision of information supplementing or replacing sensor data for the control system to act on. This is the use of AI as a decision aid so its operation may not need to meet the full rigour of a decision-making system, but must still be considered as part of the input to it. This will be critical for lethal decisions.

As a broad generalisation, AI in both types will be based on the analysis of data received during the mission from the autonomous system, its controllers and its environment. The aim is to produce the most appropriate responsive action. This response may be by either the controller or the autonomous vehicle.

We can split the functions that AI can perform into several categories based on the system's complexity:

1. Interpretation of incoming data;
2. Modelling of the system's world;
3. Predicting the future based on its knowledge of its world and the probable changes;
4. Choosing among the options it sees for affecting the future;
5. Acting or recommending actions to the operator.

The property of AI systems to improve their own performance by learning from experience is a separate problem. It can be postulated with some confidence that a weapon system will not be used if its performance after release may be different to that expected at release. Changing behaviour due to it learning from events during the mission rather than react in a way that its operator can predict is unlikely to pass an Article 36 Review as currently understood. The conclusion is that a prediction can be made that a future weapon control system could use strong AI, provided that its learning ability is frozen at a sufficiently early date before operational use that it can pass an Article 36 Review and have clear rules-of-engagement set.

The use of AI as an expert system requires a different approach. Almost by definition, we want to use the computer to provide a solution to a problem which may be stated clearly, but we do not necessarily know how to find the solution. Should the expert system provide a solution, we may not know how it reached it. The prediction here is that AI may provide expert decision aids during the targeting process, but the results must always be verified for credibility by a knowledgeable human. What 'credibility' and 'knowledgeable' mean here will, no doubt, be the subject of much debate.

Military personnel build trust of their equipment during training and exercises prior to operations. The level of trust will change during operations, depending on numerous factors including how well the equipment operates in real combat. This trust applies to all types of equipment whether hardware or software. Software decision aids are no exception, leading to the possibility, if not actual probability, that the machine result will be believed without questioning it. An engineering problem which may have to be addressed in the future is how to indicate uncertainties in results to the commander. This is a complex problem, and the commander will probably have a high workload and will be looking for certainty not uncertainty. There may need to be a threshold, below which the commander must make an assessment before acting. The legal implication is that above that value the result is completely trustworthy, in engineering terms 100% certain which is never true. A related problem is that when AI is involved, the type of error will be different to that of deterministic algorithms.

Future operations with autonomous systems will need close cooperation between humans and machines. Intelligence in the automated systems leads to operation as a team rather than the machines being simply tools which is the traditional approach. Teaming is planned be part of UK operations [6] with consequent implications for training of both humans and the AI systems. There is also research into teaming between humans and intelligent software agents for the control of multiple robots [7] in operations. There is an underlying assumption that an AI system must be able to ask the human questions and may eventually build up a level of trust in individual humans. It is also likely that tasks will be transferred dynamically between humans and an autonomous agent with associated risks and ethical concerns [8]. It is proposed here that these methods could be developed as part of the system-of-systems architectures discussed in Chapter 10. The human and agent would act as a node in the C2 chain, but the separation of human and agent/machine decisions must still be clearly separated for lethal actions. This may be a problem as there will be close interactions between the human and the agent.

One particular problem is that increasing the reliability of the automation leads to the human paying less attention and losing situational awareness. The whole approach to the design of the system and human–machine interactions must be reviewed for future autonomous systems in order to gain the full benefits and retain human responsibility [9]. Teaming techniques will apply to military decision-making [10]. They will need to have some specialised military requirements for lethal decisions so that the human is still able to accept his or her responsibilities under IHL, i.e. the human can verify the decision, based on adequate knowledge.

An AI system which learns does so by one of two types of learning:

- Supervised learning: This uses a set of known inputs; the AI system's internal weighting parameters are set and then adjusted so that the output is the desired one for these training inputs. Pattern recognition is a good example of this type of learning.
- Unsupervised learning: This does not have an external training set, but simply lets the system minimise one or more of its error functions by adjusting its internal weights by itself. This type of learning can produce networks which are good at detecting trends or patterns that were not previously visible or even expected.

Training a military AI system using supervised learning with a known, reliable data set gives the possibility of producing a system with sufficient reliability and predictability to meet IHL requirements. Defining the review process and criteria for setting rules-of-engagement will be a challenge. Training using unsupervised learning may never produce a system that will meet IHL due to its unpredictability.

IHL is clearly an issue at system and system-of-systems level. There are, however, other less-obvious problems for the procurement chain described in Chapters 4 and 5. These concern the way the contractual, technical requirements are written by the procurement agency and how the supply chain, with its set of interlinked contracts and subcontracts can demonstrate compliance to them. The system and system-of-systems may be designed assuming that the subsystems will supply results that come from predictable performance by the subsystem and its modules. If a subsystem has a function using AI with unsupervised learning, then its results may not be sufficiently predictable to meet IHL criteria. Without pre-dictability, the overall system-of-systems behaviour may not be definable, but will have to be constrained to act within technical and IHL constraints. Even if the level of unpredictability is within IHL constraints, it may be necessary to develop new methods for testing learning systems to demonstrate compliance.

The long lifetime and number of assets with their upgrades and configuration problems will bring an extra level of complexity even if the learning subsystem is frozen at defined times. Two versions of a strong AI based learning subsystem may be at the same build standard, but have used different information for post-delivery learning. As a result, they may not supply the same information to the system-of-systems even if they are given the same input parameters. Ensuring that the weapon-release commander has adequate predictability of post-release perfor-mance will be difficult, but it is a necessity under IHL. This may require new logistics tools as a simple record of Issue and Version numbers of hardware and software will not necessarily give the commander the information required for lethal decisions. He or she will need a succinct description of the performance of the system in the system-of-systems at a time when it can be understood and retained.

9.5.3   *Exploiting commercial technologies*

Civilian commercial technologies can be harnessed to deliver military capabilities that would be prohibitively expensive to develop for the limited military market. Even though the Global Positioning System (GPS) satellite constellation was developed by the DOD for military purposes, it is difficult to imagine military budgets paying for the array of portable devices that are available at any consumer electronics outlet. Digital maps have also become detailed and flexible to use due to commercial exploitation of market opportunities.

There was an early, somewhat naïve belief in the late 1990s and early 2000s that commercial technologies could be used directly in military systems. Commercial-Off-The-Shelf (COTS) was considered to be the way forward for most procurement programmes. Unfortunately many factors conspired against this. Some have already been mentioned – environmental ruggedness, products only being produced for typically less than five years, availability of components for support of in-service equipment. In addition, commercial software and operating systems have at least one significant upgrade in timescales of three to five years, with the close coupling of processor chips and operating systems.

The next stage was to change the meaning of the COTS acronym to Customised-Off-The-Shelf (COTS) and a slightly new approach with Modified-Off-The-Shelf (MOTS). The principle is to base a new military design on commercial system principles and use commercial components wherever possible. To a large extent this has worked when combined with modernised procurement processes matching this philosophy. COTS is now generally taken to mean Commercial Off The Shelf, which helps clarify that the design has unmodified, identified, commercial components embedded in a military design procured through military procurement processes.

COTS is strongly recommended by UK MOD. CM8278 [11] paragraph 25 states:

In our drive to deliver value-for-money, we will buy off-the-shelf where appropriate, in accordance with the policies set out in this paper, because this generally allows the UK to take full advantage of the cost benefits of buying from a competitive market. This approach applies to systems, sub-systems, and components.

Note that the use of existing military equipment from known suppliers of it may count as 'off the shelf' as against a bespoke product which needs a development programme. Similar guidance can be expected from other states for the same reasons as MOD give in CM8278 – less risk and cost, higher reliability, etc. The authors also recommend changes to system design to take advantage of COTS products. This is in paragraph 30, with the relevant words put in bold here:

In order to buy off-the-shelf effectively, we need to recognise these differences, take action to get the benefit of civil markets where we can (**including by simplifying potentially complex systems**), and focus our investment in research and development in those areas where the market cannot fulfil our needs or where we can influence the market effectively.

Some of the implications are discussed in Section 4.10.

One area where commercial and military developments have run in parallel and complemented each other is modular design. This can be defined as designing systems so that they are built up, as far as possible, from modules. Modules will be at the level of the subsystems or components defined in Figure 5.1 or lower. It is difficult to define a module but 'you know one when you see one'. (This is a variant on the well-known expression, I can't describe one, but I know one when I see one. This is supposed to have been used once about robots by Joseph Engelberger, the developer of the first industrial robot.) The main feature of a module is that it has sufficient defined interfaces to allow it to be replaced by another module which may have a completely different internal structure.

The combination of modular design, advanced commercial technology and responses to urgent operational requirements or needs can bring rapid upgrades to military capability. Even at system level, modules bring flexibility in upgrades. A UAV system can be designed with the concept of standardised interfaces between the UAV, the Ground Control Station (GCS) and the communication bearers. This allows the current UAV to be upgraded with a different airframe and sensor suite but still controlled with the current GCS and communication links. Similarly, the GCS can be upgraded and still use the same UAV but exploit its capabilities in a better way. The example Unmanned Air System (UAS) and its system-of-systems in Chapters 11 and 12 show how this can be approached.

9.5.4   Commercial autonomous systems

Increased autonomy in on-board functions will enable autonomous systems to operate with limited or no supervision, provided they do not make lethal decisions.

Consequently, stealthy platforms could be sent into remote areas, remaining silent until triggered by an input to a sensor to carry out more detailed analysis of its environment. For example, the autonomous system could be one or more ground vehicles which remain in cover until one of them detects a radio transmission on a military frequency band. They then switch on electro-optical surveillance sensors, carry out target indication analysis and transmit the information to a remote operator.

The fast-moving consumer market has new or improved products introduced every few months. This gives lifetimes of a few years for most processor or computer-based products and so removes the need for long-term supplies of a specific semiconductor device. This is at odds with the military need for long-term support. In principle, the problem can be solved by having modular designs in a well-specified architecture. A function can be replaced with a module which may not have the same internal components as the original one but does have an identical or equally acceptable interface specification. This is straightforward when the hardware conforms to commonly used standards. When the units are installed in difficult environments such as in a fast-jet aircraft, an armoured vehicle or a submarine, the units will probably be designed to a bespoke variation on the standard to meet space, weight, and power requirements. It is probable that an autonomous system within an AWS, whether part of the lethal decision chain or not, will have to be reviewed for its risk level and whether it is safety critical. This will add a further layer of complexity and cost on the already bespoke design variation.

There is certain to be military use of many algorithms and software developed for civil and commercial applications. It is likely that image recognition, tracking and data mining will be among them. In principle these can be assessed under Article 36 when used in the system or system-of-systems. Architectural analysis will show the role of the software in any lethal decision-making chain. The results should then be used as a basis for Article 36 reviews of individual algorithms and their implementation in the autonomous system. Particular note will have to be taken of the requirement for the use of results from the system as made, and that test scenarios are representative of the scene and sensors used.

There is an additional unwanted property of machine-learning algorithms known as algorithmic bias. This is where the results from a learning algorithm are biased in particular directions by biases in the datasets used to train them. It is usually associated with AI systems that learn from large datasets which have some form of selection-effect in their derivation. Military systems may have the opposite problem, limited datasets from actual operations or exercises. Larger datasets may be gathered from simulations, but these will be biased by the assumptions and limitations in the simulation setups. It is not yet clear how algorithmic bias will affect military AI systems or how it can be countered.

The theory of 'decisions-under-uncertainty' has been developed by economists and others as a methodology to help a human make difficult decisions. A US Army Corps of Engineers report recommended its use in decision-making for military projects although the applications given there are ones that are not related to weapons [12]. AI is well suited to decision-making with uncertain inputs so it may well be used for commercial decision-making software. Consequently, this type of product will probably be considered for some parts of the lethal-decision sequence. There will be considerable difficulties in passing an Article 36 review unless the basis of any decisions can be explained. The algorithm will also need to be successfully demonstrated with quantifiable results in realistic conditions.

There will be extensive regulation of many aspects of commercial autonomous systems over the next few years. This arises because of the commercial pressure to develop drones, autonomous cars, autonomous ships and applications of AI, etc., with public concerns about their safety. The regulatory environment will shape technical developments for commercial applications and so will spill over into the military domain with any commercial algorithms used for military applications. Regulations also drive technology, for example, effective collision avoidance systems will be needed for cars and drones. The commercial implementations which meet these regulations may well be better than any affordable military solution and so will be used in preference. However, it will be necessary to demonstrate that the software works in the differently constrained applications in military operations and training. There will need to be an upgrade plan in place for the lifetime of the platforms or systems hosting commercial autonomous systems.

Swarms of small autonomous systems are being studied for commercial applications with an extensive literature; see, for example, Reference [13]. There is a useful briefer review of swarm robotics by Brambilla *et al.* from an engineering perspective [14]. They are also coming into use in the military field, but at a much slower rate than predicted in the early unmanned system roadmaps. It is likely that they will find many applications, but their C2 structure will be critical given the low number of radio frequencies available for communication links in an operational area.

There are technical developments in many other fields such as materials and satellite communications. These not only bring new capabilities, but also render some military capabilities obsolete. For example, new materials and design software were used to give the F117 aircraft a reduced Radar Cross Section (RCS) rendering it difficult or impossible to detect with then-current conventional military radar. Radar has improved to detect targets such as the F117 making further reductions in RCS necessary but the cost to develop a more-stealthy platform such as the F22 Raptor fighter aircraft is high.

Large autonomous systems, comparable in size to current fast jets, ships and road vehicles, are under development in many countries for all domains. It is likely that there will be considerable read across from civil to military solutions, for example, Unmanned Underwater Vehicles (UUVs) are already widely used for marine exploration work with many different designs in service giving transferable experience for military UUV design. These large autonomous systems are not discussed further as their design only indirectly affects AWS issues.

The continual drive for economic efficiency in manufacturing has led to well-publicised robot production lines for large volume markets. Lower volume production methods now include 3D printing which, among other advantages, may give savings for military logistics by allowing spare parts to be made from a downloaded data file on a printer at a deployed military base far from the manufacturer. An idea of the advantages for military uses can be seen from Reference [15]. 3D printing, combined with automated and connected systems, will bring additional advantages and savings for military logistics in the next few years [16]. These will be applied to AWS in the future, probably in the same way as their use for any other military asset.

Whatever advances are made in AWS over the next decade or two, they should only be operated with human control over lethal decisions, with constrained actions. The user will also need to understand their behaviour to a level which allows them to act as authorised commanders under IHL. Opponents can be expected to use automated systems against us, but they may not be constrained by the same legal and moral frameworks. Counters to an opponent's autonomous systems must be developed by a state in order to guard against their use even if the state does not want to develop their own equivalent weapons.

9.5.5   *Decision aids or decision-makers?*

Decision aids are discussed at several points in this book as tools that make a recommendation to a human decision-maker. There is an implicit assumption that the human will then make a decision based on this recommendation, other information and previous experience and knowledge. The problem that arises is that of automation-bias. This is where a human accepts the recommendation and makes the logical decision and action from that information even if it is wrong and the human has contradictory information. The bias is seen in many fields and widely discussed [17], and even when there are more than one decision-maker in the loop [18].

One logical conclusion from this is to design decision-aids that present options, but do not make a recommendation, allowing the user to make their informed decision. This was the choice made for a decision aid for retargeting multiple Tomahawk cruise missiles when in flight towards a range of targets [19]. Operators were presented with a choice of missile/target combinations that could be chosen to execute strikes within allowed times-on-target. Tests showed that automation bias significantly reduced the number of correct decisions when the aid made a recommendation. Overall, presenting the possible choices in an intelligible way should allow faster decision-making, even without recommendations, and more time to ensure adherence to IHL in the target areas.

Another logical approach is to assess whether the automated recommendation gives a higher rate of correct decisions than the human without any automated recommendations. If this is the case, then it can be argued that the system should act on the automated decision-aid's recommendation without human intervention. This is an analogous argument to an automated response from a Missile Attack Warning System (MAWS) or sentry gun where the required reaction time is too short for a human. Any procurement authority considering such a system should take considered legal advice, and the technical team must make a very careful and thorough analysis of the evidence for 'correct' decisions. This is a good example of where decisions can only be, and must be, evidence-based.

9.6 Urgent requirements for operations

Every military campaign is likely to require equipment that is not in the current inventory. Hopefully the need is identified during campaign planning but this may not necessarily be the case, so special procedures are used. An urgent military need of this type is known as an Urgent Operational Need (UON) in the US and Urgent Operational Requirement (UOR) in the UK. The process to meet them is always fast and everyone involved tries to compress timescales to the minimum possible. There is also a need for flexibility during the programme so that the military need is not negated by unnecessary bureaucracy and over-engineering. However, there is always a minimum set of standards and processes that must be followed, even if the logic for them is not clear to all.

Details will depend on the state, but if the requirement is for any 'new means or method of warfare', an Article 36 review will have to be carried out. In addition, rules-of-engagement will need to be drawn up during the process as the introduction into service phase will almost certainly be followed immediately by deployment to theatre. It is possible that the equipment will not require an Article 36 Review but the results of the introduction into service may impose such tight restrictions on the new equipment's use that it cannot be used in that campaign. Clearly, this is an undesirable result, especially if further modifications cannot resolve the problems, emphasising the need for an Article 36 Review at the concept stage. Training programmes will be devised and introduced in parallel with the design, trials and introduction into service work so that the users will operate the equipment in accordance with IHL.

It is possible that the engineering teams involved will not be involved in, or even aware of, the initial Article 36 reviews. This makes it essential for the engineering staff involved to ask about the trials for the system's introduction into service, and the requirements for evidence that will be needed for setting rules-of-engagement.

Typical UORs/UONs could be:

- Enabling a platform to use an existing weapon that is already used on other platforms, for example, using a different bomb type on an existing manned aircraft;
- Improved sights on a tank;
- Updates to EW suites;
- Adding extra protection to infantry vehicles;
- Adding IED detection systems to a tracked robot.

These would probably not need an Article 36 review, but the following examples could:

- Adding a guided bomb to a surveillance UAV;
- Adding a gun to a tracked robot with firing instructions given by the robot's operator.

Both these UOR/UONs are for modifications to, or new combinations of, equipment that is in the existing military inventory, so it might be assumed that no further legal reviews are necessary. The argument being that the equipment has passed Article 36 Reviews, so another one is not needed. This is not true as these UOR/UONs provide a new method of warfare. The procurement authority will have to work with the legal team carrying out the reviews and the design authority and their team to ensure adherence to Article 35. The teams, between them, should generate a body of evidence which may need to be supplemented but can be used as a starting point for the Article 36 reviews during procurement and also for setting rules-of-engagement.

The combination of reduced military budgets and advanced technologies in civil applications does create a military need to introduce selected civilian products into operations under UOR/UON procedures. This trend is likely to continue in the future. Examples can arise in asymmetric warfare where Red is using devices such as mobile phones or small drones against Blue. The only possible Blue response may be to use civilian technologies to counter them. If these are likely to generate target identification leading to a lethal effect, then Article 36 reviews should be considered.

Clearly, the rise in civilian applications of autonomous systems and software is likely to lead to military imperatives to utilise them. In principle, there is no difference between the UOR/UON processes and those used in any procurement programme for commercial technologies as discussed in Section 9.5.

9.7 Countering autonomous systems

9.7.1 Physical attack

Large military unmanned systems may be smaller than the equivalent manned system, but have similar characteristics, such as manoeuvrability, speed and weapon payload[1]. This means that they will be vulnerable to the same countermeasures as their manned counterparts. It can be anticipated that the countermeasure and counter-countermeasure cycle will continue as various nations and organisations develop new applications of suitable technologies. The requirement for an AWS to have communication links with its commander for any lethal decision does render them susceptible to attacks on the link's physical infrastructure.

Small UAVs, commonly called drones, are now a popular consumer product. There are various initiatives around the world to regulate their use and register ownership but, given the large number of both manufacturers and delivered drones, it is probable that they will be used in asymmetric warfare against states in many ingenious ways. A closely related problem is that of drone incursion into flight paths at airports. The Federal Aviation Authority (FAA) has a research programme looking for solutions with trial programmes underway [20]. This, and similar programmes, are likely to provide solutions which can be transferred into military applications.

---

[1]This may seem a surprising statement as the autonomous system is not constrained to support a human with all their needs. However, in order to deliver the same weapon load, the range, payload and stealth requirements are similar so the need for fuel and power will be similar to a manned system. There will also be requirements for space, weight and power for the intelligence needed to replace the human.

Physical destruction will be an option as a defensive measure. The system specifier should also consider two results from a kinetic counter-drone system: the end-point of any projectile which fails to hit the target; and the landing point(s) of the debris from the successful hit on the target. The danger to civilians and civilian infrastructure should be minimised in the same way as for any weapon system. One of the systems chosen by the FAA for its trials programme uses an RF transmitter to interfere with the drone's C2 system rather than using a kinetic strike [21], probably for this reason.

Civil defence systems against drones will be subject to domestic law, but the use of such systems by military forces in an operation may be subject to IHL. Design requirements for any military system should be written in consultation with relevant legal authorities who will also need to be involved in the trials programme at delivery as a minimum.

9.7.2   Cyber-attack

Cyber warfare and cyber-attacks are now recognised threats to all networks and computer systems. Even a lack of direct connectivity may not be sufficient defence, as was shown by the STUXNET attack on Iran. Therefore cyber-attacks must be considered in the design of any automated system. We will assume here that indirect methods of attack such as exploiting a lax data-handling procedure are beyond the scope of this book. It will also be assumed that cyber warfare defences will be an integral part of any AWS design.

It is shown in Chapter 10 that the commander must be in control of the AWS and that it must have predictable behaviour both in following instructions and if there is a failure in communications with it. Now consider the simple AWS shown in Figure 9.1. This comprises an automated weapon controller connected to a weapon by a C2 datalink, the automated controller being connected to one or both

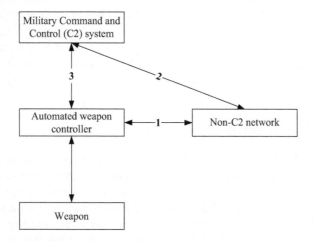

Figure 9.1   Networked automated weapon system

of its superior node and an external network. It can access information from either the C2 chain (Link 3) or the network (Link 1) or both. The superior node accesses the network through Link 2:

- If there is no connection between the automated controller and the network (Link 1 is absent), all information comes via the C2 system and so has to pass existing validation procedures before it is acted on. If there is not a human in this link, there will be robust processes which meet Article 36 requirements.
- If there is no connection between the military C2 chain and the network (Link 2 is absent), the commander can only monitor or control the information that the autonomous system accesses through the system itself. This may not be an insurmountable problem if the design takes this into account, but it may make it impossible to detect a cyberattack on the network giving corrupted data to the weapon system. This is likely to be considered a significant risk in any review and mitigation measures must be in place.
- If the C2 connection (Link 3) is broken but the weapon system still stays connected to the network, it may be possible for the C2 system to monitor what the weapon system is doing in response to the data it receives from the network. It may be possible for the weapon system to be monitored and controlled via the network, but this may not fully meet the requirement for the system to be always under control. It will also open up a vulnerability to cyber-attacks.

It must be possible to design a system with sufficient internal capabilities to be judged as robust against the above risks, but it is reasonable to say that it will be complicated and demand sophisticated error detection methods and strict limits on its allowed actions with low authorised power.

The conclusion to be drawn from the above arguments and similar ones is that the cyber drivers on autonomous weapon systems will be for them to be as self-contained as possible as far as targeting is concerned. This may reduce the AWS' capabilities to the on-board ones rather than exploiting extra data and off-board processing power available over the network. The ability of the controller to predict actions of the system when acting on its own will remain as a fundamental requirement, even if this is limited to knowing they no longer have the anticipated control. In extremis, the AWS may be assumed to be under the control of Red and designated as hostile so it can be destroyed by Blue forces.

9.8   Non-kinetic AWS

Cyber warfare has been mentioned in several places, but cyber weapons are not covered in this book as their legal status is not clear at the time of writing. Any engineer asked to work on them should ensure that the programme has the correct legal authorisation. They should then check the legal framework for the state where they are working. An operational analysis exercise such as that in Chapter 4 should be carried out, or they should ensure that one has been carried out so their work does not breach the law as interpreted in their country.

The term 'non-kinetic weapons' covers a broad range of offensive weapons, usually exploiting one or more parts of the electro-magnetic spectrum[2]. They do not necessarily directly cause death or injury to a person, but can do so either directly or indirectly, so they should meet the general principles of IHL. Regardless of wavelength, there are three underlying precepts:

1. There is a clear command chain from the high-level commanders to an authorised operator;
2. The weapon system will be able to differentiate between military and civilian targets;
3. There is a clear delineation between human and automated decision-making functions.

The first precept should be true for all military campaigns. If the operator can see the target area directly or indirectly, using sensors, then the second premise is fulfilled. The designer must ensure that the third is true, both by design and verified by trials. Without evidence confirming the precepts for the system, the weapon either cannot pass an Article 36 review or will have very severe constraints placed on its use.

A high-power laser beam replacing a missile or other ordnance will need to meet similar requirements to a traditional kinetic weapon. There will need to be a legal ruling on the applicability of the Protocol on Certain Conventional Weapons (CCW). This prohibits the use of weapons whose purpose is deliberate blinding and must be used to place design requirements in the system specification.

It is likely that hyper-sonic missiles will become widespread. Both Russia and America are developing them according to newspaper reports [22,23], so the response time of the defence system needs to reduce appropriately. One important driver for laser weapons is that they can deliver a lethal force to a target in a much shorter time than a conventional projectile. However, this only reduces the delivery-time part of the defence systems' response. The other main time is from detection to the decision to fire the weapon. If this is made by a human there will be their response time which is at least 200 ms if the response is made immediately. It is clear that there will be pressure to automate the identification process and couple this directly to the laser-weapon's 'fire' command. This application of identification algorithms makes them part of an AWS so there will be a clear need for full Article 36 reviews of the whole system including the identification algorithms.

9.9 Future operational analysis

9.9.1 Problems for traditional operational analysis

Traditional operational analysis, as described in Chapter 4, to find counters to a new threat allows the analyst freedom to start with a blank sheet of paper to find an ideal solution. Existing assets are then inserted into the modelling when the desired

---

[2]The breadth of use of the spectrum for all types of operations defensively and offensively can be seen in: DOD 2016, Joint Doctrine Note 3-16, *Joint electromagnetic operations*.

Blue capability is specified. There will always be a place for this approach, but reducing budgets, long platform lifetimes and the rapid, sometimes unpredictable, developments in civilian technologies bring a need for a different approach to developing capabilities. This is especially true for autonomous capabilities where the civil market is driving autonomous system and AI developments in all fields. It should be noted that there are legal questions being raised about responsibilities for some of these civil applications. Analogies with the legal position for weapons are discussed in Chapter 13.

Long platform lifetimes have already been mentioned several times and it is clear that platforms will need regular upgrades to their capabilities. Additional pressure will come from demographic factors, potentially leading to a reduction in the number of personnel in most nations' armed forces. Clearly enhancing mission systems to increase platform capability and reduce human workload will become increasingly attractive.

COTS has been discussed in Section 9.5 and the reasons for the pressure to use it wherever possible. In order to take advantage of COTS products, there will need to be changes to operational analysis and system design as COTS products will almost certainly need modifying before they can come into military service.

It can be foreseen that a military user will request an assessment of the military utility of a particular suite or programme that has demonstrated autonomy in the civil market. The operational analysis will then have starting questions such as:

1. What extra military capability will this specific, or similar, product bring to our armed forces?
2. Will this capability bring any advantage over likely adversaries, and if so what will be the benefit in likely operations?

In answering these questions, there will have to be a fairly detailed system-of-systems design concept. With long-lifetimes for platforms and other equipment, many systems and capabilities will be defined for known time periods. The analysis will have to include the capability constraints set by the combination of the existing materiel and predictions of the opportunities and limitations of civil software when used for military purposes. It may then become a way of specifying the timescales for upgrades to other subsystems so that the new software can be incorporated to give more autonomy.

The original problem of countering a new threat then has the starting point of a set of capabilities at all dates into the future. The question becomes one of finding a counter using variations on the then 'current' capabilities and identifying the required modifications to existing or new COTS equipment. In the same way that a library of potential hostile threats can be built up, a parallel library of counters to them can be established and invoked when needed. The library could also be used in making choices between upgrade paths.

9.9.2   *Operational analysis and system concepts with COTS modules*

Hardware has practical problems with mechanical interfaces, environmental requirements, etc. Software, however, is often seen as portable and an easy way to upgrade a military capability. This may be true for some requirements. Stock control for logistical support and project management are obvious examples where COTS products on standard computers usually offer superior performance at a lower cost than specially produced military systems.

Commercial pressures are leading to rapid advances in automation in all fields including the militarily-relevant ones of decision aids, decision making, data analysis and imagery. This is not the place to predict the likely advances over the next few years, however there are sets of questions that should be asked about the software and algorithms that engineers may be asked to insert into existing systems to enhance autonomy levels. These questions are those that apply to software to enhance the automation in the targeting process. The answers will be part of the evidence used at the Article 36 Reviews. There will be others of a more general nature concerning stability of the supplier base, security of the processor, etc., which are not covered here.

9.9.2.1   Questions about algorithms

- Who owns their intellectual property?
- Does the algorithm have a known mathematical basis?
- Is the mathematical basis known to the weapon system designer?
- What are the assumptions in the mathematical expression of the algorithms?
- Can the algorithms be modelled so that the software implementation can be verified?

9.9.2.2   Questions about the software

- Is the software safety critical?
- What language is the software written in?
- Who owns the intellectual property?
- Will the software or data be highly classified? If so, will there need to be a self-destruct mechanism on any processor or database in a projectile that may fall into hostile hands?
- Has the supplier developed and written the software themselves?
- Is the software under configuration control by the supplier(s)?
- How robust is the software to operating system upgrades?
- What documentation will be supplied with the software?
- Will the customer have access to the code?

Software documentation tends to be one of three types:

1. Non-existent except as sales leaflets;
2. Leaflets plus contractually binding statements when delivered;
3. Detailed handbooks in soft or paper copy.

Only Type 3 is likely to be acceptable for weapon control.

When these questions are answered, it should be possible to write down the limits on the acceptability of the software for military use. One critical question is whether software produced to civil standards, possibly with uncertain provenance, can be given the required power for legal use in a targeting chain or decision aid for a lethal decision. If not, there will be severe limits on its use, or it may not be acceptable for use in targeting.

The acceptability limits can define a set of options for military applications. Each option may be considered to be a module which can be inserted into a suitable place in the system-of-systems. The existing system-of-systems architecture will place limits on the locations for the new module, as well as the hardware requirements for the module. The authorised power (See Panel 3.1 and Section 10.9) that can be assigned to the module will need special attention and may vary depending on is location.

The operational analysis is then an analysis of the benefits of each option in a range of representative scenarios. Instead of asking about new capabilities to meet a capability gap, it addresses the two questions in Section 9.9.1. This approach may be likely to become one of the future areas for operational analysis that will need development of new techniques to assess cost, risk and benefits.

9.10 Future changes in International Humanitarian Law (IHL)

9.10.1 Introduction

Currently, there is no specific international law for autonomous systems in warfare, whether AWS or Lethal Autonomous Weapon Systems (LAWS). The Geneva Conventions, API and APII, apply, but there is also discussion about the applicability of the Martens Clause to LAWS and whether specific new treaties are needed. Several groups and international organisations, including the UN have proposed new treaties and bans which are discussed in Section 9.10.3.

There is a wider concern with controlling 'robots' which we can take to mean autonomous systems and the ethical codes they should follow. There are numerous books on the topic and it is certain that human rights laws and regulations will follow [24–27]. These will have an impact, whether directly or indirectly by influencing technical advances due to market forces.

9.10.2 The Martens Clause

The Martens Clause was proposed by Professor von Martens, the Russian delegate to the 1899 Hague Convention II [28], and it was included in the *1907 Hague Convention IV Respecting the Laws and Customs of War on Land*. It was a compromise about the status of resistance movements in occupied territories, but has become one of the accepted principles of IHL [29]. Despite this, there is no single agreement on its interpretation. The original statement was:

> Until a more complete code of the laws of war is issued, the High Contracting Parties think it right to declare that in cases not included in the Regulations adopted by them, populations and belligerents remain under the protection and empire of the principles of international law, as they result from the usages established between civilized nations, from the laws of humanity and the requirements of the public conscience.

Its importance to IHL is seen by it appearing in modified forms in subsequent treaties, notably in the Preamble to APII:

> The high contracting parties ... Recalling that, in cases not covered by the law in force, the human person remains under the protection of the principles of humanity and the dictates of the public conscience, ... Have agreed on the following. (Then the articles of APII follow.)

The principles are included in API as general principles. Most of these are given in Appendix A7 and are included as part of the systems engineering analysis in Chapters 10, 11 and 12.

9.10.3   Potential new treaties

There was considerable public debate about the role of weaponised UAVs in Iraq and Afghanistan and calls to ban them [30], with meetings of international lawyers to debate the issues involved [31]. There is agreement that LAWS are a potential new type of weapon and meet the terminology of API Article 36 as a 'new means or method of warfare'. What is not clear is whether existing IHL is sufficient, or whether a new convention or protocol is needed.

The UN has formally started discussions about LAWS and their place in IHL. The mechanism used in 2016 was an International Committee of the Red Cross and Red Crescent Review Conference of the High Contracting Parties to the CCW, i.e. the nations which have signed the CCW. A Group of Governmental Experts[3] (GGE) on emerging technologies in the area of LAWS was established. This was followed in 2017 and 2018 by GGE meetings under UN auspices.

Reports about the GGE meetings and inputs to them are available [32]. The GGE considers that the main issues to be addressed are:

- Characterisation of the systems under consideration in order to promote a common understanding on concepts and characteristics relevant to the objectives and purposes of the CCW;
- Further consideration of the human element in the use of lethal force; aspects of human-machine interaction in the development, deployment and use of emerging technologies in the area of LAWS;
- Review of potential military applications of related technologies in the context of the Group's work;
- Possible options for addressing the humanitarian and international security challenges posed by emerging technologies in the area of LAWS in the context of the objectives and purposes of the Convention without prejudging policy outcomes and taking into account past, present and future proposals.

International treaties take many years to negotiate and agree so, even if the GGE decides that a new treaty or protocol under CCW is necessary, it is unlikely that there will be immediate effects on engineering design.

---

[3]Experts in this context include lawyers, diplomats, military officers and lobby groups as well as academics and technologists.

References

[1] National Science Foundation. A roadmap for U. S. robotics. From internet to robotics. 2016 edition.

[2] US Department of Defense. *Unmanned systems integrated roadmap FY2013-2038*. Reference number 14-S-0553; 2013.

[3] UK Ministry Of Defence. *Future character of conflict*. 2010, updated December 2015. Available from https://www.gov.uk/government/publications/future-character-of-conflict [Accessed on 23 April 2018].

[4] UK Ministry of Defence, *Strategic Trends Programme Future Operating Environment 2035*; 2015, pp. 23-25. Available at: https://www.gov.uk/government/publications/future-character-of-conflict [Accessed on 24 April 2018].

[5] Odell L.A., Wagner R.A., and Weir T.J. *Department of Defense use of commercial cloud computing capabilities and services*. Institute for Defense Analyses, Paper no. P-5287, 2015.

[6] UK Ministry of Defence, *Joint Concept Note JCN1/18, Human-Machine Teaming*, 2018.

[7] Chen, J.Y.C., and Barnes, M.J. 'Human–Agent Teaming for Multirobot Control: A Review of Human Factors Issues'. *IEEE Transactions on Human-Machine Systems*, 2014, vol. 44(1): pp. 13–29.

[8] Nothwang W.D., McCourt M.J., Robinson R.M., Burden S.A., and Curtis J.W. 'The human should be part of the control loop?' IEEE Conference Resilience Week (RWS), 2016, pp. 214–220.

[9] Endsley M.R., 'From here to autonomy: Lessons learned from human–automation research', *Human Factors*. 2017, vol. 59: pp. 5–27.

[10] Christensen J.C., and Lyons J.B. 'Trust between humans and learning machines. Developing the gray box'. *ASME. Mechanical Engineering*. 2017, vol. 139(06): pp. 9–13.

[11] UK Ministry of Defence report CM8278, *National Security Through Technology: Technology, Equipment, and Support for UK Defence and Security*, February 2012.

[12] Schultz M.T., Mitchell B.K., Harper B.K., and Bridges T.S. *Decision making under uncertainty*, U.S. Army Engineer Research and Development Center, Report no ERDC TR-10–12, 2010.

[13] Ying T., *Swam intelligence*. Three volume set. London: IET; 2018.

[14] Brambilla M., Ferrante E., Birattari M., and Dorigo M. 'Swarm robotics: a review from the swarm engineering perspective'. *Swarm Intelligence*. 2013, vol. 7: pp. 1–41.

[15] The web page: http://www.tandemnsi.com/2016/02/additive-manufacturing-holds-promise-for-military/ [Accessed 19 January 2019] is a report on a US Army and University of Maryland meeting about futures for 3D printing in military application.

[16] A review can be found at the UK Ministry of Defence website: https://www. contracts.mod.uk/do-features-and-articles/three-key-developments-trans-forming-military-logistics/ [Accessed 19 January 2019].

[17] Mosier K.L., and Skitka L.J. 'Human decision-makers and automated decision aids: made for each other?' Chapter 10 in Parasuraman R., Mouloua M. (Eds) *Automation and human performance, theory and applications*. Lawrence Erlbaum Associates Inc. 1996, Reprinted 2009 by CRC Press.

[18] Skitka L.J., Mosier K.L., Burdick M., and Rosenblatt B. 'Automation bias and errors: are crews better than individuals?'. *The International Journal of Aviation Psychology*. 2000, vol. 10(1): pp. 85–97.

[19] Cummings M.L. 'Integrating ethics in design through the value-sensitive design approach'. *Science and Engineering Ethics*. 2006, vol. 12(4): pp. 701–715.

[20] FAA news item on 1 July 2016; *FAA Expands Drone Detection Pathfinder Initiative*; Available at https://www.faa.gov/news/updates/?newsId=85532 [Accessed on 25 April 2017].

[21] Wade A. 'US to trial British anti-drone system at airports'. *The Engineer*. 2 June 2016.

[22] See for example The Australian 11 March 2018, available at: https://www. theaustralian.com.au/news/world/russia-claims-to-have-tested-undetectable-supersoninc-missile/news-story/9d13262571b5baee4bbff587d1bf58de Accessed on 25th April 2018].

[23] See, for example, The Cable 1 March 2018. Available at http://foreignpolicy. com/2018/03/01/pentagon-official-says-u-s-hypersonic-weapons-research-underfunded/ [Accessed on 25 April 2018].

[24] Wallach W., and Allen C. *Moral machines. Teaching robots right from wrong*. Oxford: Oxford University Press; 2009.

[25] Lin P., Abney K., and Bekey G.A. (eds). *Robot ethics. The ethical an social implications of robots*. Cambridge, Massachusetts: The MIT Press; 2012.

[26] Bostrom N. *Superintelligence, paths, dangers, strategies*. Oxford: Oxford University Press; 2014.

[27] Calo R., Froomkin A.M., and Kerr I. (eds). *Robot Law*. Cheltenham UK: Edward Elgar Publishing; 2016.

[28] Ticehurst R., 'The Martens Clause and the Laws of Armed Conflict'. *International Review of the Red Cross*. 1997, No. 317. Available at: https://www.icrc.org/eng/resources/documents/article/other/57jnhy.htm [Accessed on 30th April 2018].

[29] Roberts A., and Guelff R. *Documents on the laws of war*. Oxford: Oxford University Press; 3rd Edition, 2000. pp. 8–9.

[30] Asaro P. 'On banning autonomous weapon systems: human rights, automation, and the dehumanization of lethal decision-making'. *International Review of the Red Cross*; 2012, vol. 94(886): pp. 687–709.

[31] International Committee of the Red Cross. *International humanitarian law and the challenges of contemporary armed conflicts*. Report No. EN 32IC/15/11, 32nd international conference of the Red Cross and Red Crescent, 2015.

[32]   The GGE has produced reports on its meetings and several nations have produced 'position papers' giving their views on LAWS and IHL. Some of these are available at the UNOG site: https://www.unog.ch/, entering GGE into the Search box. Some are 'Restricted' and not openly available. The main papers are at https://www.unog.ch/__80256ee600585943.nsf/ (httpPages)/7c335e71dfcb29d1c1258243003e8724?OpenDocument#_Section2 [Accessed 30th April 2018].

Chapter 10

Systems engineering applied to International Humanitarian Law (IHL)

10.1 Introduction

The four basic tenets of International Humanitarian Law (IHL) stated in Section 7.6 are military necessity; humanity; proportionality; and distinction. They are used here to derive engineering requirements that are applicable to Autonomous Weapon Systems (AWS). Architectures are used to examine the most important functions in autonomy and help identify the problems faced in making technical developments compliant with IHL. The problem can be stated simply as, 'what authority must an automated system have to act autonomously and be compliant with IHL?'

Chapter 7 introduces the legal background to the use of weapon systems. Section 7.10 gives the engineering requirements which follow directly from specific treaties and articles within them. The legal requirements for precautions in targeting are outlined in Chapter 8 with some of the problems to be addressed for more automated weapon systems given in Section 8.6. Sections 7.10 and 8.6 only address a subset of the requirements for a new weapon; they do not look at the performance of the complete weapon system and the automated functions within it which are the subject of this chapter.

The analysis and systems engineering techniques presented in Chapter 4 are applied in this chapter to the Geneva Conventions and Additional Protocol I (API) discussed in Sections 7.5 and 7.10. The overall context of weapon system procurement is discussed briefly, then the Geneva Convention requirements are derived from API.

A system architecture is needed to describe the human and machine decision-making processes involved in lethal attacks[1] as well as the systems or components which deliver the lethal effect. The 4D/Real-time Control System (4D/RCS) architecture [1] is chosen and used to derive generic requirements for subsystems within the architecture. These are applied to a littoral scenario in Chapter 11.

The concept of authorised power is introduced in this chapter. This gives a useful method of both limiting the behaviour of the weapon system and giving a

---

[1]The word 'lethal' is used in this book to describe any action which may involve harm to humans. This may be indirect in that the target is inanimate, such as a building, but humans may be close enough to suffer harm from the weapon's effect. It is usually assumed that the weapon is kinetic, i.e. there is a release of explosive energy, but other mechanisms may be used such as kinetic energy from high-velocity non-explosive projectiles, or directed energy weapons.

basis for system tests to demonstrate adherence to API. When used for weapon systems, it is based on the legal and regulatory requirements mandated by IHL and not based on one particular branch of applied ethics.

10.2   Context

The system considered in this chapter is a weapon system used in a military campaign for targeting specific objects. Targeting is a process for making the decisions about weapon release during a mission. For the purposes of this book, the process starts after the target list is issued (see Figure 8.1) and ends shortly after impact. It is implemented using a weapon system which may be distributed across several platforms. It is postulated that the weapon system can operate with a considerable degree of autonomy and the systems engineering problem is to see how much of the system can be made autonomous and remain compliant with IHL.

Target information comes from a range of sources, many of which may be remote from the weapon commander, supplying stored and real-time information into the weapon-release decision. This means that even if an autonomous targeting system is possible, it will still have to fit into the existing military infrastructure of the nation using it.

This chapter makes the same assumption as the majority of legal discussions about AWS – that it is used to attack pre-defined targets as part of a wider campaign. There is no explicit discussion of an AWS having a self-defence system such as the Missile Attack Warning System (MAWS) in Panel 3.2. However, the analogies between the release of weapons and countermeasures are close as there must be clear identification of a threat and a target. In the case of the MAWS, there may be a choice between a lethal and non-lethal response as well as considerations of the military necessity of one autonomous system defending itself against another one when no human life is directly involved.

The systems engineering process starts with the context of the force structure as well as the AWS's likely operational uses. It is highly likely that much, if not all of the nation's Command and Control (C2) infrastructure will already be described using an architecture framework such as those discussed in Section 4.2.4. These frameworks include definitions of the command chain, the communication interfaces and bearers supporting it. If available, the architecture framework can be used to set the system context in a rigorous manner. If not, it will be necessary to develop something similar.

Once the context has been described, there are at least two approaches which can be taken to the procurement of such a system. These are described in Sections 10.2.1 and 10.2.2.

10.2.1   Bespoke solutions

When a weapon system is procured for a specific platform such as a ship, requirements and specifications must be written for that particular platform. This will be essential for those parts of the system which are integrated into the platform's systems such as its C2 and fire control system. Clearly, national and international standards will be used wherever possible. The weapon system may itself be a complex weapon system[2] which needs constant updating from the targeting system after weapon release.

A weapon system will be included in the original platform procurement contracts, probably as a bespoke system even if it uses existing subsystems for either economy or commonality reasons, or both. The platform system design will have an architecture that will be described using tools chosen by the prime contractor, which may be specific to that company.

It is unlikely that the weapon system will only have interfaces with the host platform. Target information will come from external as well as internal sources even if its communications are routed through the platform. Interfaces between the weapon system and the rest of the nation's infrastructure will need to be defined. The external infrastructure will probably have existing architecture descriptions using a standard framework.

Every new platform and its weapons will have been reviewed under API Article 36 with any necessary legal requirements and testing carried out at the time. System changes will need to be reviewed to decide if a new Article 36 review is needed. An intention to increase the automation of the targeting system makes it almost certain that a new Article 36 review will be needed.

The problem faced by the specifiers and designers of upgrades to increase automation will be identifying those functions which can be upgraded, and are within the available financial and technical budgets. The existing hardware will place constraints on locations and processing capabilities, the AWS will demand specific types of information, whilst the platform may need updates from new external information sources such as unmanned surveillance systems. The complete system will be constrained by the available space, weight and power.

Bespoke upgrades will be necessary for upgrades to a major platform, but an early decision will be whether there should be commonality with other automated weapon functions being developed for other platforms or from research programmes. The design will then have to integrate these other concepts into the platforms' infrastructure.

10.2.2   Automate each function

Another approach to autonomy is to identify the functions in the targeting process and make them more autonomous by a programme of upgrades. It is explained in Section 5.4.1 that consideration of the autonomy level of critical functions is an approach favoured by some international lawyers for assessment against IHL criteria. Modern distributed systems are also described in terms of system functions so it seems advisable to base the design on automating functions, concentrating on the 'critical functions' involved in targeting and lethal decisions.

---

[2]This use of the adjective 'complex' to describe weapon systems in this book is not to be confused with the term 'Complex Weapons'. These are strategic and tactical weapons reliant upon guidance systems to achieve precision effects. Complex Weapons include the missiles that provide the terminal effect and the weapons systems (but not the platform) that fire them. This definition is taken from the agreement between the Government of the United Kingdom of Great Britain and Northern Ireland and the Government of the French Republic Concerning Centres of Excellence Implemented as part of the 'One Complex Weapons' Sector Strategy, dated, 24 September 2015, Paris.

Implementing autonomy through functions is a feasible way to introduce results from research programmes into new assets and systems.[3] The approach should then enable incremental upgrades to be made as technology evolves. Although designers of a new system have the advantage of freedom of choice in approach, there will be many constraints on them. Space, weight, power, timescale and financial constraints are self-evident, but there will be others, including interoperability with other targeting systems and weapons already in the nation's inventory. The API Article 36 reviews should then be integrated into the design process.

This approach, using functions, should be considered for upgrades to existing systems as well as new ones, even when they are bespoke. Defining functions, their boundaries and interfaces will be an important part of the initial feasibility stage. The results should also be reviewed to see if they show that there is a maximum autonomy level for that function in the existing asset and its system-of-systems.

There is a high risk of unexpected behaviours when the new functions interact with existing ones. Rigorous testing will be difficult due to likely 'wicked problems' discussed in Section 4.2.2. This is where a problem is found, but it may be impossible to reproduce the same test conditions, or where an attempt to fix the problem only changes it rather than removes it. The project risk register should include this type of problem as they occur with all upgrades, but an autonomous function's interactions with other systems may be more complex than normal, especially if there is any degree of learning in the function.

10.3   Requirements derived from the four IHL principles

10.3.1   General principles

The general principles of IHL relevant to weapon systems are stated in API Article 35, quoted in Sections 7.11.2 and 7.11.3. Article 36 then requires that new means of methods of warfare be reviewed for breaches of IHL not just API, so our generic requirements must cover all IHL.

The four basic principles underpinning IHL and API Article 35 are military necessity, humanity, proportionality and distinction. It is convenient to derive system requirements under these headings although different aspects of them are in several articles in API and other treaties. States also usually use these headings when stating their interpretation of IHL in military manuals [2,3]. The derivation is carried out in the next four sections.

The relevant API articles are included in Appendix A7 of Chapter 7.

10.3.2   Military necessity

Force can only be used if necessary to achieve a military objective. It follows that force cannot be used unless there is a military objective, and that if the objective has been achieved, then further force cannot be used.

---

[3]The provisos would include that the Technology Readiness Level (TRL) of the research is at least 5 in the context of the nation's assets in the likely operational scenarios.

It also follows that the force must be under control. This is achieved by a C2 chain of nodes which are authorised entities as defined in Panel 1.1 and 3.1. Each node may be human or machine and may have its own internal communication network. Every node will have a defined place in the command chain and the sole node who or which is authorised to give instructions to it. The humans will know their authority and the limits on their freedom of action from training and operational procedures. The automated, machine nodes must have their authority defined, and limits set on their behaviour for their defined place in the command hierarchy. These must be an inherent part of their design and set by their requirement specifications.

This concept can be extended to cover an AWS which may itself be an authorised entity but with less authority than a human. The authority level will be closely related to the rules-of-engagement for the operation. Rules-of-engagement are currently interpreted by a human targeteer for any weapon release decision, mainly based on requirements from the proportionality principle. A fully autonomous targeting system must be able to interpret them consistently to the same standard, without human intervention. If it cannot make these decisions consistently, the weapon system cannot be considered to be under control.

The communication links for commanding a node may be different to those for controlling it, and may both be different to those used elsewhere in the C2 chain for technical and/or security reasons. Link integrity must be monitored to ensure that all commands are received correctly otherwise the node may be out of control. If the communication links do fail at times other than planned interruptions, there must be contingency plans in place which will require interpretation for the particular circumstances at the time.

Panel 10.1 gives a set of requirements for an AWS derived from consideration of the factors given in this section.

Panel 10.1:   Requirements derived from the necessity principle for an AWS

1. The AWS is a node in a C2 system which is part of a legitimate military force;
2. The node commanding the AWS (its superior node) shall have authority to send it on a strike mission;
3. The AWS shall have communication links with its superior node in the C2 chain;
4. The AWS shall know when the integrity of its communication links is compromised;
5. There shall be workable contingency plans in place if the communication chain is compromised. The AWS shall be capable of choosing the appropriate plan;
6. The AWS shall have clearly defined authority to act, and defined responses if it cannot act within its authority;
7. The AWS shall be able to interpret rules-of-engagement based on pre-existing information and on updated data during its mission.

10.3.3   *Humanity*

This principle restricts the suffering and damage to the level actually needed to achieve the military objective. Decisions under humanity can be made by humans before the AWS's mission starts. If there are no updates, the autonomous weapon system becomes like the early cruise missiles which could only be fired against isolated targets with no non-military objects in the target area.

This chapter assumes initially that the AWS will have sensors, access to off-board information, and will have to make humanity and proportionality judgements itself. The general case will also assume that the system must search for defined types of target in a cluttered environment and make its own decision about whether to release its weapon or not. It may itself be the weapon.

There is a requirement in API Article 57.2 (b) that an attack must be cancelled or suspended if it becomes apparent that an object is not, or is no longer a military objective, or that the loss of civilian life will be excessive in relation to the antici-pated military advantage. Suspending or cancelling an attack in the light of new information means that the nodes must receive information about the target area during the attack and be able to act on it. Clearly there will be a cut-off time (t_0) after which it is impossible to call off the attack. Conventionally this is weapon-release time but even relatively simple guided missiles or guided bombs can be steered away from their target to a safe area for impact. Highly automated weapon systems will have more complex options so that t_0 will need to be defined for each system.

Panel 10.2 gives a set of requirements for an AWS derived from consideration of the factors given in this section.

Panel 10.2:   Requirements derived from the humanity principle for an AWS

1. The AWS shall be able to recognise all symbols, radio signals or other means of identifying objects or areas which are the subject of special pro-tection, such as medical facilities;
2. There shall be clear criteria set before the attack starts to identify the target and that it is legitimate;
3. The AWS shall have methods to check that the military personnel in the target area have hostile intent;
4. There shall be a defined time, designated here as t_0 after which it becomes impossible to prevent the weapon hitting the target;
5. The AWS shall have a method to confirm up to time t_0 that the target is still a legitimate object to attack;
6. The AWS shall be able to take suitable action to minimise its lethal effect if the target is no longer a legitimate one;
7. If the AWS can choose the lethality of the weapon it releases, it shall use the minimum compatible with achieving the military objective.

10.3.4   Proportionality

Proportionality is arguably the most subjective principle to apply. The concept is that the losses caused by an attack must not be excessive in relation to the antici- pated military advantage. There is considerable overlap with the humanity principle and the two sets of requirements derived here also overlap but any overlap can be dealt with in deriving the requirements for a specific system.

Panel 10.3:   Requirements derived from proportionality principle for an AWS

1.   The AWS shall not attack any area unless it has been specifically tasked to attack a designated target or target type in that area by its authorised commander;
2.   The AWS database shall have a set of areas, locations or other identifiable objects which it cannot attack unless specifically authorised to do so by its commander explicitly overriding the protections for this set. The set shall be capable of being updated during a mission;
3.   The AWS shall not release a weapon unless it has a single, designated target which it can be reasonably expected to hit or damage;
4.   The AWS cannot attack a designated target unless it has a means of dis- tinguishing it from other objects in the target area;
5.   The AWS shall know or calculate the trajectory of the released weapon and the likely differences between the aimpoint and the impact point;
6.   The effects of the released weapon shall be limited and known by the AWS;
7.   The AWS shall either be capable of calculating the effects of all its weapons on the designated target's area or passing all relevant information to its comman- der and not releasing any weapon until instructed to do so by its commander. All considerations are to include the likely differences between aimpoint and impact point;
8.   The AWS shall have clear quantified criteria from its commander on the allowed collateral damage. If the target area does not meet these criteria, the information shall be passed to the commander and no weapon can be released until subsequently authorised by the commander.

Minimising civilian losses has become increasingly important as weapon accuracy has increased. Kinetic effect can now be delivered to an accuracy of a few metres, often with the result that only one weapon is needed to achieve the military advantage. This, combined with high-quality surveillance of the target area, means that a reasonable estimate can be made of military and civilian casualties arising from the attack. The estimate has to include an allowance for the likely distance between the aimpoint and where the weapon will actually hit. This distance will be a known weapon parameter (The Circular Error Probable – CEP) and can be incorporated in the AWS' algorithms. The balance between losses and advantage is made by judgement in the context of wider strategic and political considerations.

Proportionality is expressed legally through API in two places. Article 51 extends a general protection of the civilian population against the dangers of military operations and gives several rules which must be observed. Articles 57 and 58 give precautionary measures and state that *constant care shall be taken to spare the civilian population, civilians and civilian objects.* The proportionality principle itself is given in API Article 57, paragraph 2(b).

It is not the role of engineers to make proportionality decisions or, by extension, to design systems that have implicit criteria for the decision built into the algorithms. Some of the rules, such as the prohibition of reprisals are based on intent which cannot be assessed by current technologies. Expert Artificial Intelligence (AI) systems may be proposed to make proportionality decisions but, as discussed in Section 9.5, AI learning systems in weapon systems cannot be anything other than a decision aid for the foreseeable future.

The approach taken here is to write generic system requirements that put the rules of API into engineering terminology and ensure that all the relevant information is presented to one human, the autonomous weapon system's commander. The authorised human either makes the proportionality decision or sets quantifiable criteria so that the automated system can make the decisions under clearly defined conditions.

Panel 10.3 gives a set of requirements for an AWS derived from consideration of the factors given in this section.

10.3.5   Distinction

It might appear obvious that opponents should be identifiable as such so that they can be attacked. The converse is also true, that your own forces are identifiable. Military uniforms have a long history in fulfilling this principle. The distinction between two sets of opposing forces is not the only distinction required under IHL. There is the requirement to distinguish civilians and civilian objects from military ones as well as from opposing forces. A further complication is objects and infrastructure which have dual uses, i.e. a facility can have both civil and military uses at the same time. Power distribution networks are an obvious example.

API Articles 53–56, 59 and 60 state objects, areas or aspects of the environment that cannot be attacked in a conflict, some of which will have distinctive symbols to proclaim their protected status. Articles 61–67 cover civil defence units and activities which should also have distinctive emblems. Protection of medical facilities and staff is defined in Articles 8–31 although much of this is not directly relevant to autonomous weapons. The requirement to identify the type of object is included under humanity and in Panel 10.2 as they are designated for humanitarian reasons.

The distinction principle is potentially the easiest to put into engineering terms. There are many types of sensor available, with object recognition algorithms being a well-established field. It should be noted that precision in terminology and interpretation of claims for performance is an important issue. Panel 8.3 *The four meanings of ATI* is but one example. The IHL requirement is very simple to state, but specific requirements will depend on the automated system, the likely operational scenarios and the acceptable recognition accuracies for each category of object in those scenarios. It should also be noted that there is a requirement to identify civilian objects as well as military ones.

A general problem for system designers comes from modern asymmetric warfare. It can be very difficult to identify hostile combatants who are disguised as, or actually are members of the civilian population. The Geneva Conventions and their associated Additional Protocols assume that the conflict is between two armed forces of a similar type, but the same requirements are applied to both types of conflict. This has led to increased use of intelligence information in the targeting process as well as long-term surveillance with automated processes and tools for interrogating large data sets. The principle remains the same, identifying hostile forces and distinguishing them from non-combatants; the method of achieving it has evolved with the threat.

Panel 10.4 gives a set of requirements for an AWS derived from consideration of the factors given in this section.

Panel 10.4: Requirements derived from the distinction principle for an AWS

1. The AWS shall be able to distinguish between civilian and military personnel;
2. The AWS shall be able to distinguish between military and civilian objects;
3. The AWS shall be aware of all protected locations, objects and areas in its operational area;
4. The AWS shall be able to identify all protected areas near its intended target or targets;
5. The AWS shall confirm that a designated target remains a legitimate military target until time t_0 and abort the strike if the target is no longer legitimate.

10.4 Developing the architecture framework

Once the context and the system-of-systems are known, its architecture can be used to derive requirements for a specific AWS. This is the most general approach although it may appear to have a large overhead 'cost' in developing the architecture. However, this cost should be recovered by savings later in the programme.

The weapon system and/or platform hosting it will be integrated into the wider national or coalition force structure which will be supplying real-time information on the operational and target area. The interfaces and data flows will need to be specified. This is more comprehensive and rigorous if the architecture used for the weapon system has common features with those describing the rest of the system-of-systems.

It is clear from the necessity requirements in Panel 10.1 that the source and validity of the data and the integrity of the network are critical. The composition and structure of the force will depend on the campaign, and the operation will be conducted within that structure. Therefore it is good practice, if not essential, that the automated weapon system is described using a recognised architectural framework. Interfaces and data flows can then be verified for compliance with requirements even if the force structure is not the one envisioned when the automated parts of the weapon system were procured.

Chapter 4 explained how architectural views are used to describe integrated and flexible system-of-systems. Data about the system-of-systems and systems are held in a repository in a consistent format which is amenable to architecting. The system-of-systems and systems are entities but using different architecture views allows someone to answer a specific question about them. The user draws up the views which show the relevant systems, interfaces and data flows in the context of their question.

The specific question to be addressed here is to derive Article 35 requirements for a proposed generic AWS. At this stage the system must be kept at a conceptual, but representative level, recognising that a fully autonomous system is probably not realisable. When general requirements are derived, it is possible to apply them to derive a more detailed architecture for a particular weapon system in the context of likely operations. The question becomes one of identifying the targeting functions within the weapon system, especially the 'critical' ones and the subsystems which together perform these functions. The AWS requirements can then be derived from the Geneva Conventions and API, and deciding where human intervention is a mandatory requirement for the system.

We are considering a relatively small part of a defence infrastructure so it can be assumed that most of the operational analysis has been carried out, as described in Section 4.8 and illustrated in Section 11.2. There will also be architectural descriptions of part, but probably not all, of the infrastructure around the weapon system. It is reasonable to assume that the architecture has a meta-model. A meta-model based on the MODAF one is shown in Figure 10.1. The specific problem that we are addressing is to find the systems engineering requirements for an autonomous targeting capability that meets the IHL capability requirements given in Panels 10.1–10.4.

Figure 10.1 sets the context for strike operations which include all activities after the issuing of the approved target list. The target locations are not necessarily known before the operation starts. The system-of-systems is identified as the 'Solution Architecture' with 'Strike' as its activity. 'Autonomous Targeting' is then the specialised activity within it, provided by a 'Logical Architecture' which delivers 'Autonomous Targeting' through an 'Autonomous Node'. The 'Autonomous Node' is the AWS which has a set of 'Functions' used to carry out the roles in the strike mission.

Further architecture views can be developed to describe the targeting system on its platform and in a representative scenario. This is the preferred approach when there are questions concerning the use of a specific system in likely operations. The particular views chosen will depend on the question being asked and the architecture of the system under consideration. This chapter derives generic requirements for autonomous targeting derived from IHL, so an alternative approach has to be taken here which does not rely on a specific weapon system implementation. A standard architecture model is used that can be applied to an organisation, a human or a machine as the aim includes exploring the boundaries between human and machine.

Increasing automation in a set of functions means that the description of the node performing the function must be valid for both human and machines. As technology evolves and automation increases, more human functions will be transferred to the machine. The requirements on the quality of the solution and decision will remain constant whether made by human or machine.

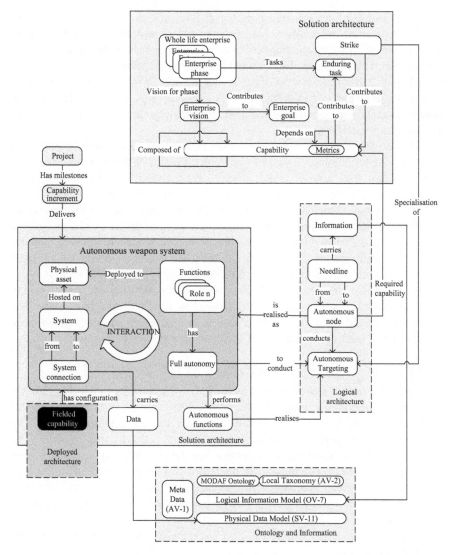

Figure 10.1   Meta model for an AWS (Based on the MODAF M3 meta-model, UK MOD Crown Copyright)

10.5   Generic weapon system architecture

10.5.1   Architecture requirements

Once the context is set, and we have a set of requirements, the next problem is to find a suitable standard architecture to describe the autonomous weapon system. Figure 10.1 defines the context and Panels 10.1–10.4 set the IHL capability requirements for an AWS. There is not a single 'correct' architecture, instead one has to be developed which allows us to derive systems engineering requirements for design and test purposes. We need the architecture that answers our problem, which is to increase the number of autonomous functions performed by the system, with an aim of replacing the human if this is possible. This is the same as asking how we can replace human decision-making with automation.

Chapter 2 discusses human decision-making and shows how it can be represented as a three part cognition model: awareness, understanding and deliberation. Several variants were mentioned and there are probably many more that have been developed for specific purposes.

The requirement that we have is for an architectural description based on the three-part cognition model which can be implemented as a human or as a machine node. It must also describe a Real-time Control System (RCS). The 4D/RCS standard [1], is a widely used model that meets our needs. It is not the intention here to develop a detailed model using it, instead, the three-part structure will be used so that function and system requirements for an AWS can be drawn up in a recognised framework. It is then possible to derive more detailed models of subsidiary nodes using these requirements. If the reader wishes to do this, there are useful explanations of applying the 4D/RCS standard in References [4] and [5].

Provided that there is internal consistency in the 4D/RCS description of the AWS, the model should be able to fit into the chosen architectural framework. The interfaces with the system-of-systems need to be defined in an unambiguous way and care taken over the data needs of the AWS and system-of-systems.

10.5.2   The 4D/RCS reference model (NSTIR 6910)

This reference model was developed by the US Army as part of their programmes to develop autonomous land vehicles. It provides a conceptual framework for implementing intelligent vehicle systems and a reference model architecture. The standard also gives engineering guidelines for its application to army vehicles. The concept covers intelligent RCSs structured into levels with response times from milliseconds to weeks. The model will be used in this chapter to specify autonomy requirements in terms of human cognitive functions. The engineering guidelines in the standard are for one application so they are not directly applicable to the general problem but they are illustrative.

The reference model is made up of a hierarchy of nodes (authorised entities) which are defined as:

> An organizational unit of a 4D/RCS system that processes sensory information, computes values, maintains a world model, generates predictions, formulates plans, and executes tasks.

The nodes all have the same basic structure, shown in Figure 10.2 which is Figure 4 in Reference [1]. Each node can be implemented as a machine-based controller or its functions can be performed by a human or group of humans.

The definitions of the five functional elements in a node given in the standard are:

> **Value judgement** is a process that computes value, determines importance, assesses reliability and generates reward or punishment.

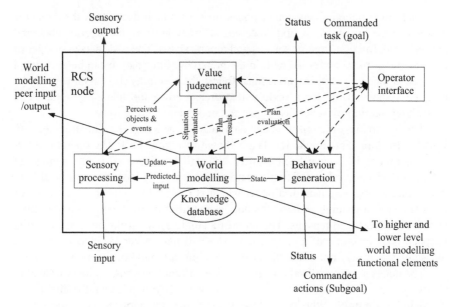

Figure 10.2   A 4D/RCS node (taken from NSTIR 6910)

Sensory processing *is a set of processes that operate on sensor signals to detect, measure and classify entities and events and derive useful informa-tion about the world.*

World modelling *is a set of processes that construct and maintain a world model. A world model is defined as an internal representation of the world.*

Behaviour generation *is planning and control of actions designed to achieve behavioural goals. A behavioural goal is defined as a desired state that a behaviour is designed to achieve or maintain.*

A **knowledge database** *contains the data structures and the static and dynamic information that together with world model processes form the intelligent system's world model.*

The model was developed for both human and machine-based decision-making and has the advantage that it allows a clear separation of the two. It assumes a physical instantiation of the autonomous system, but it can have wider applications [6].

The key advantage of this node definition is that the node is based on human cognition: awareness comes from the Sensory Processing; understanding comes from the World Model; and deliberation is carried out by the Behaviour Generator sup-ported by the Value Judgement component. A node can make predictions about the world and its effect on it. The node can then take action based on those predictions.

The lowest level nodes in the hierarchy have very little freedom of action, the standard gives examples of these such as servos driving vehicle wheels. The nodes at the level above these are called primitives with a response time of less than one second. Subsystems are the level above this. All levels in a military hierarchy can be described as nodes using the definition given above.

Each node interacts with other nodes in ways which depend on their relative position in the hierarchy. A node receives instructions from its superior node and passes up information from its sensors and reports status. The node acts as a superior node to one or more nodes below it in the hierarchy. Information can be exchanged between all nodes, but instructions must follow the hierarchy. It is assumed that the required response time for a node will decrease down the hierarchy – the node controlling steering must act faster than the node planning the route.

The hierarchical organisation can have many layers from headquarters level to vehicle effectors. Three node levels are shown in Figure 10.3. The operator interfaces of Figure 10.2 have been omitted and will be discussed in the context of authorising weapon release. The standard assumes that the operator will be able to interrogate almost any part of a node, which is reasonably valid for the driver or commander of a ground vehicle; it is not necessarily true for other assets for either technical or workload reasons. The pilot of a single seater strike aircraft in hostile airspace is an example of a commander who will have a very high workload and whose platform may not allow direct intervention into parts of the avionics suite.

The nodes at a given level all have equivalent functions, responding on steadily shortening timescales and more restricted horizons as you go down the chain. The command chain links to the lowest level through the Behaviour Generators. Sensor information flows between the Sensor Processors. The architecture has 'hierarchical levelling' which means that the nodes at a given level all have comparable

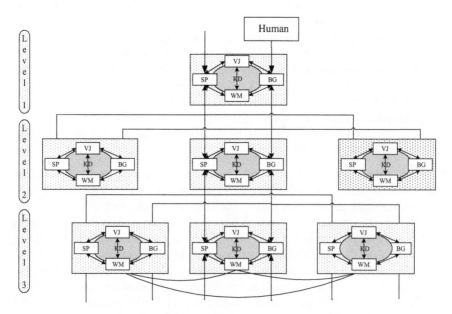

Figure 10.3   A 4D/RCS node hierarchy (links between WMs at the second level omitted for clarity)
   [Key: Value judgement (VJ), Behaviour generator (BG), Sensory processing (SP), World modelling (WM), Knowledge database (KD).]

response times to instructions and external changes. The Knowledge Database is partitioned between node levels to match their hierarchical levels. This allows the higher-level nodes to re-plan the mission phase in real time whilst the lower nodes continue to plan for, and react to, rapidly changing local circumstances.

The Behaviour Generator controls the actions of the node as it carries out its role in response to commands from its superior node. The commands are executed as tasks which the Behaviour Generator interprets as one or more goals and a set of activities to achieve the goal(s). These may be divided into subtasks for that node or delegated to lower-level nodes. It has a scheduler which sets timelines for completion of the tasks. The task may be to deliver a particular result such as identify an object in an area, or it may be to reach and maintain a steady state such as a constant frame rate whilst surveying an area.

Each Knowledge Database shares information across its tier level via the World Models at its level. The Knowledge Database links providing the sharing across a Level have only been shown at Level 3 in Figure 10.3 for clarity. It also has an information repository with configuration control.

10.5.3   Approach and definitions

The meta-model in Figure 10.1 provides the starting point as it shows the AWS in the context of its system-of-systems and the military operation (Strike). The AWS block in Figure 10.1 can be described as a 4D/RCS node as shown in Figure 10.2. The human command inputs are into its Behaviour Generator at Level 1. The Sensory Processing inputs are the intelligence and other relevant data available at the time of issuing the command. They will be updated during the mission.

The AWS must deliver the set of functions which perform the n roles in Figure 10.1; the functions being delivered by one or more nodes within the AWS. By definition of a military C2 structure, these nodes must be in a hierarchy as shown in Figure 10.3. The AWS is the Level 1 node which can be decomposed into several Level 2 nodes which are themselves decomposed into Level 3 nodes.

The aim is to establish the IHL requirements for the weapon system in terms that apply to the cognitive functions in the Level 1 node which is initially assumed to be human. The Level 2 nodes may be human or machine, the aim of this analysis being to understand their functions and the requirements on them if they are all implemented as autonomous functions. Section 10.8 looks at the issues if the Level 1 node, the AWS, is implemented as a machine under human supervision.

The functions identified by lawyers and discussed in Section 5.4.1 are target acquisition, tracking, selection and attack [7]. These assume a particular type of mission which has a relatively long timescale for continued observation of one or more targets, followed by selection of the target to attack. The Observe,Orient, Decide and Act (OODA) process described in Section 2.7 with discussion of automating it in Section 8.5 is more general as it does not make these assumptions. The relationship between the two is not exact, but for most purposes it can be taken as:

- Observe starts before and includes target acquisition;
- Orient includes tracking;

- Decide is equivalent to select;
- Act is attack or weapon release.

Therefore, we look at the functions underpinning the phases of OODA, their activities and how they can be expressed in an architecture.

Several terms have been used earlier in this chapter and are given in Panel 10.5 for consistency and clarity in comparing engineering terminology with the legal usage of the word:

Panel 10.5:   Definitions of terms used in deriving architectural requirements

A *phase* is a time-based section of a mission. There may be several roles performed during a phase. OODA describes four phases of a targeting mission.

A *role* is the part played by the weapon system in the system-of-systems. Surveillance is an example.

A *function* is one of the activities performed by the weapon system as part of its role. Target identification is an example. It is not the same as a *cognitive function* used in decision making.

Several *tasks* make up a function. 'Make a radar image of objects within a specified distance from a location' is an example.

A task comprises an *activity* and a *goal*.

Definitions of targets from STANAG AAP-06 are given in Panel 8.1, but it is sensible to add to these by defining non-targets as objects which are not to be attacked due to the nature of their use, treaty obligations or identified as being of limited or no military value.

10.6   Architecture requirements

10.6.1   *Roles and functions during each phase*

When the weapon system receives a targeting command to search for a target type in an area, but not necessarily at a known location, the Behaviour Generator will plan the mission as far as it can. The time for the Observe phase will not be known, but there will be fuel and survivability constraints on it. There will be generic plans for the remaining phases which can be used for planning purposes and will set some limits on the Observe phase. The Level 1 Behaviour Generator will assign tasks and their timescales. Figure 10.4 shows the OODA phases of a mission with the main roles during each phase and the information needed for the start of each phase. The functions are carried out by a node, with one or more nodes carrying out a role.

The systems, subsystems and primitives which carry out the roles of Surveillance, Monitor Localised Scene and Guide Weapon will depend on the specific domain and platform(s). There are fundamental differences in the engineering solutions for sensors,

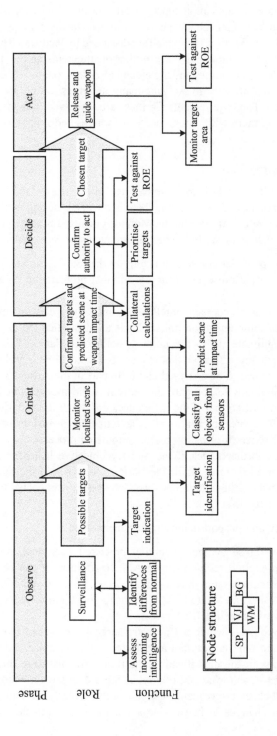

Figure 10.4 OODA process with roles and functions performed during a targeting mission. Each function is carried out by a node which has the structure shown in the panel

[Key: Value judgement (VJ), Behaviour generator (BG), Sensory processing (SP), World modelling (WM), Knowledge database (KD).]

weapon guidance and communication links for air, sea and land capabilities. The possible level of automation for each will also be different. However, there are generic requirements derived from Section 10.3 which are discussed in Sections 10.6.2, 10.6.3, 10.6.4 and 10.6.5. These requirements place further requirements on the data which must be in the Knowledge Databases and are given in Section 10.6.6.

The role Confirm Authority to Act has many activities which are currently performed by humans. The requirements for these are discussed in Section 10.6.4. Approaches to meeting them with autonomous subsystems and functions are given in Section 10.8.

10.6.2   Surveillance

The aim of the Surveillance role is to examine areas of air, sea or land looking for objects that may be targets, and then tagging them in the database and world models. They are examined more closely during the Monitor Localised Scene role. Military necessity requirements lead to the following requirements for Surveillance:

1. The system must be under control, so it follows that only designated areas are to be examined for the presence of targets and that protected areas are recognised as such;
2. The role must include accessing intelligence and other relevant information available over its communication links. This is to ensure that the target can be confirmed as legitimate up to the point t_0 defined in Panel 10.2. Network delays and data latency will mean that in practice the last time is earlier than t_0. These delays will need to be analysed and form part of the Article 36 Review;
3. There must be methods to compare the current world model with the initial model and criteria set so that if the differences are large, the superior node is informed. This is to prevent irrational behaviour due to misinterpretation of sensor and intelligence information by the automated processes;
4. Necessity forbids further attack if the military objective has been achieved. This leads to a need for a capability to perform analysis on the target scene after weapon impact if the autonomous weapon system is expected to perform a second attack if the first fails.

Distinction, humanity and proportionality requirements lead to:

1. An ability to recognise symbols denoting protected personnel, objects and areas. This is an explicit requirement in API. In addition, it can be expected that the system should be able to recognise military vehicles and combat dress used or worn by Blue and White forces. This is to give them the expected level of protection from attack in IHL;
2. An ability to tag objects which are potential targets. The set of likely targets will almost certainly be larger than actual targets due to noise and clutter as well as limitations in the classification algorithm. At this stage this is not a problem if the set is reviewed and revised as more data are gathered;
3. The system must check target identification results regularly up to time t_0. This is to ensure that changes in knowledge about the target and the target area

during later phases of OODA continue to confirm that attack with a lethal weapon is legitimate;

4. The world model must include some method to designate objects and personnel that are civilian and potential dual use objects;
5. An ability to update current world models with all available information at regular intervals;
6. The system must either identify areas where there will be little or no collateral damage if hit by the weapon, or have such places in its database. These are to act as safe areas if the weapon has to be redirected after release, or it cannot self-destruct.

It is highly likely that there will be objects which do not meet pre-set criteria as military objects but are not clearly civilian. These have to be classified as unknown, and treated as if they are military and hence potential targets. They will be passed to the Monitor Localised Scene role for further examination and must be classified later as targets or non-targets.

10.6.3 Monitor localised scene

This role is to take the unknown and indicated targets from Surveillance, identify them more accurately and tag civilian objects and personnel in the target areas. The monitored scene must be sufficiently large to include objects such as vehicles which may come close to the targets at the time of weapon impact.

The necessity requirements will be the same as those for Surveillance. The distinction, humanity and proportionality requirements will also include those given in 10.6.2 but there are additional requirements:

1. There must be a method to relate the observed scene to that expected and make a coherent situational analysis of it. This is a world model. It must contain time-related information to allow tracking of moving objects;
2. The system must be able to make predictions about the scene at weapon impact time;
3. Target indication becomes identification so there must be image and object recognition algorithms which can be applied to all relevant objects in the target scene(s) with previously agreed confidence levels for the positive identification of military objects;
4. All objects classified as unknown must be subjected to review with target identification processes and reclassified as civilian or non-target if they do not meet the criteria for classification as Red military objects;
5. The role must be able to recognise military objects which are part of its own force.

The humanity requirements in Section 10.3 mandate that the sensors continue to monitor the target area during the whole of the Decide phase of OODA. This is so that scene changes can be assessed and confirm that new, perhaps higher resolution information, still confirms that the object is the legitimate target as identified. There will probably be detailed differences in the scene surveillance procedures.

Identification of the non-targets will become more important for calculations of likely collateral damage.

10.6.4   Confirm authority to act

This role will have all of the requirements in Sections 10.6.1–10.6.3 plus:

1. Defined actions that it is authorised to take without reference to its superior node;
2. Defined actions that it must refer to its superior node and receive authorisation before taking them;
3. Defined actions that it can take subject to defined criteria. These criteria may come from calculations made by the node;
4. The role must include performing collateral damage calculations for every potential target. This will require blast range information for its weapons and some rules about damage levels to object types such as brick or wooden buildings, tents, buses;
5. The ability to calculate aim-point and impact-point errors for the weapons it can release;
6. The ability to divert the weapon to a 'safe' impact point and t_0 for each of these points if it cannot self-destruct;
7. Criteria based on rules-of-engagement for the criteria-based authority. This also places a requirement on the mission planners to provide rules-of-engagement in a way that can be interpreted unambiguously by an automated system without human assistance;
8. There is a need for a set of fail-safe actions at several points in the activities making up this function. These will need to be specified in detail for every implementation of an automated function in any weapon system;
9. The node must report all weapon-release decisions to its superior node;
10. There must be a method for the superior node to monitor lethal decisions and inject new data if necessary.

Implementation of the Confirm Authority to Act role will depend on the domain and assets. It is examined in more detail in Section 10.8 as it contains the roles and functions which rely most on human judgement and have difficult problems associated with automation, i.e. proportionality and humanity. Distinction is in Surveillance and Monitor Localised Scene; military necessity is an initial precept for the Strike node as the target list gives targets which are already authorised, based on human judgement at a more strategic level. Although it is an initial precept, it will need to be considered again as part of the final weapon-release decision.

10.6.5   Release and guide weapon

This makes the assumption that the AWS is a complex system with a separate weapon to release, such as a GPS-guided bomb or a missile with its own sensor which locks on to the target that it is given. If the AWS is the weapon and can autonomously interpret a scene and identify its target then this section describes the role executed by the human commander. If the AWS releases a weapon which can select its target, then the weapon must also execute its own 'mini-OODA loop' which must pass an Article 36 Review.

Implementation of this role will be very domain specific, but the following generic requirements can be derived:

1. Maintain guidance of the weapon as long as possible;
2. Continuously monitor the target area up to and, if possible after weapon impact;
3. The system must be able to redirect the weapon to hit a 'safe' contingency area if one exists, the target is no longer legitimate, and the redirection is possible.

10.6.6   Knowledge database contents

The Level 1 Knowledge Database has common elements across all functions. Its partitioning, if any, will depend on the application and implementation in that system. The database must have at least the following information:

1. Data on the symbols denoting personnel, objects and areas of special protection in a form suitable for all the sensors in the system;
2. A map of sufficient resolution for the weapons in the system;
3. Recent and updated copies of the force's recognised air, sea or land picture, depending on which environment the system is operating in;
4. Coordinates and extents of all protected areas;
5. Initial world model which was preloaded at the start of the mission;
6. Current world model;
7. Criteria for unacceptable differences between the current and initial world models;
8. Map of operational area with civilian and dual use objects tagged;
9. Distinguishing features of military objects and personnel;
10. Distinguishing features and behaviours of objects and personnel that make them potential targets;
11. Agreed criteria for accepting indication of military and unknown objects;
12. Agreed criteria for accepting identification of military objects;
13. Actions the node is allowed to take;
14. Rules-of-engagement expressed in terms that can be unambiguously interpreted by the node for the scenes it is likely to encounter;
15. Rules-of-engagement-based criteria to set the system's authority level;
16. Criteria for its authority based on laws and rules that are not part of the rules-of-engagement;
17. Actions the node can take without authority from its superior node (Its authorised power);
18. Actions the node cannot take without authority from its superior node (Every action except those explicitly stated in its authorised power.);
19. Set of fail-safe plans for contingencies;
20. Areas with no non-military objects in them to act as alternative fail-safe aimpoints;
21. Blast damage data for its weapons.

Clearly this list is not comprehensive, but it gives a minimum set of information which is required for an autonomous system. Even with this information set, it is not clear that all functions in the weapon system can be automated.

There is an assumption that the data in the Knowledge Database is not corrupted during the mission. Meeting this assumption is part of the overall cyber-security policy and implementation of the Blue forces.

It may be possible to write requirements for putting rules-of-engagement in machine-intelligible form and unambiguously stating criteria so that the system can have a high level of automation and hence autonomy. However, these will be very difficult to turn into algorithms that are acceptable to legal authorities. If this is achieved, the software implementation will have to have a very high level of integrity and reliability for making lethal decisions. Risk analysis may make it necessary for the algorithms and software to be defined as safety-critical before any use in operations.

10.7   Cognitive function requirements

The requirements from Section 10.6 can be merged and system requirements derived for each of the cognitive functions in Figure 10.2. These are given in Panels 10.6–10.9[4]. The requirements in the panels are indicative of those needed for a procurement contract but should not be taken as a definitive set.

Section 8.6 discusses issues for automating targeting processes with seven capability restrictions on an AWS in Section 8.6.3. These should also be included in any considerations of this section for an AWS, although most of them are included in the requirements below, or could be expected to be included in a proper engineering design process. They are also included in the discussion in Section 10.8.

Panel 10.6:   Sensory processing (SP) requirements

1. The SP must have the ability to recognise symbols denoting protected personnel, objects and areas;
2. If the SP has its own processing capability, the outputs must be fully compatible with the WM;
3. The SP must have the ability to classify objects as military, civilian or unknown;
4. The SP must have a target indication capability. This capability may also include indicating objects as unknown if there is a reasonable level of doubt about their nature;
5. The SP must have the ability to tag objects as military objects and if they are potential targets;
6. The SP must have a capability to identify targets to previously agreed confidence levels for objects in the approved target list for the operation;
7. The SP must be able to continuously monitor the target area up to, including, and, if possible, after the weapon impact.

---

[4]The 4D/RCS standard uses the abbreviations SP, WM and VJ for the functions in a node and these are used in the panels.

Panel 10.7: World modelling (WM) requirements

1. The WM must have a world model with time-related information and a capability to track moving objects;
2. The WM must be able to make predictions about the scene at weapon impact time;
3. The WM must have methods to compare the current world model with the initial model and have criteria for the differences that trigger reference to its superior node or to a human in its command chain;
4. The WM must have an initial world model to act as a starting point to interpret the SP outputs;
5. The WM must be able to access intelligence and other relevant information available over its communication links;
6. The WM must update its current world model with relevant and available information at regular intervals;
7. The WM must have a method to relate the observed scene to that expected and make a coherent situational analysis of it;
8. The WM must include some method to designate objects that are civilian and potential dual use objects;
9. The WM may need a capability to perform analysis on the target scene after weapon impact and the ability to confirm the success or failure of the attack.

Panel 10.8: Value judgement (VJ) requirements

1. The VJ must have criteria for its authority derived from the extant rules-of-engagement;
2. The VJ must have pre-set criteria for the decisions the weapon system is authorised to make and act on;
3. The VJ must either identify areas where there will be little or no collateral damage if hit by the weapon, or have such places in its database;
4. The VJ must review the identification of all objects classified as unknown with relevant new data and reclassify them as civilian if they do not meet the criteria for reclassification as military objects;
5. The VJ must have the ability to calculate aim-point and impact-point errors for the weapons it can release;
6. The VJ must be able to perform collateral damage calculations for every potential target.

Panel 10.9:   Behaviour generator (BG) requirements

1. There must be a method for the superior node to monitor lethal decisions under consideration by the BG and inject new data if necessary;
2. The BG must report all weapon-release decisions to its superior node;
3. The BG must have defined actions that it is authorised to take without reference to its superior node;
4. The BG must have defined actions that it cannot take without referring to its superior node and receiving authorisation before taking them;
5. The BG must have designated areas to be examined for the presence of targets;
6. The BG must not authorise attacks on protected areas, objects or personnel;
7. The BG must have a range of fail-safe actions that it will take if necessary;
8. The BG must maintain guidance of the weapon as long as possible;
9. The BG must be able to redirect the weapon to hit a contingency area, if one exists, when the target is no longer legitimate and the redirection is possible.

10.8   Issues in moving from automation to autonomy

10.8.1   Approach

The question to be addressed here is, 'how many functions in a weapon system can be made autonomous and still meet IHL?' The context has been set in Section 10.2 and 10.4, a first level architecture proposed in Section 10.5, with requirements in Section 10.7. It can be seen from Section 10.6.4 and Figure 10.4 that the node performing the Confirm Authority to Act role will have the majority of the activities which are currently carried out by a human. Any autonomous decisions made by the weapon system's node in this role will be based on information generated during the previous roles, so we must look at the reliability of the functions generating this information. This is necessary as a human will understand, and compensate for, shortcomings in these functions based on their experience as well as the stated performance. There will also be the information in the Knowledge Database and the rules-of-engagement as interpreted by the Behaviour Generator.

The Behaviour Generator will configure the weapon system for the Decide phase so that the IHL and rules-of-engagement criteria for weapon release are met. The domain-independent information flows are shown in Figure 10.5, the grey boxes denoting the Level 1 node's cognitive functions. The difference between an automated and an autonomous system is whether the Behaviour Generator's 'Release decision' can be made with or without human intervention, i.e. the node's Authority Level and the reliability of its Value Judgement cognitive function.

The tasks within the functions in Figure 10.5 may be delegated to Level 2 and lower nodes if necessary. If the cognitive functions are performed by humans, the same humans can carry out these tasks, configuring themselves as a set of Level 2 nodes. Similarly, the Level 2 nodes may be implemented on the same hardware in

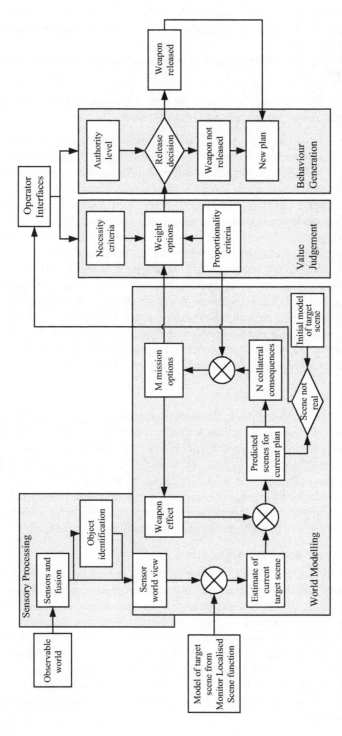

Figure 10.5   Information flows in the weapon system during the Decide phase. The grey boxes are the Level 1 4D/RCS node's cognitive functions

the weapon system as the Level 1 node. It should be clear that nodes at successively lower levels operate with shorter timescales and have less authority and freedom of action than higher ones.

It is useful to introduce the concept of *authorised po*wer here. It was defined in Panel 1.1 and 3.1 as:

> The range of actions that a node is allowed to implement without referring to a superior node. No other action being allowed.

At the level of the 4D/RCS primitives and servos, a node can only act in one way in response to a command. However, when more than one response is possible, the node is only authorised to instruct subordinate modes if they can all act without human intervention. The concept can be applied to the weapon system and nodes below it, its cognitive functions or functions within them. Authorised power will be used in the analyses in the rest of this chapter and succeeding ones.

The configuration of the weapon system as a Level 1 node will have been different during the preceding phases, but still under control of the Behaviour Generator. The AWS's responses to inputs whilst performing its roles in the preceding phases, are the subject of Sections 10.8.2–10.8.5. It is probable that a nation's armed forces will have guidance, if not actual rules, about confirmation of a target's legitimacy using several information sources. If possible, the organisation of the Level 2 node's tasks and the integration of their outputs into the Level 1 node should be compatible with this guidance. This will then enable IHL-compliant targeting to be an integral part of the operational use of the technologies.

10.8.2   Automating Observe

The observe phase is carried out in the surveillance role using functions analysing intelligence and observed information. It can be relatively straightforward in cases such as air-to-air combat, but very difficult in others, such as hitting a specific road vehicle with a small weapon in a cluttered environment. The requirements in Panel 10.6 can be implemented at some level with current technologies. The problem that arises is the reliability of practical identification algorithms and how to compare this with legal criteria. The techniques are different for targeting in each domain, giving wide differences in applications and inferences about target types.

10.8.2.1   Target and non-target types

The air-to-air and surface-to-air environments will have a limited number of object types to identify and there are a myriad of radar and electro-optic techniques for identifying them. It can also be expected that the recognised air picture will be available over a secure network such as Link 16.

The naval domain also has a limited number of object types with large military platforms being fairly easy to identify both from sensors and intelligence data. Smaller objects which may deliberately be designed to confuse are more problematic. Fast Inshore Attack Craft (FIAC) are the classic example of these as they can look like many of the large numbers of small craft operating at any given time in and around harbours.

Fixed ground targets such as communication towers, buildings, military bases and airfields are on most maps or can be inserted from surveillance information prior to any attack. The IHL problems arise if they can be used for civilian purposes as well as military ones. Consequently they may not be a legitimate target at the time of the strike operation. Mobile ground targets range from large mobile missile launchers down to individual road vehicles which may be deliberately concealing themselves in civilian traffic.

The information required to identify an object as a legitimate target will depend on the domain (land, air or maritime, both on and below the surface) and its type. This affects the weapon system's World Model and the information in Figure 10.5. As an example, ground targets have been split into categories for further examination below. The division is not precise and needs to be reviewed when considering an actual system being supplied under contract. Similar lists of target types can be readily drawn up for the other two domains.

1. Fixed ground targets

    Fixed ground objects' designation as targets, non-targets or dual-use objects can be made before the mission, and put into the initial world model with their category. The World Modelling function can be programmed to use the sensors and off-board data to monitor dual-use objects to detect parameters such as number of people entering or leaving it, or radio signals emanating from it, to confirm or change its status as military or civilian. The legal question would then be whether this change of status should be confirmed by a human, external Signals Intelligence (SIGINT) or left to the automated functions in the weapon system. System performance from analysis and trials will provide evidence for the review and also as guidance for setting rules-of-engagement.

2. Large vehicles

    Large vehicles generally have identifiable features. Some features such as tracks will indicate but not prove military use. Other features such as the multiple windows of a bus will be indicative of civilian use. Mobile missile launchers will have distinctive features from the missiles, but there will need to be trials proof of an automatic identification algorithm's ability to distinguish between them and a large fuel tanker. Off-board data can be used as part of the assessment, for example, detection of radio transmissions from the vehicle as part of a hostile C2 chain.

    It can be postulated that it will be possible to identify some large vehicles with sufficient precision for an automatic weapon release to be considered, but this is not taking into account the collateral damage that weapon impact may cause.

3. Tracked vehicles

    Tracks are used by construction and agricultural vehicles as well as military ones, so their detection by, for example, radar will only be indicative of military use. The architecture should ideally include identification by a different algorithm, for example, one looking for gun barrels or missiles as well as tracks. The vehicles behaviour may also be significant, for example, if it is accompanied by a radar or communication vehicle.

4. Small vehicles

Identification of an object as a small vehicle should be straightforward, all small vehicles could be categorised as possible targets for further examination. The problem then becomes one of identifying a specific type of vehicle among many hundreds of types. Unless the identification is of an explicitly military type, for example, a mobile anti-aircraft gun, there are many reasons why it should not be attacked, even if there is a belief that hostile forces are using it. Changing the object's category from small vehicle to that of legitimate target will almost certainly require mission-specific rules-of-engagement.

5. Tracking moving objects

The world model includes time information from sources such as sensors that can track moving vehicles and people over useful timescales. This enables inferences to be made about the intent of the people or the driver of the vehicle. The behaviour of the object under surveillance may fit the anticipated behaviour of a hostile combatant and allow its indication as a possible target. Any attempt to make an identification as a target by an algorithm would need careful discussion with legal advisers.

6. Non-military objects

The pragmatic approach to identifying non-military objects is to apply the following rule: anything that has not been identified as a target or non-target to the agreed confidence level for the operation is a non-military object and tagged as 'unknown'.

Some types of non-military object can be identified by the same algorithms used for target identification. It is unlikely that all types will be able to be identified. Current identification techniques require a database of signatures or templates that are matched with the sensor outputs. The consequence is that all non-military objects must be considered in the collateral damage calculation even if they are not identified as a specific type of vehicle.

7. Safe areas

These are areas which do not have any non-military (i.e. civilian and unknown) objects in them. They become alternative aim-points for the guided weapon after its release should the original target area become illegitimate due to unforeseen changes such as the appearance of a civilian object near it.

Identification algorithms have used the technique of matching key image features to those in a database accessed during the image processing. The recent developments of image recognition using machine-learning algorithms, a subset of AI, brings a new technique to object identification, but does not remove the need for a large set of target and non-target information. This will be used for training the functions carrying out the identification rather than for direct comparisons. Both techniques will only be as reliable as the available databases of targets in military scenarios. (Learning systems are discussed in Section 10.10.)

Summarising the above, it is possible that a relatively small number of target types may be reliably identified as legitimate targets automatically. The weapon system's sensor suite and processing should be supplemented by, or include, inputs from external sources over the host platforms' datalinks. This may increase the probability of correct identification as a target.

The Sensor Processing function does not include collateral damage calculations. It cannot authorise any weapon release although it does provide the scene-related information for these calculations. Consequently, its reliability and accuracy should be considered in reviews of the AWS' critical functions.

The design and test processes will produce confidence levels for indication and identification of objects as a particular class or type of object. The confidence levels are highly likely to become part of the data used by one or both of the Value Generator and Behaviour Generator cognitive functions. The values will be used to generate the list of possible targets for further analysis. Therefore, they have become a part of the targeting decision process.

10.8.2.2 Authorised power for Observe phase

The authorised power of the Observe phase will be those of the Sensory Processing cognitive function and are:

1. Confirm fixed ground objects as ones already on the pre-loaded map;
2. Indicate and tag objects as possible targets if the confidence level of identification is above an agreed value;
3. Identify and tag objects as a specific type if the confidence level of identification is above an agreed value;
4. Tag objects as legitimate targets if they have been identified and meet agreed criteria from the Knowledge Database;
5. Tag objects as non-targets if the confidence level of identification is above an agreed value;
6. Tag all other objects as unknown;
7. Search for, and identify areas which have no non-military objects in them to act as alternative 'safe' aim-points after weapon release;
8. Pass indicated and identified target lists to other cognitive functions.

The definitions and limitations will be clear to the engineers implementing new targeting algorithms. There are risks with reuse of existing algorithms; they may have been produced as decision aids and have less precise definitions, or had less rigorous testing than required for autonomous operation. It is essential that all limitations are captured in documentation which will be available to military as well as technical staff, such as military handbooks. They should be made available for the Article 36 reviews and for future reference when rules-of-engagement and upgrades to autonomous target selection are considered.

10.8.3 Automating Orient

10.8.3.1 Automating Monitor Localised Area Role

This role has two tasks: carry out more detailed surveillance of the smaller areas around the indicated targets, identified targets and unknowns; and to classify the objects as either civilian non-targets or as military targets. The resulting set of targets is then passed to the Value Judgement function for decisions about actions. The original operational analysis and system design should have ensured that this set is small. If it is, continued detailed surveillance and application of target identification algorithms can be carried out in a short enough time to meet the military requirements of the attack.

The first step must be for the Behaviour Generator to take the list of possible targets, estimate the times for identification, for detailed surveillance of the areas around the targets, and activities for collateral damage calculations. There will be a balance to be struck between numbers of identifications, areas to be surveyed and collateral calculations. The reference time will be the time between starting the Orient and Decide phases allowed in the mission plan.

There must be a contingency plan if the time required is longer than the planned time. This could be to prioritise some types of indicated targets according to pre-defined criteria, or to refer to a higher node or human. Again, as in the Observe phase, the algorithms become part of the targeting decision process. This is not a technical problem if the possible-target list is a decision aid and presented to a human for decisions, but could be a workload problem. If autonomy is considered for Orient, the Article 36 review will have to identify the algorithms to be reviewed and the criteria for their acceptance. Again, information will have to be documented in a suitable form for legal and military reference when rules-of-engagement are drawn up.

10.8.3.2   Predict scene at impact time function

The format and structure of the world model will be specific for the domain and weapon system under consideration. However, some elements will be in common. Those coming from the requirements listed in Section 10.6.4 are a minimum set. The world model will be built up by a process similar to the one shown in Figure 8.3 and discussed in Section 8.5.4.

It is fairly straightforward to build a model of the scene at the time of the main observations. It is possible to build up a knowledge of the local 'patterns of life' for some scenarios which both improves predictions and highlights deviations from normality. However, the scene will evolve even without any influence from the assets in the strike force.

Predicting scene changes will be problematic even with knowledge of the velocity vectors of the moving objects. There will be errors due to measurement uncertainties in the sensors which can be calculated but there will be additional uncertainties due to changes in the speed and direction of the objects in the scene which may not be predictable to an external observer. It may be possible to put upper limits on their later positions due to their manoeuvrability.

Updates from all surveillance assets and sensors will be required, their timing will be part of the system design requirements. The errors and uncertainties may be partially compensated by reducing the scope of the world model to the area around one or two objects with increased surveillance of them. Military needs for continued surveillance of potential threat areas may make this impossible.

This function and its implementation will be specific to the weapon system and system-of-systems. It will have to be tailored to the likely operations, their scenarios and the weapon system.

The world model will be used for selection of non-targets for collateral damage calculations, and hence part of the proportionality decision. Therefore its provenance and reliability will need to be part of the Article 36 review. Given the nature of the uncertainties, its performance will need to be understood through its use in trials and military exercises, so there will need to be an on-going dialogue between users and suppliers.

10.8.3.3 Authorised power for Orient phase

The authorised power in the Orient phase will include those in Section 10.8.2. Additional powers will be:

1. Choose suitable world model for analysis of target scene;
2. Decide what information is needed during the Orient phase to identify targets and plan its collection;
3. Implement agreed contingency plan or plans if there is insufficient time for full scene analysis;
4. Make predictions of scenes at future dates and compare with sensor data;
5. Log errors in world model based on measurement uncertainties;
6. Identify scene changes outside the predictions from measurement uncertainties and include in world model;
7. Produce predicted maps of areas around individual targets at weapon impact time. These will cover the area affected by the weapon and classify all objects in it;
8. Refer decisions to superior node if the agreed criteria about the scene are not met.

10.8.4 *Automating Decide and Act*

The node's behaviours in these two phases are dominated by the elements shown in the Value Judgement and Behaviour Generator cognitive functions in Figure 10.5. They make decisions and take actions based on the inputs from the world model predictions. It should be clear from the discussions in this chapter that if this phase is conducted by humans using their judgement and the automated inputs are treated as decision aids, then compliance with Article 35 should not be a large problem. This is not the case if the inputs come from autonomous functions and this phase will be executed autonomously, i.e. no human input to the decisions made by the weapon system.

It should be clear from Sections 8.5.3 and 8.5.4 that it is possible to automate many routines within the functions in the Observe and Orient phases. It may be possible to express rules-of-engagement in a way that can be interpreted unambiguously by an algorithm for some of the routines which would make them compliant with IHL. Both sections have shown that these routines will be part of the targeting process and must be reviewed under Article 36.

Sections 8.5 and 8.6 also discussed the problem of errors in the outputs from the preceding phases. Some can be expressed in standard statistical parameters generated by the measurement algorithms. However, some errors such as predictive errors due to interactions between objects in the scene will be very difficult to quantify. Progress in the civil autonomous vehicle industry may provide solutions in the longer term, but they may not be applicable for military situations.

All the errors and uncertainties come together in the Value Judgement cognitive function. Their types and reliability will depend on the domain and system implementation. However, some general guidelines for the node's authorised power can be made here which will be applied in Chapter 11.

Rules-of-engagement can be set that define the type and numbers of non-military object which can and cannot be damaged by the weapon's effects. These parameters can then be put into the Knowledge Database as proportionality criteria. The node's authority level in the Behaviour Generator will then examine the quantified errors and uncertainties in the accompanying data about the objects in each target area. If there are no unknowns, then it may be possible to make a proportionality decision to release or withhold the weapon.

Military necessity criteria are set during the preparation of the target list, so if the system is able to monitor the list, either adding or deleting items during the mission, then some necessity criteria are met automatically. Ensuring the system is under control should be part of the continual monitoring of the integrity of the weapon systems C2 and data links and their bearers. Again, it will be possible to derive parameters to allow automatic checking of the necessity criteria.

When an air-launched laser-guided weapon is released for hand-over to a forward air controller, the aim-point is replaced by the controller's capture-basket for the weapon, with contingency plans if the controller fails to take control.

Assuming the weapon system is responsible for the weapon up to impact, the system will need to monitor intelligence information and the area around the aim point up to time t_0 so that it can guide it to a pre-determined safe aimpoint entered in the Knowledge Database from earlier phases.

It must be assumed that there will be reviews of all release decisions, so it is essential that the system keeps a record of all data used at the time. This gives a requirement for configuration control of the Knowledge Database and the C2 status up to and including the weapon's impact time.

10.9   Authorised power for a Level 1 node

The Behaviour Generator cognitive function in Figure 10.5 has an authority level which has the information needed to make the release decision. This will be all the criteria which must be met by the data used in the Value Judgement decisions in weighting options. It is essentially the summation of the authorised powers of all the functions used by the system in the mission and is the basis of the definition of authorised power as defined in Panel 1.1 and 3.1. The Level 1 authorised power for any autonomous system defines the actions which it can take. In any given application area, it is based on, at least, the ethical, legal and regulatory requirements mandated for that application, and not based on one particular branch of applied ethics.

Defining the parameter set that must be included in the node's authorised power will be one of the most important considerations for the procurement authority in placing any contract for a highly automated weapon system. The allowed values of error and uncertainty for fully automated or autonomous operation will be the subject of detailed Article 36 review. Only nodes that pass the review to deliver autonomous functions can be on separated from the human functions. This is consistent with the requirement for clear separation of human and machine functions in the architecture and its implementation.

The authorised power and authority level of each system will depend on the details of its implementation. More detailed discussions are given in Chapter 11.

10.10   Learning systems

It is likely that AI learning systems will be considered for functions in an automated weapon system. They may be considered suitable for target and non-target identification activities, and the subjective Value Judgement activities. There are both technical and contractual difficulties which must be considered as well as the IHL-based cognitive function requirements in Section 10.7. The discussion here should be supplemented by the ones in Sections 3.4 and 3.5.

Any learning system requires a large reliable dataset for training. This will need to be representative of actual mission conditions including noise and clutter. Generation of an unbiased reliable dataset will be a large task. Some nations may have large datasets for target identification, but the algorithms' reliability for classification of civilian and unknown objects will need to be verified to the same level as that for military targets. Similar considerations will apply to any algorithm used for assessing proportionality criteria, but the training data may be more difficult to find or generate.

Procurement contracts must have test criteria that depend on repeatable results in similar conditions. This may not be possible with a learning system that adapts its results on the basis of new data. The alternative is to freeze the algorithms at the point of contractual handover from supplier to customer. If this is the case, then the system no longer learns and it can be tested and evaluated as any other firmware and software system. The acceptance criteria will then include specifications for that issue of the network and algorithms and the training data.

Article 36 reviews would be in place in the procurement process, but would need to take special note of the AI part of the system. Any further software upgrades from additional learning would probably need to be the subject of further Article 36 reviews except minor ones in object identification and classification.

References

[1]   Albus J., Huang H-M., Messina E., *et al. 4D/RCS Version 2.0: A reference model architecture for unmanned vehicle systems*, NISTIR 6910, National Institute of Standards and Technology, Gaithersburg, MD, USA, 2002.

[2]   UK Ministry of Defence. *Joint Services Publication 383 (JSP383), The joint service manual of the law of armed conflict*, 2004, with amendments of 21 May 2014.

[3]   US Department of Defense. *Law of war manual.* December 2016. Available from https://www.defense.gov/Portals/1/Documents/pubs/DoD%20Law%20of%20War%20Manual%20-%20June%202015%20Updated%20Dec%202016.pdf?ver=2016-12-13-172036-190 [Accessed on 14 May 2017].

[4]   Albus J., Barbera T., and Schlenoff C, 'RCS: An intelligent agent architecture'. *Intelligent Agent Architectures: Combining the Strengths of Software Engineering and Cognitive Systems*, AAAI Press, no. WS-04-07 in AAAI Workshop Reports, 2004.

[5]   Schlenoff C., Albus J., Messina E., Barbera T., Madhavan R., and Balakirsky S. 'Using 4D/RCS to address AI knowledge integration'. *AI Magazine.* 2006. Vol. 27(2): pp. 71–80.

[6]   Gillespie A.R., and Hailes S. In preparation.

[7]   International Committee of the Red Cross and Red Crescent (ICRC) Report 4221 *Autonomous Weapon Systems: Technical, Military, Legal and Humanitarian Aspects*, Expert Meeting, Geneva, Switzerland, 26–28 March 2014, Part III, page 62.

Chapter 11

Systems engineering for a new military system

11.1   Introduction

Earlier chapters in this book have discussed concepts and general principles that apply to the development and procurement of autonomous systems. Their systems engineering implications only become clear when applied to a specific problem. This chapter and the next one take an illustrative example of a hypothetical new air-to-ground targeting capability. The chapter is fairly detailed in order to show the range of complex issues and problems that arise in introducing a new system into an existing infrastructure, in this case a military one. Similar complexities will arise in most domains and must be solved by their systems engineers.

Chapters 8 and 10 show that targeting and the release of lethal weapons are the most difficult areas of International Humanitarian Law (IHL) for engineers and users. Section 8.5 shows some of the problems in automating the Observe, Orient, Decide and Act (OODA) targeting process.

This chapter considers solutions to the targeting problem which maximise the level of autonomy in critical functions and the limits which come from Articles 35 and 36 of API in the Geneva Conventions. It follows the logic of Chapters 4 and 5 to find a solution based on current in-service technologies. The capability requirement is turned into a system that can be integrated into a nation's military forces. This requires both satisfactory technical performance and demonstration of compliance with IHL through the mechanism of Article 36 Reviews[1].

The application of most of the principles is dominated by the contractual issues in procuring the systems, subsystems and modules which meet the engineering requirements, with many problems only arising during system integration.

Article 36 Reviews are considered at two programme points although there will be more. These are chosen to show how the legal issues demand clauses in one contract chain right down to the specialist company delivering algorithms which are key to the IHL distinction principle.

Section 11.2 shows how the problem is analysed and the capability defined so that it can be delivered by an Unmanned Air Vehicle (UAV) system integrated into the nation's and coalition's existing Command and Control (C2) infrastructure. The UAV system is the system of a UAV, its ground station, the control and data links between the two and its operators.

---

[1]See Sections 7.11 and 8.6 for the explanations of Articles 35 and 36.

Section 11.3 gives the solution that will be the subject of contracts awarded after a tendering process.

Section 11.4 shows how the system-of-systems delivering the capability is divided into systems with procurement contracts based on technical, military and financial considerations. For simplicity, the solutions are initially based on current automated technologies. The critical areas for the necessary Article 36 review are discussed in Section 11.5 in the context of representative contractual arrangements and how the necessary technical evidence can still be supplied by the engineering teams.

The example is for the air domain, but very similar processes will be followed for new land and maritime capabilities. Generic problems, for example the unaffordability of the perfect solution and integration into the existing coalition command structures, will be the same. Maritime and land procurements are also constrained by existing commercial arrangements for the supply and maintenance of almost everything that interfaces with the new equipment. This potentially leads to non-ideal supply chains which could give difficulties in delivering adequate technical evidence for Article 36 reviews unless explicitly addressed.

Chapter 12 covers the problems when significantly more automation and autonomy are introduced into the targeting chain, triggering detailed Article 36 reviews. These reviews drive many of the decisions made during the procurement process and the testing and trials of the new systems, so that adequate evidence can be presented at the reviews. The procurement programme is constrained by the contractual decisions made in Chapter 11. Chapter 12 examines autonomy in all critical functions and whether they can or cannot meet Article 35 and 36 requirements.

11.2   Analysis of the problem using Chapter 4

11.2.1   The state's problem

The process starts with the (fictitious) national government assessing likely threats and possible reactions to them. It is assumed that the state will be one partner in a coalition of several nations. Political assessments of likely opponents and initial military planning show that it will be necessary to capture and occupy land held by an opponent or alliance with an equal level of military capability. The opposing nation has a coastline and navy, so land, sea and air capabilities must be assessed.

Military staff decide that all campaigns will necessitate the coalition's forces (Blue[2]) making an opposed landing on the shore of the hostile nation (Red) near an urban conurbation. Blue will not have air dominance, so their air assets must be capable of operating in contested airspace. Red will defend territory with air strikes and the use of naval and ground forces close to the landing area. Figure 11.1 gives a schematic representation of the scenario.

The state's military staff are very concerned about a new weapon system that will be deployed by Red and which may be supplied to other potential adversaries

---

[2]This book follows the conventional terminology of Blue for own assets, Red for hostile and White for the UN and Non-Government Organisations (NGOs).

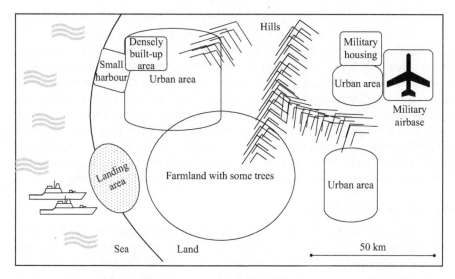

Figure 11.1   Littoral invasion scenario

by its manufacturers. This is a very accurate missile with a range of 50 km fired from mobile launchers. It represents a significant threat because the missile has a very low radar cross section and Infra-Red (IR) signature. Further refinements include the ability for the missile, when in flight, to introduce manoeuvres to confuse counter-measures. These make the missile very difficult to detect after launch and almost impossible to counter with a kinetic response whether guided or unguided.

The tracked launcher is about 7 m long and can travel at speeds up to 80 km/h. It is anticipated that the launchers will be hidden in both urban and rural areas, appear briefly to launch their missile(s) against the Blue forces, then either return to their hide, or move to a new location. Intelligence information shows that Red has a large stock of missiles, but few launchers, making launcher destruction a high priority.

The launcher can fire up to four missiles against different targets in less than 1 min. It does not need to stop to fire, but must limit its speed to less than 15 km/h and be on a straight, smooth track for 30 s prior to release and continue on the same track during release. A second multiple launch may be made within a low number of minutes of the first or, more likely, a few tens of minutes later from a new site. The launch direction is anywhere within ±90° of the line to the target. There can be obstacles such as high buildings on either side at missile launch but no obstacles within a few metres of the launch path. The launcher has the capacity to jam or spoof Global Positioning System (GPS) signals in its immediate vicinity. This precludes the use of a GPS guided bomb from a weaponised surveillance UAV even if it has the capability to predict, at bomb release time, the GPS coordinates for the moving target at the time the bomb reaches the ground. There is a high risk of unacceptable collateral damage if the launcher is attacked in an urban area with any weapon that cannot be guided onto the target. The launcher is not armoured and can be destroyed using a small air to ground anti-armour missile already in the nation's inventory.

Blue's only current way of countering the threat is to use high-altitude-long-endurance surveillance UAVs which are operated from remote UAV ground stations, and call in a fast-jet to attack it with whatever suitable ordnance it is carrying. This is unlikely to be effective as most fast-jets will only be in the operational area if they are already on a mission which may be higher priority than attacking a missile launcher. Also it is likely that they will be carrying GPS-guided ordnance, increasing the chance of unacceptable collateral losses, especially with likely restrictions on weapon use in crowded urban areas. Blue anticipate that Red will recognise these restrictions and use urban areas for launch sites increasing the justification for a new capability.

11.2.2   Required capability

This is stated to be the capability to detect and destroy a launcher over the whole area where it will be able to attack Blue forces. It is highly desirable to destroy a launcher before it fires its missiles but this is not essential. The strike capability can be delivered by fast-jet or weaponised UAV, but the number of available fast-jets with suitable weapons is limited.

An additional required capability is to be able to detect and attack other types of military vehicles under similar conditions. This is so that the system delivered by the procurement contracts is not so specialised that it can only be used against one target type.

The capability includes being effective for operations in day, night and all-weather (DNAW[3]) conditions. The anticipated littoral operations will be in areas where cloud cover can be anticipated for a significant number of days every year, so DNAW capabilities are given a high weight in all the analyses.

11.2.3   Initial operational analysis

The question to be answered is whether Blue can reduce their losses from the Red mobile missile launchers to an acceptable level without losing too many surveillance assets. This is posed as the Sponsor Problem in Figure 4.1 and the NATO Code Of Best Practice process is followed, as described in Chapter 4. The analysis starts with existing results from campaign planners using war-gaming techniques based on the assets currently available or anticipated to become available to both sides.

The balance of air power leads to the conclusion that the majority of manned aircraft operations by Blue will be:

- Limiting the effect of Red air strikes on the Blue littoral forces. This will mainly be by air combat;
- Air strikes on Red ground and sea forces;

---

[3]The strict definition of DNAW is Day Night All Weather. The different environmental conditions in DNAW operations usually require different technical solutions. 'Night' implies infra-red sensors and 'All-Weather' implies high-resolution radar or the ability to operate below low cloud and with infra-red sensors.

- Air strikes on Red airbases;
- Suppressing Red's air defence systems which will be a combination of local weapons and a larger integrated one comprising surveillance and tracking radars supporting a network of Surface to Air Missile (SAM) systems.

It is unlikely that sufficient manned aircraft will be available to patrol effectively for the missile launchers. The Red air defence capabilities will force Blue's large manned area-surveillance aircraft to stand off and be fully occupied supplying information to landing-area commanders about Red air strikes and naval threats.

11.2.3.1 Current Blue assets

The starting point of the analysis is to list the relevant assets that could be used to provide the desired capability. These are given in Table 11.1. The most cost-effective solution will be to provide the capability by reassigning or modifying current equipment and capabilities. Procurement of a completely new weapon system can be expensive and take many years to come into service.

The nation's armed forces will have both dedicated data links and one or more secure networks. The assumption made here is that the campaign will be conducted with the forces acting as part of a coalition. Most assets will be part of a coalition

Table 11.1   Current Blue assets that can be used to provide the capability

Asset	Role	Control and data distribution	Comments
Satellites	Regularly updated results from surveillance of operational area	Coalition intelligence centre. Data distribution over coalition network.	Requires very light or no cloud cover over target area.
High-altitude-long-endurance UAVs	Continuous surveillance of operational area	Remote ground station. Data distribution over coalition network.	Requires very light or no cloud cover over target area unless equipped with synthetic aperture radar.
Medium altitude surveillance UAVs	Local surveillance for limited durations	Re-locatable ground station. Data distribution over coalition network.	Satcomms link between UAV and ground station. May fly below cloud cover, but then vulnerable to ground defences.
Tactical UAVs	Tactical ground strikes using on-board missile with homing head	Re-locatable ground station. Data not normally distributed beyond local commanders.	Uses direct radio link between UAV and ground station. Can operate at low levels below cloud cover.
Fast-jets	Ground and air attacks of Red assets	Pilot and air command centre. Information passed over voice and Link 16 communication network.	Ordnance mainly GPS and laser guided weapons. Small homing-head guided missiles also used.

linked with a system such as Link 16. This is the communication network used across NATO with most assets being connected to it. A useful overview can be found in Reference [1].

If the capability is needed urgently, modifications to existing equipment may be the only solution and met using Urgent Operational Requirement (UOR) processes[4]. The UOR deliverable will still have to be reviewed under Article 36, and rules-of-engagement criteria rapidly and rigorously identified.

There is an assumption that Blue will have continuous surveillance of the operational area from satellites and high-altitude-long-endurance UAVs. The surveillance UAVs operate at altitudes above most, but not all of Red's surface to air missiles, and have no defensive aids. These are operated from remote ground stations and the information passed to the relevant local command centre. There is some level of target indication in this information but legitimate targets must be confirmed before a strike can be authorised. Fast-jets and weaponised UAVs may be cued to these potential targets by local commanders, but rules of engagement will require that there is a confirmatory identification before weapon release. The confirmation may be by the attacking aircraft or from another nearby surveillance asset. The fast-jets keep out of Red's air defence missile range for as long as possible, then come in quickly for ground strike or air-to-air combat and pull out rapidly when their mission is completed.

The disadvantage of using the high-altitude-long-endurance UAVs is that their main mission is to give oversight of the contested beachheads and large scale movements of Red forces. This means that they will not be available to cover the large areas where the missile launchers can hide. Their electro-optical sensors[5] will be unable to give results if there is cloud cover. Synthetic aperture radar sensors, if fitted, can give high-resolution images of limited areas of the ground. However, this requires straight and level flight for a period of tens of seconds or minutes, which may not be acceptable in contested airspace.

Blue have two types of UAVs in their inventory suitable for this task. Both have one dedicated ground station for each UAV in use at the time. The first is a medium-altitude UAV, mainly used for surveillance under the command of a ground station which can be located anywhere as it uses a Satcomms link to the UAV. It is designed for flight times of several hours at medium altitudes but is not very manoeuvrable. It can be role-fitted with an optical, IR or synthetic aperture radar sensor. As with the high-altitude-long-endurance UAVs, the raw sensor data is sent over a dedicated datalink to the ground station who distribute it to the relevant local commanders. The UAV has been designed assuming air dominance in its operational area which will not be the case in this scenario. Its design does not include a defensive aids suite and its poor manoeuvrability limits its usefulness as a weapon platform for difficult targets such as the missile launcher. Some of Blue's

---

[4]UORs are the UK equivalent of the US Urgent Operational Need (UON).
[5]The term 'electro-optic sensor' covers visible and near infra-red wavelengths, but often is taken to mean a sensor that responds to visible wavelengths.

inventory of these UAVs can be fitted with a GPS-guided bomb, but these are not acceptable for use against the missile launchers.

The second type of UAV is a tactical one. It is smaller and designed for tactical ground strikes. It can operate at higher speeds and lower altitudes than the surveillance one. It has a smaller airframe and has a shorter endurance of a few hours, but has been designed assuming that it will operate at low altitudes. Its principle weapon is a small air-to-surface missile with a homing head and a range of a few kilometres against targets such as armoured vehicles. The ground station must be located within a few tens of kilometres of the UAV to remain in line-of-sight radio contact with it as it does not have a satellite link. The missile's homing head is cued on to its target by commands from the operator in the ground station over its dedicated command link. They are usually operated as tactical weapons under the land force commander to support ground troops.

Both types of UAV are capable of launch and recovery from simple facilities such as wide roads. Neither have any on-board defensive aids or Link 16 for direct communication with other assets. Both have dedicated ground stations with a direct control and video link to their UAV. The surveillance UAV's ground station is transportable, but is usually located at a remote base and distributes its information over the operational network. The tactical UAV ground station has to be located locally due to its need for line-of sight comms. Its data is not normally distributed beyond its ground station and local commanders as its role is to destroy hostile assets in support of local troops.

11.2.3.2   Initial analysis results

The first results from the operational analysis are the limitations on the ability to destroy the missile launchers with the use of current assets. These are:

- The number of UAVs that can operate in any area is restricted to the number of operational ground stations. Very few ground stations will be available locally due to competing pressures in supply ships to the beachhead and very limited space on board ships or air command aircraft if the UAVs are to be operated from them;
- The available radio spectrum for communication links will also limit the numbers of UAVs that can operate in any given area. (All communication links will be in heavy use during this operation due to the number of assets and rapidly changing scenarios);
- The use of optical or IR sensors monitored by the UAV operator gives a low probability of detecting any missile launcher as they only appear briefly. This gives a requirement for rapid repeat surveillance of likely areas, increasing the number of surveillance UAVs for acceptable coverage;
- The use of UAVs on patrol in shared airspace will restrict the operational freedom of fast-jets due to the probability of collisions;
- There is a, probably acceptable, 'friendly fire' risk of loss of both types of UAV due to artillery or naval fire from Blue forces against Red forces;
- The surveillance UAVs have no self-defence capability rendering them susceptible to losses from common Red anti-aircraft weapons;

- The tactical weaponised UAVs have no defensive aids and are susceptible to many Red threats due to their operation at low level. Their vulnerability and the number required to cover the necessary area make them unacceptable as the sole method of detecting missile launchers;
- All UAVs have no Link 16 communications capability as their use until now is for surveillance in segregated airspace, with data passed to other units via their ground stations.

11.2.4   Refining the operational analysis

The next step in the operational analysis is to look at possible new capabilities that will increase the effectiveness of operations to destroy the missile launchers. The analysts will need to liaise closely with military and technical advisors during this phase. The analysis will also be guided by the nation's long-term defence procurement programme and realistic budgets. Although not a direct part of analysing a likely scenario, the business case will be stronger if any new assets, or modifications to the existing ones, will improve other capabilities for other missions and increase flexibility for the nation's armed forces.

It is decided that the design and build, or purchase of a fleet of weaponised UAVs that can operate with high survivability in hostile airspace is not affordable when compared with other requirements in the state's defence inventory. Consequently, solutions must be based on maximising the use of available and affordable assets.

The first approach to the analysis is to increase the size of the existing fleet of medium-altitude surveillance UAVs. These will monitor activities over the whole area where the missile launchers can attack Blue forces. Their role is to identify the missile launchers when they leave their hide and call up a strike by a tactical UAV from the land forces. The number of UAVs required has to include predicted attrition rates due to Red action as well the number required to survey the areas with a medium to high probability of the presence of a missile launcher.

The operational analysis then examines a range of technical changes that can be made to the current assets and C2 infrastructure to improve effectiveness. An initial list of measures of merit will be compiled to assess relative effectiveness of different changes. Eventually this evolves into the ones discussed in Section 11.2.7. The main conclusions from the analysis include:

- Surveillance using optical or IR sensors will require low altitude operation if there is cloud cover, or close-in observations are needed to confirm the target. Operation at altitudes of a few hundred or a thousand feet will give the UAVs some screening from the Red area surveillance radars which are set back from the coast to give better defence against Blue air or missile strikes;
- Potential operational areas are likely to have extensive cloud cover and rain for many days every year, sometimes with cloud cover persisting for several days;
- An alternative to low level surveillance is fitting available medium-altitude surveillance UAVs with a synthetic aperture radar capability;
- Red may attack Blue UAVs by the use of SAMs. Losses are mitigated to some extent as the Red air defence system will have difficulties dealing with both

Blue's general fast-jet air support to ground operations and multiple surveillance UAVs. Defensive small arms and machine gun fire by Red may also be effective in some circumstances;

- Multiple surveillance UAVs will need to operate over Red territory, the likely number is around three, assuming that intelligence information or military assessment will reduce the likely operating area for the missile launchers;

- The number of surveillance UAVs exceeds the capacity of any local shipborne or airborne asset to host their ground stations;

- Tactical UAV ground stations will be brought into the operational area the early stages of the invasion as their normal use is support ground troops. The extra requirement for this capability is for the ground statin to be located as far forward as possible in order to have line-of-sight communications with the UAV when operating tens of kilometres into hostile territory;

- Human–machine interfaces will be critical due to the anticipated high workload on the operators with multiple UAVs of two different types, the tactical and medium level surveillance ones, having to cooperate closely;

- Launcher detection using current target indication algorithms is difficult due to the low Technology Readiness Level (TRL) of the algorithms, and the trade-offs between a low false alarm rate and high detection probability. A supplementary method would be to use a missile launch detector triggered by the launch flash of the missile. These are part of some defensive aids suite suites;

- When cued, the tactical UAV will carry out a close-in missile strike. Final confirmation of the target and weapon release will be by the tactical UAV operators under most circumstances. There is the possibility of the surveillance operators remaining with 'eyes on' the target and commanding a weapon release from a tactical UAV, but this is unlikely;

- It is not essential for the launcher to be hit before the first missile launch as Red has a limited number of these assets, so a high kill probability of the launcher is more important than a kill before launch;

- The surveillance UAV will require some form of defensive aids suite to reduce losses to an acceptable level;

- If it is not possible to fit an adequate defensive aids suite to the surveillance UAV, the alternative is to fit a different engine and stand-off at higher altitudes until cued in to search a specific area;

- Surveillance UAVs will be lost, but the loss rate is acceptable with three UAVs airborne at any time as it will still ensure destruction of all Red missile launchers over an acceptable time;

- The tactical UAV should have some defensive aids fitted, but this may not be possible due to the limited size of the airframe;

- Existing tactical UAVs cannot achieve high altitudes, so will have to stay airborne in relatively safe medium and low altitude areas whilst awaiting strike instructions. (Safe areas for a UAV will have higher risk levels than is acceptable for a manned aircraft.) Even there, they will require some level of defensive aids suite which may release chaff or flares, or cause rapid manoeuvres. Consequently, its autopilot will need to be much more capable than its current one;

- The data rate for live video feeds from airborne UAVs in the operational area exceeds the capacity of the available datalinks by a large factor. The suggested technical change is to use Link 16 to pass highly compressed video feeds from the UAVs to all UAV ground stations and area commanders. The UAVs will need to be fitted with Link16, on-board image recognition and data compression software to provide this data distribution. The single direct video link available for the launcher-hunting operation will be connected to the tactical or surveillance UAV and its ground station as appropriate at that stage of the mission;
- There is only one C2 link available for all three surveillance UAVs, so it will have to time-share between them. Therefore, the UAVs will require an increased level of autonomy to maintain their stations;
- Red will try to jam or interfere with all of Blue's communication links. Continuous enhancement of link security is not part of this capability requirement, but operational effectiveness will set requirements for an ability to continue with the mission during periods of lost communication links;
- It is unlikely that Blue will be able to achieve acceptable levels of hits on the missile launchers in the few minutes that they will be visible. This can be overcome if there is a Ground Moving Target Indication (GMTI) and Track (GMTT) capability[6] on the surveillance UAV which can track the launcher in cluttered conditions. This gives a capability to destroy the launcher after its return to its hide.

The operational analysis process will involve several iterations before reaching a stable and acceptable solution. It ends with the three outputs given in Section 4.8:

1. A description of the military capabilities that deliver the desired improvement in force effectiveness (Section 11.2.5).
2. An outline technical requirement for the systems delivering the capabilities (Section 11.2.6).
3. Tables of each of the measures of merit, quantified if possible or scored for comparison in various scenarios (Section 11.2.7, but without any scenarios).

These are only outlined here as the detail will be specific to a particular nation or coalition, its assets and those of the hostile nations. In practice, they will be more general than implied by the conclusions above as the solution must have wider applicability than this scenario. Additionally, the suppliers must be allowed some scope for negotiation and to offer new technologies which may give better performance than the assumptions in the original analysis.

11.2.5   Unmanned air system capabilities

The UAV system is defined for the analysis as a UAV, its ground station, the control and data links between the two, and its operators. For simplicity, the airfield

---

[6]GMTI and GMTT are usually associated with radar systems. Note that radar engineers and users call every object detected by the radar a target, whether it is a military target or not. It should be possible to avoid ambiguity by the context.

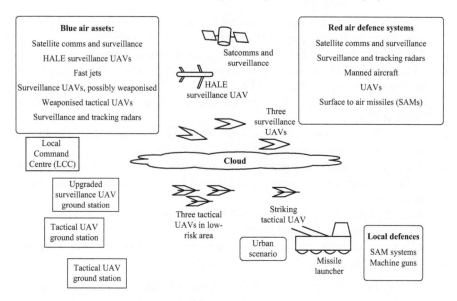

Figure 11.2   Scenario, missile launcher strike in littoral operations

systems required for its launch and recovery and handover to its ground station are left out of this analysis. The desired capabilities resulting from the operational analysis will be presented as a document and supporting evidence to senior decision-makers. It will give the required capabilities for the system and changes to the surrounding infrastructure. In practice several operational scenarios will be considered, but for simplicity, only one is shown diagrammatically in Figure 11.2. This type of cartoon is used for operational analysis work and is often used as one of the views for the architectural analysis using one of the frameworks given in Section 4.2 and discussed in Section 11.2.6 below.

The document will include the following capabilities, but with more detail and quantification:

- The ability to survey continuously areas extending to the 50 km missile range beyond the Blue front line;
- Surveillance and tactical UAV assets to have self-defence capabilities to counter SAM attacks;
- Tactical UAVs that can be deployed with their ground stations to remote air-bases for launch and recovery;
- Missile warheads to be small in order to minimise collateral damage near the target, but large enough to penetrate likely tactical hides for the launchers;
- The ability to fly in contested airspace without live video feed to operators;
- The ability to fly in response to sensor inputs but without direct commands from the ground station;
- Day, night and all weather capability;

- Continuous indication to operators of the location and movement of tracked vehicles including the missile launcher in the surveyed area;
- The ability to track indicated targets in urban and rural areas;
- Live video feed of an area 250 m^2 around the target area to the operator to be provided after target indication (The resolution will be expressed as a rating on the military version of The National Imagery Interpretability Rating Scale (NIIRS) [2,3].);
- The ability to destroy military vehicles detected by the surveillance assets in a timeframe of a few minutes, preferably less than 3 min;
- The ability to have both types of UAV under the command of the air or ground command centre, with command being switched between them occasionally, as required;
- A set of requirements from Through-Life Capability Management (TLCM) considerations[7] such as interoperability with specified other systems, software maintenance and upgrade policies, upgrade to weapon types and coordinate systems;
- Upgrade strategy, plans and indicative costs.

The capability to meet Article 35 requirements is implicit rather than explicit. Section 11.5 discusses the reviews in detail for this application.

11.2.6   Architectural analysis

The procurement authority and their technical advisers will jointly examine the operational analysis results and determine if there is a requirement to modify existing architectures. The architecture of the UAV system and its larger system-of-systems will be expressed using one of the frameworks in Section 4.2. If changes are necessary, the procurement authority will propose one or more architectures for the required numbers of both types of UAV, the ground stations and the current coalition C2 architecture. The UAV system architecture will need to include the computer and intelligence components – Command, Control, Communications, Computers and Intelligence (C4I).

External intelligence data will be passed to the operators in all the ground stations. The location of the algorithms for target indication and identification will be split between the UAV and the ground station. The enhanced autopilot necessary for defensive manoeuvres will override the flight path ordered by the operator, so the division of authority and routing control between the UAV and ground station will need to be reviewed and probably changed fundamentally.

The starting point is the architecture which would be used for this mission without this extra capability. This is shown in Figure 11.3. Note that this is the C2 architecture and not the connectivity architecture which will show all the communication channels in use. The overall command role is initially the maritime commander as the air assets will be deployed in support of the naval and ground forces during the landings and establishing the beachhead. At some stage, the overall command will move from naval to ground commanders. This is not relevant to the air forces as they are always under

---

[7]These are given in more detail in Section 4.11 and Appendix A4 of Chapter 4.

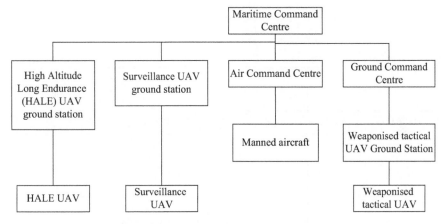

Figure 11.3   Existing C2 architecture

the control of the air command centre. The exception would be when command of a strike aircraft for weapon release is passed to a forward air controller. Normally, the strike aircraft are in packages of a few aircraft with their Package Leader having a direct radio link to the air command centre.

Currently, the flight path and other tasks of the surveillance UAVs are controlled[8] by their operators who will be located out-of-theatre. The operator is under the command of the maritime command centre through the general coalition network which provides the link between the air command centre and the ground stations. There are two data links: the link providing the piloting capability for the operator to fly the UAV and the video link providing the surveillance data which is distributed by the ground station over the coalition network to authorised recipients.

The C2 chain for the tactical UAV system has a similar architecture to that for the surveillance UAV system except that it is currently under the control of the ground command centre as it is used as a tactical asset responding to rapidly changing local circumstances.

The launcher–hunter capability will require very close coordination between the surveillance and tactical UAV controllers. The conclusion is that there must be a local command centre for both surveillance and tactical UAVs executing these missions. This is also consistent with the requirement for the tactical UAVs to be based nearer to the operational area than the surveillance UAVs. This leads to the proposed architecture shown in Figure 11.4. Although the threat from the launchers is to ground or naval forces, the initial clearance for use of both types of UAV will be by the air command centre. This is because there will be extensive air activities for many purposes over the whole operational area. Air coordination will

---

[8]Control here means that the pilot or operator gives the UAV a set of waypoints and the UAV chooses its own flightpath to reach its destination. The human acts in a supervisory role, deciding the objectives and type of flightpath required.

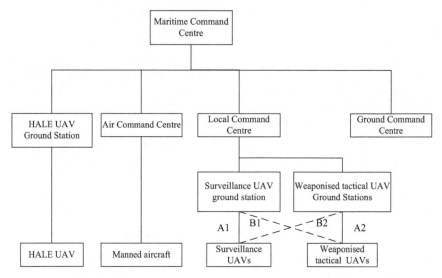

Figure 11.4   Proposed C2 architecture
[Links B1 and B2 will replace links A1 and A2 for certain phases of operation.]

be by the air command centre using their recognised air picture[9]. As the operation evolves, it may be better to transfer the local command centre to be under the command of the ground command centre.

The architecture shown in Figure 11.4 can be used as a basis for the competition stage of procurement, recognising that some bids may offer different architectural changes. The bids are likely to have various proposals for the use of the available communication channels. There are STANAGs for the interfaces between UAVs and ground station as well as Link 16; the appropriate ones will be specified in the invitation-to-tender for each subcontract.

11.2.7   System-of-systems effectiveness

The problem faced by the analyst is to decide the measures of merit for this capability and their weighting for decision-making during the procurement. Initial lists will be drawn up as soon as the sponsor question is posed; extensive revision of both their definition and importance is a normal part of the analysis process. It is assumed here that the measures of policy effectiveness will have been assessed and used in deciding that a littoral operation is required. Maximising the use of existing assets as part of both procurement and military policy will also have been included at that stage.

---

[9]The recognised air picture is a controlled central database of all known aircraft in the operational area with data about each one. It is maintained and kept at the air command centre with Blue aircraft and controllers having restricted access to the relevant parts.

The main criterion for this example is the rapid neutralisation of the missile launchers. This can be made one of the measures of force effectiveness and will flow down into the other measures of merit. However, this is too limited as the analysis has been carried out for a fairly specific scenario, with current technologies. Additional important factors to be considered are:

- Flexibility of the proposed systems to be used to provide other capabilities;
- The potential for capability upgrades by inserting new technologies;
- All aspects of reliability;
- Supportability.

This is effectively consideration of TLCM, which must appear in the dimensional parameters if nowhere else.

Tables 11.2 to 11.5 give example tables of measures of merit. These are restricted to the subset relating to automation. There are two types of UAV, the tactical and surveillance types, so there will be two sets of Tables 11.4 and 11.5, one for each UAV type. A full set of measures of merit will be more exhaustive including many others such as platform stability for image and synthetic-aperture-radar map resolution which are not included here.

The parameters and criteria will be rated in importance, and tenders from bidders will be assessed against them. Ideally these will be quantified so that tender compliance can be unambiguously assessed. Usually this is not possible for some measures of merit which have to be expressed and measured qualitatively. In these cases, analysts and programme managers often use a Red, Amber, Green (RAG) approach, a qualitative methodology which is not discussed in detail here. The principle is that for a given implementation of a system, the value assigned to a measure of merit or to each project risk, or other parameter, can be classified as red, amber or green. Green is acceptable, or even exceeds the requirement, amber may be acceptable but needs further consideration, red is unacceptable. Sometimes blue is used to denote clearly exceeding the specification.

Table 11.2   Measures of force effectiveness

Measure of force effectiveness	Assessment criteria
Night performance	• Quality of IR imagery in different light conditions
Poor weather performance	• Quality and interpretability of radar data • Quality of IR imagery using different wavelengths
Number of surveillance UAVs	• Area which can be covered by each UAV • Time taken to image area with adequate resolution • Reduced time on-station due to increased payload of Link 16 and defensive aids • Attrition rate due to Red's air defence system

(Continues)

Table 11.2 (*Continued*)

Measure of force effectiveness	Assessment criteria
Response time between target identification and Blue weapon hitting target	• Response times of each link in surveillance and decision chains • Time for tactical UAV to reach target area • Estimated time between launcher leaving hide and firing for that location • Estimated time for launcher to return to hide or go to new location
Number of operators required for the use of UAV fleet	• Data rate from surveillance UAV fleet • Number of tactical UAVs
Number and location of ground stations	• Effectiveness of one operator controlling multiple UAVs
Fusion of UAV information with other intelligence data from other sources	• Availability of other authorised intelligence • Timeliness of other intelligence information
Ability to distinguish between military and civilian objects	• Assessment by simulation and modelling of test scenes • Relative performance of radar and electro-optical imagers
Ability to track identified targets	• Track accuracy • Number of targets that can be tracked simultaneously • Track before identify ability
Ability of tactical UAV to operate from rudimentary airbases	• Ability to operate from tarmac roads • Ability to operate from grass fields

Table 11.3 Measures of C2 effectiveness

Measure of C2 effectiveness	Assessment criteria
Optimal location of UAV system in coalition C2 chain	• Position of UAV system node in operational area C2 architecture • Access to timely intelligence data from other sources such as electronic surveillance measures
Possibility of passing weapon targeting and release decision to and from forward air controllers	• Ability to link to forward air controllers
Ability of surveillance and tactical UAV to maintain station whilst not under direct operator control	• Air space management system's ability to keep aircraft separation for specified time • Sense and avoid capability • Auto-response from defensive aids suite if UAV under attack • Continuous maintenance of imagery transmission during automatic flight periods
Ability of surveillance and tactical UAV to continue mission whilst data and command links jammed by hostile emitters	• Capability of automated systems on UAV to make decisions other than those in previous row
Quality of data and information presented to operator	• Workload criteria based on human–machine interface quality and psychological factors • Human ability to interpret radar data

Table 11.4   Measures of performance

Measure of performance	Assessment criteria
Logistics chain needed for repairs and replacements	• Speed of response to replace battle losses • Long-term cost of support • First, second or third line repairs and maintenance
Target indication algorithms	• Criteria for indication • Probability of correct indication • Probability of false alarms • Speed to make indication
Target identification algorithms	• Criteria for identification • Probability of correct identification • Probability of false alarms • Speed to make identification
Number of images that can be transmitted to operator in real time	• Image compression factors • Available bandwidth under combat conditions with multiple users
Resolution of real-time images received at ground station	• NIIRS criteria • Difference between Link 16 imagery and direct video feed • Interpretability of radar information • Location of UAV's target position on direct feed imagery
Distribution of target information	• Speed of distribution to C2 nodes for targeting • Speed of distribution to weapon operator • Reliability of target information supplied to weapon operator
Probability of hitting designated target	• Guidance accuracy of weapon • Coordinate transfer accuracy from image to weapon aim point • Missile homing head performance

Table 11.5   Dimensional parameters

Dimensional parameter	Assessment criteria
Modular design	• Number of modules • Overlap of individual module capabilities with military decision-making nodes • Ability to replace modules in service • Likely capability enhancements by upgrading individual modules • Commonality of modules with current in-service modules/systems

(Continues)

Table 11.5   (Continued)

Dimensional parameter	Assessment criteria
List of interface standards to be used between subsystems	• Minimise the number of standards used • Absence of system-specific standards and definitions
Standards to be used between system and other coalition and national materiel	• Proportion of standards that are already in use by nation and coalition partners • Security of national information in coalition operations
Data link requirements	• Number of links • Bandwidths • Ability to use the existing bearers • Security when disrupted • Data latency
Lifetimes of platforms and infrastructure	• Comparison of lifetimes with other assets that system relies on
Component reliability	• National criteria on mean time between failures (MTBF)
Probability of correct target identification	• Human success rate using targets in diverse range of test scenes
Probability of false alarms	• Human false alarm rate using targets in diverse range of test scenes
Ability to identify civilian people and objects	• Human success rate using diverse range of test scenes
Image compression	• Acceptability of imagery from surveillance UAVs during their real-time transmissions
Ability to track a target	• Results from models and tests of test scenarios
Electro-optic sensor resolution	• Ability to differentiate between military and civilian personnel and objects
Synthetic aperture radar map resolution	• Ability to differentiate between military and civilian personnel and objects in radar images of resolution supplied by the synthetic aperture radar sensor

11.2.8   Pre-competition activities by the procurement agency

11.2.8.1   Contract review

The existing infrastructure and contracts will be reviewed by the procurement authority to see what changes are needed and/or possible. The aim is to understand the contractual implications which come from the measures of merit and higher-performance specifications which must be achieved at an affordable cost.

A note on terminology: The consortium, company or division of a large company responsible to the defence ministry for delivering or maintaining a major platform, system or service is known as a prime contractor, called Prime in this book. The prime contractor usually has a chain of suppliers down to specialist suppliers of specific technologies. These suppliers are often known as second, third (etc.) tier suppliers. The simplification is made here that all equipment and services not produced by the prime are produced by lower tier companies called Tier in this book.

The existing facilities and contracts are:

- Existing C2 infrastructure with prime contractor Prime(C2). This company has experience of successfully integrating new platforms into the nation's and coalition's C2 infrastructure;
- Satellite communications between UAVs and their ground stations supplied by a service contract on a prime contractor delivering Satcomms, Prime(Satcomms);
- Link 16 infrastructure between all assets in the operational area, run by the state's armed forces with contracts on Prime(C2);
- Surveillance UAVs, supplied by company Prime(Surv UAV) who also act as prime contractor for all installation, maintenance and modification work on the platform;
- Ground stations capable of commanding two surveillance UAVs simultaneously with two operators. The ground stations are supplied and maintained by company Tier (Surv Ground Station);
- Deployable tactical UAVs supplied with their dedicated ground station by a prime contractor called Prime(Tactical). The interface standards to external systems are the same as used for the surveillance ground stations. They are owned and maintained by the state's armed forces, with a series of contracts on Prime(Tactical).

The analysts and technical advisers examine how to add the desired capabilities to the current infrastructure at minimum cost. The result is that the minimum number of UAVs required to be maintained on station during the littoral operation will be four tactical and three surveillance UAVs.

11.2.8.2 Analysis

Operational analysis, including modelling and simulation, shows that the seven UAVs can be operated by three operators, two for continuous control of the tactical UAVs and one monitoring the outputs from the surveillance UAVs and other surveillance and intelligence assets. This has the important proviso that control of any of the seven airborne UAVs can be assigned to any operator. This is to ensure that when a surveillance UAV operator has a target indication, he or she can switch to a video link for the area. Their workload is considerably reduced if their ground station has some level of automated target indication capability.

The surveillance UAV operator then provides a continuous data stream to the tactical ground station used for targeting. The tactical UAV operator may need to take direct control of the surveillance UAV radar sensor, if fitted, or analyse on-board data directly. He or she will then use the same video displays and remain under the same local commander. It is also decided that both types of UAV must be able to operate with the same C2 structure as the current UAVs. The operational analysis includes detailed simulation of the possible human–machine interfaces and operator workload using tools such as those outlined in Chapter 2.

The analysis also showed that the current number of UAVs, ground stations and operators will be fully occupied with support to the beachhead and immediate area. Therefore, it is decided that the capability to seek and destroy the missile-launchers will necessitate procurement of new equipment, not just upgrades to current materiel. However, the policy decision is made that the new equipment must be upgrades to existing designs, with maintenance of interface standards and minimal retraining of personnel.

The procurement decision is to procure three more ground stations for the new capability. (Two deployed to the operational area and one spare.) Cost and operational compatibility reasons necessitate a repeat purchase on the original supplier, Prime (Tactical). Some modifications will be needed as there will be more interaction with the surveillance UAV and ground station than in the existing design. During contract negotiations, company Prime(Tactical) offers a compliant ground station based on the old design and to upgrade the existing ground stations to the new capabilities. This is accepted, but as two contracts, one for the new ground stations and an extension on an existing contract for maintenance of the existing ground stations to upgrade them to the new standard.

The DNAW requirement using electro-optical sensors necessitates low-altitude operation with increased exposure to shoulder-launched SAMs and unacceptable attrition rates. The initial solution is to fit a defensive aids suite to both UAVs, as well as the Link 16 unit needed for C2. This gives a major problem, the space weight and power requirements for a defensive aids suite and Link 16 are stretching for both UAV suppliers, Prime(Surv UAV) and Prime(Tactical). Special equipment can be developed, but their power budget will significantly reduce the time-on-station for both UAVs.

High-altitude operation with a very limited defensive aids suite capability gives an acceptable attrition rate for the surveillance UAV with some reduction in performance. Consequently, a sensor with a synthetic aperture radar mapping capability from high altitude is recommended as the compromise solution. This gives DNAW performance with a resolution of a few centimetres. The defensive aids suite requirement is left as an option in the invitation-to-tender. Link 16 is still considered to be essential as the surveillance UAV must be fully integrated into the communication structure of the operational area. Link 16 is a line-of-sight system, but the UAVs' Link16 outputs can be relayed from the operational area communications infrastructure to other remote locations.

There is still a problem with the tactical UAV. It has been designed for medium and low-altitude operation in uncontested airspace. The design is based on the assumption that it will always have a direct line-of-sight video link from its electro-optic targeting sensor to its operator. A consequence is that it has a limited-performance autopilot and no space on its airframe for a radar antenna. It is decided to keep this basic design so the weapon operator can see the target area. It will continue operating at medium and low altitudes but will need:

- A defensive aids suite to achieve acceptable survivability;
- A more capable autopilot;
- Beyond-line-of-sight communications so that it can operate at low levels and at ranges of many tens of kilometres which put it beyond line of sight from the ground stations.

The choice of subsystems within the defensive aids suite will have to be the subject of an operational analysis study using knowledge of anticipated threats.

A fall-back option is to use a new engine to allow higher altitude operation whilst on patrol and drop down for the target strike when cued. There will still be a need for a upgraded autopilot, but with different specifications.

11.2.8.3 Technical advice

The technical advisers recommend that limited processing power on the surveillance UAV makes target indication rather than target identification software an achievable objective for on-board analysis. It is proposed that the radar gives a Link 16 output with locations of indicated targets to accompany the much-compressed radar images. Consequently, there may be a need to use the wide bandwidth video link to access extra radar data for target identification and confirmation by the surveillance ground station. The target indication software will also be installed at the ground station for reference purposes. The division of processing power between the UAV and ground station along with the allocation of the Link 16 and video link bandwidths is included in the deliverables from the suppliers allowing them flexibility for their solutions.

A radar study concludes that an existing radar module can be fitted to the surveillance UAV, replacing the current optical sensor. It has a synthetic-aperture-radar mapping capability, but needs more transmitter power to give the desired range when at medium altitudes. It also requires more processing power to support on-board target indication algorithms and radar data compression. However, synthetic aperture radar offers the same map resolution independent of range, so the surveillance UAVs can deliver a surveillance capability at the higher altitude but with no electro-optic sensor. The military advisors consider that this enhanced capability will increase operational capability in poor weather for many other scenarios so extra business cases are made to add radar and Link 16-capabilities to all the existing surveillance UAV assets. The video link is by Satcomms and the preference is for the radar data to use the same link.

There are several disadvantage with the radar sensors: the images are not necessarily easy to interpret; synthetic aperture radar takes a longer time to survey a given area than optical sensors; the radar mapping requires straight and level flight to form the images, so the flight path is predictable for Red's air defences; the radar transmitter can be used by an anti-radiation missile. Due to these factors, the military staff place a requirement that if weather conditions allow, the surveillance UAVs for this operation must be role-fitted with their original optical sensor and not have the radar capability for those flights.

The surveillance UAVs will be the same as the current fleet, with the exception of the radar sensor replacing their electro-optic sensor and adding a Link 16 module, so there will be minimal extra maintenance costs. (The procurement agency may make the business case for fitting Link 16 as standard on all UAVs of this type.) The current operator training courses will need enhancement to include training for Link 16 and radar image interpretation as there will be no optical imagery coming from the UAV. The extra radar power will not overload the power supplies or cabling, but will reduce the time-on-station. The trade-off between radar surveillance and reduced time on station is considered to be acceptable by the military authorities. The UAV will still require its sense-and-avoid capability which is provided by a simple forward-looking electro-optical sensor. These do not have adequate resolution for surveillance of target areas.

Trials are carried out simulating rapid cuing of strikes with the tactical UAV against targets using a direct video link of its optical sensor to the operator. These show that even with no defensive aids suite the risk of loss is low in the time between cueing, strike and return to station. As a result of the trials results, the defensive aids suite requirement can be relaxed from the initial requirement for high-risk areas to those for the lower risk loitering areas. The electro-optic sensor is not the same as the one fitted to the surveillance UAV so it is not possible to fit radar to the tactical UAV. This leads to a requirement for an interface between the radar and electro-optic sensor imagery so that both can be compared directly by one operator.

Following the revisions to the system-of-systems necessitated by these pre-competition activities, the proposed architecture must be validated by the process described in Section 4.13.

11.3   The system-of-systems delivering the capability

11.3.1   The C2 infrastructure

The IHL issue is that the weapon is under control and that the rules of engagement are followed. This makes the C2 architecture one key issue. The C2 architecture in Figure 11.4 meets the overall needs and it, or an equivalent one, must be implemented. This is more of a military policy issue rather than a technical one, especially having a command centre for this capability.

The links for the seek-and-destroy operation will have to fit into the coalition's communication architecture. There will be at least one network between and within coalition members' forces and command centres. Link 16 will connect all assets in the operational area in addition to any other direct links. There will also be other dedicated links in the operational area. Spectrum management will be an issue that should be arranged in the preparations for the campaign, but there will always be tactical issues as well as interference from Red's counter-measure systems.

Figure 11.5 shows the connections that will be required when one tactical UAV is attacking a missile launcher that is under surveillance by one UAV with a radar sensor and one with an electro-optic sensor. The weaponised, tactical UAV and ground station assigned to the strike will be connected by the direct video link and Link 16. The remote surveillance ground station will receive its Link 16 surveillance data via the satellite link. The three remaining tactical UAVs are on holding flight-paths in relatively safe areas. The surveillance UAV that is not part of the strike operation may be in a holding area or may continue surveillance of other areas, the ground station operator relying on a target indication cue from it for attention.

The design and analysis of the connections and networks must consider the missile launcher as a data source. Speed and reliability of distribution of this data to the operator of the strike UAV is an important measure of performance as this underpins the weapon-release decision.

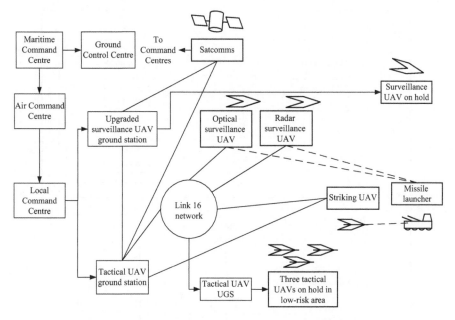

*Figure 11.5   Connectivity diagram for weaponised tactical UAV strike
on a missile launcher*

The first technical issue is ensuring that all the interfaces between nodes and assets meet the same standards. This apparently simple statement covers considerable integration effort as there is always some level of interpretation of a technical standard. This can lead to problems which must be solved at the integration stage if not before.

The second technical issue is that the design must allow transfer of command from maritime to land command centres without a break or a period when it can receive commands from both or neither as its superior node. The solution will be a mixture of technology and operational procedures to ensure that the change is effectively seamless and that no information is lost.

11.3.2   Changes to UAV ground stations

Currently the surveillance and tactical UAV systems each comprise one UAV and a dedicated ground station. Both systems have been designed with interfaces in accordance with STANAG 4586, as shown in Figure 11.6, but have different designs for the most equipment in the ground station.

There is more space available for changes to the surveillance ground station as it uses satellite communications, and is located away from the operational area, giving the option of using a larger or extended cabin. The surveillance ground station must be modified to work over Link 16 in the operational area, using a satellite to connect to the tactical Link 16 network. It will be too far from the tactical UAVs to use their line-of-sight data links to control them. However, the technical advisers consider that

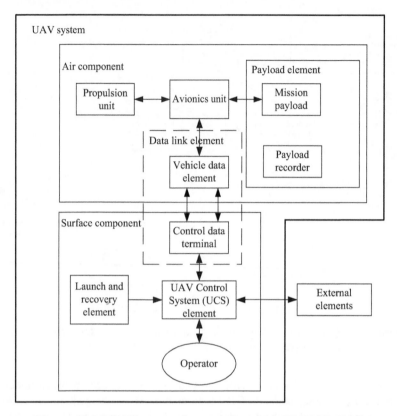

Figure 11.6   UAV system elements (based on STANAG 4586)

developments in automation will eventually allow remote operation of the tactical UAVs using link 16 and other coalition networks. Initially this would be to supervise them when in safe areas whilst the weapon operators are conducting strike operations.

Space is limited in the tactical ground station which must be transported to be near the operational area, so modifications to this ground station should be kept to a minimum. Despite this, it is decided to make the local command centre part of the tactical ground station. There are two reasons for this. The first is to facilitate handover of control from maritime to ground command centres. The second is to ensure that the tactical UAV operator will be on the tactical Link 16 network with the most recent and reliable data when conducting the strike operation. This decision increases technical complexity and the amount of equipment in the ground station but is made mainly because of the legal advice following an initial Article 36 review when the outline system-of-systems, utilising existing materiel, had been completed.

Both ground stations will need to be fitted with Link 16 and have an interface over it with the defensive aids systems fitted to both types of UAV.

The decision to operate more than one UAV from one ground station means that there will have to be extensive changes to the human–machine interface and data analysis facilities. The analysis of the compressed data and target indication cues from the surveillance UAVs will be a new capability for the ground station, with consequent human–machine interface and workload issues.

11.3.3   Changes to UAVs

It is decided that both UAV types will be retained as their different airframes and engines make them the preferred choice for the separate roles of surveillance and low-level strike.

Both UAVs require Link 16 and defensive aids suite to be fitted. The defensive aids suite should include a missile-launch detection capability on at least the surveillance UAV. It can act as both part of the self defence capability and also as a cue to look for a missile launcher, any launcher being a potential target. Depending on the responses to the tendering process, the defensive aids suite may trigger the release of chaff or flares, or the UAV may take evasive action and a new route to complete its mission.

The surveillance UAV can be role-fitted with a synthetic aperture radar which is an area surveillance technology. It will be necessary to have both radar and electro-optical sensors available for the same platform although only one is fitted for any given mission. The radar must be fitted with a moving ground target indication and tracking capability allowing it to follow potential targets. The missile launcher will have a distinctive motion prior to launch, making this a potential radar 'fingerprint' to identify it as well as any unique features in its radar return due to its shape.

On-board target indication software is needed for the times when the operator is unable to continuously monitor the sensor outputs. The procurement agency's technical advisers will have to examine the trade-offs between the specifications of the software, the two sensor types and the link capacity to download data.

Both UAVs will require the design and fit of new autopilots. These will be different for each UAV type. They must be capable of operation for periods with no operator input, as at present, but they must also be able to respond in an appropriate way to defensive aid system inputs.

Operations in military airspace that is shared with other aircraft usually require that the aircraft, or UAV is in a Link 16 circuit. This makes fitting a Link 16 terminal to the UAVs essential for either UAV to operate in the area on seek-and-destroy missions.

11.3.4   Development of tactics

The missile seek-and-destroy mission is not a fundamentally new approach to the problem, the differences being due to the use of a weaponised UAV system instead of a manned aircraft. However, new tactics will need to be developed by military advisors as it will potentially involve the use of UAVs in the same operational area as manned aircraft. These will need to be evolved in parallel with the procurement process as there will be a mutual dependence between the technical solutions and operational considerations.

Maximum flexibility is achieved if one of the surveillance UAVs is always fitted with a radar module. The main reason is that it gives two methods of indicating and identifying targets and gives an all-weather GMTI capability allowing stand-off tracking of the missile launcher before and after launch and locate its hide. The permanent availability of radar also allows for changes in weather conditions.

The evolution of tactics must be planned carefully so that the results give a suitable data set for both the Article 36 reviews and the setting of rules of engagement for the range of likely operations. Clearly the planning must use trials based on a much wider set of scenarios than the one given in this chapter.

11.4   Delivering the capability

11.4.1   The procurement problem

The activities described in Sections 11.2 and 11.3 show that the C2 architecture can be modified to incorporate the new capability and measures of merit derived for the system-of-systems. Seven UAVs are required in the operational area, but this will require an order for a larger number due to attrition in any operation, maintenance downtime, reserves and any requirement from other operations carried out at the same time. Calculation of the total requirement, including support facilities is not considered here as this will be specific to each nation and is only indirectly dependant on the level of autonomy in the system.

The next problem is to decompose the capability into systems and subsystems which can be procured. The individual systems, interfaces and communication links must be technically complete and have unambiguous specifications so that the system-of-systems will operate correctly. They must also be capable of being designed and delivered by one company or consortium against these specifications with clear delivery criteria that will generate payments at specific times (milestones).

A large systems house can be expected to bid to deliver a capability but this is essentially the procurement agency passing the whole procurement problem to a commercial partner. There may be reasons why this is not acceptable, either commercial, political or military. The specifications for the systems and subsystems must be written so that a range of companies can bid for those items that match their particular expertise. It is also likely that the procurement agency will want a particular supplier's product in the final system, regardless of who wins the final contracts. Although complicating the systems engineering problems, it is a fairly well-known situation.

It is clear that the system-of-systems delivering the capability will evolve, with more automation being introduced. The procurement managers can see that with increasing levels of autonomy in the targeting system, demonstrating adherence to IHL will be essential. Without adherence to IHL, which is a state responsibility, the system-of-systems cannot be used operationally to deliver weapons onto targets.

The Article 36 reviews will require clear visibility of the principles and processes used in the automation of the target identification process. The core algorithms for these come from specialist suppliers who are well down the supply chain. These suppliers will be concerned about commercial confidentiality and their competitive advantage in the market. This is an example of the need for specific contractual conditions about balancing the release of intellectual property for analysis with the number of trials required to demonstrate compliance.

It can be seen from the above discussion that the requirement will be met using a set of contracts based on the agency's division of the work into work packages. Of necessity, there will need to be extensive discussions between the agency, its military and its technical advisers. These may not always agree and compromises will have to be reached and budgets reassessed for affordability. Effectively, the procurement will be based on 'design to cost'. This is normal for military procurement. Governments assign an overall budget for defence which is then divided among the various programmes and commitments.

The procurement agency, in this case, decides that the most cost-effective solution is to place contracts using the existing equipment, logistics and trained staff where practical. An Article 36 review will be necessary at the key procurement stages. Autonomy will be introduced using a small fraction of the budget to raise TRLs and introduce these technologies when budgets allow. Article 36 reviews will be planned for the introduction of more automation using the results of technology demonstrators.

Regardless of the policy decisions about contract awards, the plans must include verification of the performance of the system-of-systems as well as individual systems and modules. Performance and behaviour verification are essential parts of the integration plan but may need to be a separate work package. The work content will be specific to the project and contract structure but must follow the guidance given in Sections 1.5, 5.4.4, 5.4.5 and 13.8.9. The outputs will be an essential part of the evidence presented to the Article 36 review. They should include the actual performance of the capability providing the 'new means or method of warfare'.

11.4.2   System-of-systems requirements

The requirements will probably be a set of documents running to several hundred pages with considerable detail, including all the relevant STANAGs and other standards that the equipment must meet. Interface specifications for internal and external connectivity will be stated.

The procurement programme will involve:

1. Modifying the fleet of surveillance UAVs to fit Link 16;
2. Modifying the fleet of surveillance UAVs for control by the tactical ground station links;
3. Modifying the fleet of tactical UAVs to fit Link 16;
4. Upgrades to the surveillance UAV autopilot;
5. Upgrades to the tactical UAV autopilot;
6. Fitting defensive aids suite or a different engine to the tactical UAV;
7. Reduction of the infra-red and radar signatures (This could place restrictions on the number and types of antenna used on the UAV.);
8. Radar with mapping and target tracking capabilities to be fitted to surveillance UAVs;
9. Develop target recognition algorithms that can identify moving and/or tracked vehicles. This is to be installed on the surveillance UAVs for target indication purposes;
10. Develop target identification algorithms which will be compatible with the indication algorithms. These will be for use at the ground stations;
11. Create a capability for a surveillance UAV to pass target coordinates directly to the nearest tactical UAV;

12.   Contracts for support and upgrades;
13.   New training programmes and training equipment.

It is almost certain that the bidders will include current prime contractors with their preferred subcontractors. The nation's procurement authority will make a policy decision about responsibilities for integration and the role of the primes in delivering the capability. The options for integration being either one of the winning suppliers, such as the tactical UAV manufacturer, to be prime and responsible for integration, or for a separate 'systems house' to act as prime.

11.4.3   The procurement process

The procurement process has to divide the work into packages. In general, these are competed where possible, followed by contract negotiations and contract award. The authority must consider the integration of the contract deliverables into the nation's infrastructure during this process and draw up an integration plan.

The procurement agency decides that feasibility studies will be necessary to:

1.   Examine the trade-offs for the tactical UAV system between fitting defensive aids and improved flight performance with a new engine;
2.   Specify the upgrades required for the autopilots in both UAVs;
3.   Agree the principles of target indication and identification;
4.   Assess the amount of image processing that can be performed on the UAV for target indication and identification;
5.   Assess the performance of a mapping radar with moving target capabilities that can be fitted to the surveillance UAV;
6.   Investigate methods to reduce the radar, optical, IR and acoustic signatures of both UAVs, but especially the tactical one. It may be possible to carry this out in collaboration with other users of these UAVs;
7.   Investigate options for remote C2 of the tactical UAVs by the surveillance ground station.

11.4.4   Integration plans

Delivery times, availability of test ranges and planned In-Service-Date (ISD) for various levels of capability play an important role in setting an integration strategy. After considering these factors among others, the order of integrating the various deliverables is planned to be:

1.   Fit Link 16 into a surveillance UAV and tested on test ranges;
2.   New standard tactical UAV ground station, tested with current surveillance UAV ground station in current military C2 structure for operations in military and civil controlled airspace;
3.   Surveillance UAV with Link 16 trialled in military airspace with new ground station;
4.   Mapping radar tested by radar supplier on test aircraft using direct data links and test ranges;

5. Target indication parameters chosen;
6. Tactical UAV flights with new ground station;
7. Installation of a defensive aids suite or a new engine in a tactical UAV;
8. Target indication software installed in radar on a surveillance UAV;
9. Target identification algorithms installed in surveillance ground station;
10. Target identification information over Link 16 to the ground station and confirmation of target identification to the tactical ground station.

A simplified illustration of this integration is given in Figure 11.7. There will need to be some complex interactions between contractors during the programme to ensure that the interfaces and data flows between subsystems are correct and understood. Prime(C2) will be carrying out extensive verification and validation modelling of the capabilities of the combinations of subsystems. The results will be used to refine requirements and shape the plans for introduction into service. The initial operational capability will be less than full capability, so that military doctrine and tactics can be established, based on the actual systems that will be delivered.

The delivered system will have changes to the UAVs' autopilots and flight control systems. This will need to go through a certification process by the nation's aviation authorities before the UAVs can fly in their airspace. This may require a sense-and-avoid capability for civil airspace if not for operational areas where the level of acceptable risk is higher.

11.4.5 Work packages and contract award

The procurement agency decides to split the project into the work packages given in Table 11.6. After the invitation-to-tender, down-selection process and contract negotiations, the contracts are awarded to the companies shown in the last column of the table. The contracts will be awarded on a value-for-money basis for the new capability. However, the supply-chain structure for this set of contracts will affect the award of contracts for the upgrades for more autonomous operation discussed in Chapter 12.

11.4.6 Work Packages 1 and 2, integration into C2 structure

The problem faced by the contractor Prime(C2) is implementing Figures 11.4 and 11.5 with deliverables from the range of suppliers in Table 11.6. Prime(C2) must deliver a capability which is fully compatible with the coalition's C2 systems, and meeting the state's operational philosophy. The contractor will be supplied with accurate information about the current and planned C2 systems. The state will be highly likely to operate in a coalition with other states with joint C2 systems. A typical example of a coalition C2 structure is the NATO Air Command and Control System [4].

The starting point for the work in these work packages is Figure 11.3. This structure will be a small part of the coalition's command structure as there will be large numbers of ships, aircraft and ground forces involved in any campaign or operation. The architecture of Figure 11.4 looks, and is, very similar to that in Figure 11.3 as far

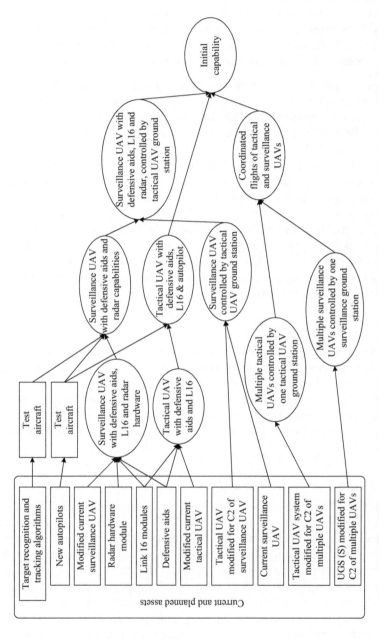

Figure 11.7 Integration plan for missile-launcher search-and-destroy capability. L16 denotes Link 16

Table 11.6 Work packages

WP	Content	Supplier
1	Integration and industry side of introduction into service	Prime(C2)
2	Subcontract from Prime(C2) for integration of UAV systems into military C2 structure	Different division of Prime(C2)
3	Supply of Surveillance UAVs with upgraded autopilot, Link 16 and synthetic aperture radar target identification	Prime(Surveillance UAV) plus subcontractors
4	Supply of tactical UAVs with Link 16, defensive aids suite and upgraded autopilot or new engine and flight-control system following study of trade-offs	Prime(Tactical) plus subcontractors
5	Study on operator workload and resolution of human–machine-interface issues for ground stations. Includes close liaison with military advisers	Specialist company or in-house technical experts
6	Supply of new tactical ground stations (Upgrade of the existing ground station to the new standard)	Prime(Tactical)
6a	Upgrade of surveillance ground station for new role with tactical UAVs	Subcontract on Prime(Surveillance UAV) from Prime(Tactical)
7	Algorithms for radar target identification and tracking on surveillance UAV. Results and compressed image to be passed over Link 16. Implementation on UAV and at ground station	Tier(Radar) with subcontract to a specialist radar algorithm company
8	Technology demonstrator for modular role-specific surveillance radar for UAV	
9	Setting to work of surveillance UAV system with Link 16 prior to entry into service, with subcontract on test range operator	Prime(Surveillance UAV)
10	Setting to work of tactical UAV system with Link 16 and defensive aids suite prior to introduction into service, with subcontract on company who operates the nation's test range(s)	Prime(Tactical)
11	Supply of direct link to forward air controllers	Prime(C2)

as UAV control is concerned. The problems arise when the requirement for interactive interchange of radar data between all UAV operators and both types of ground station. The aim being for one tactical UAV operator to have control of the surveillance UAV with live streaming of its electro-optical sensor output or real-time direct access to the radar data, depending on the role-fit sensor.

Analysis of the operators' workloads in the first analysis of a realistic scenario shows that they are unacceptably high. This includes simulations using the current ground stations and radar maps from trials. Gathering high-quality radar map data necessitates that the surveillance UAV flies straight and level for several tens of seconds or minutes which is a high-risk flight path[10]. Red is able to attack the UAV during this phase of its mission. The consensus is that one operator must remain in control of the surveillance UAV, and no other, so that optimum evasive action can be taken, whether by the defensive aids suite or by the operator, and maximum radar data gathered. This operator will also be analysing the data for the indicated presence of missile launchers and cueing further investigations. Initial checks will include deeper investigation of the radar data, followed by cueing the tactical UAV for a strike action against the target.

The tactical UAV operator will receive some radar data and the coordinates of one or more targets. As they will not be collocated with the surveillance UAV operator, the tactical UAV operator will need access to the radar data to form an opinion about how they will execute the strike operation. The legal review has concluded that it is unacceptable for the strike to start simply on the evidence from radar alone due to limitations in the performance of the algorithms. The tactical UAV operator must plan the strike to minimise collateral damage. It is possible that they may require more radar data. This could be by access to the data stored on the surveillance UAV or requesting a new, higher-resolution synthetic-aperture-radar map of a small area around one indicated target.

The internal architecture of the surveillance UAV system is in accordance with STANAG 4586. This is shown in Figure 11.5 which assumes that control of the sensor suite may need to be under the control of a ground station different to the ground station flying the UAV. STANAG 4586 has Levels of Interoperability (LOI) and in its Annex B, Paragraph 2.3.1 states that:

> LOIs have been defined for discrete levels of functionality and are independent of one another. In other words, the control of a sensor (within LOI 3) is independent and separate from the control of the air vehicle (within LOI 4).

Each of the subsystems in Figure 11.6 has an associated STANAG, including STANAG 4607 for radar data.

The procurement authority and Prime(C2) decide that the anticipated technical problems can be overcome and it will be possible to integrate the capability into the state's C2 system. Provided that the capability remains under the control of the state's command centres, the legal reviews stipulate that it can be used with the state's likely coalition partners. This has implications for the coalition, as the nation will offer a capability complete with command centre, not a set of systems which can be integrated separately into the coalition's force.

---

[10]A straight, level flightpath is the ideal one for making synthetic-aperture-radar maps of the ground. This is not a problem when there is no threat to the aircraft. In principle, other flight paths could be followed but this is at the expense of large complications in the radar signal processing and making the radar have an unacceptably low TRL for this programme. Even if techniques for non-linear flight paths are developed, the UAV will still have to fly a predictable flight path so the basic survivability problem is eased but not removed.

Table 11.7   Assessed TRL of radar technologies

Technology	TRL
Radar hardware	High
Mapping algorithms implemented on ground	High
Mapping algorithms implemented on UAV	Medium
Storage of radar data on UAV	Medium
Send data to ground over video link	Medium/High
Select data to send over Link 16	Low
Target indication algorithms	Medium
Target identification algorithms	Low

11.4.7   Radar work packages

Although the radar work packages are of relatively low financial value, their importance will become clear in the discussions about proving adherence to Article 35. This section applies the concepts in Chapter 5 to the work packages relating to the radar and target identification processes. These are Work Package 7 and Work Package 8 in Table 11.6.

The procurement authority examines the available radar technologies and assesses their TRLs as applied in the context of this programme. (As this is an example, the TRLS are only given high, medium or low values, and not values of 1–9.) The results are given in Table 11.7.

It is considered that the mapping algorithms and data storage technologies, although at Medium TRL, can be developed under the procurement contracts with acceptable risk. However, given the importance of target indication and identification, their TRLs must be increased with a technology demonstrator programme as quickly as possible. The demonstrator programme will have an early deliverable of requirements for the processing power required to include target indication algorithms in the UAV. The quality of the maps and target indication and identification results will be an important part of the technical evidence at the Article 36 reviews.

It is also decided that the final demonstrations will be on the project's radar not just in a laboratory environment or trials aircraft. The actual radar hardware and UAV must be used in the final stages of the demonstrator. Two radar units will be procured by government and supplied to the contractors. The procurement agency will set up part of Work Package 2 so that the trials radar can be integrated into a surveillance UAV at an early stage in the programme.

WP8, the technology demonstrator, is awarded, after competition, to a specialist radar company to:

- Identify features of missile launchers that will be visible in the radar maps from the surveillance UAV;
- Derive algorithms to indicate locations of these features in a radar image whilst the launcher is static, and whilst it is moving;
- Derive algorithms to identify missile launchers and the constraints on their interpretation by an operator;

- Show that a mapping radar can be installed on the surveillance UAV using the same mechanical and electrical interfaces as the electro-optical sensor;
- Demonstrate through flight trials that radar target indication data can be sent down over Link 16 and compressed images over the video link. This will initially be with a trials aircraft and then on the surveillance UAV;
- Provide radar data from the surveillance UAV ground station that uses standard protocols, so a range of radar image-processing firms can undertake Work package 7, the identification algorithms;
- Provide options for pre-processed radar data that can be made available to the specialist company that wins the contract for Work Package 7, the identification algorithms. The data is to be provided using a standard data protocol acceptable to the procurement agency's technical advisers.

11.4.8 WP3 (Part), upgrade surveillance UAV autopilot

In common with many UAVs, control of the surveillance UAV is via a satellite link, giving a need for an autopilot. There will be link delays giving delays in the UAV's response to commands. Also the pilot's knowledge of the UAV's flight path and intentions will lag by this time plus the time for him or her to understand any changes.

In this example, we assume that the autopilot is fairly sophisticated and can choose, and fly, a route between waypoints. It will have the ability to land autonomously at pre-planned airbases. This means that it has full control over all aircraft flight control functions. It will also be safety-critical, with the engineering implications that follow (multiple redundancy subsystems, strict design controls, specialised operating system, etc.). The procurement agency decide that the cost for a new specialist company to change the autopilot would be unacceptably high, so negotiations start with Prime(Surveillance UAV) to upgrade the autopilot software.

The chosen defensive aids comprises a Missile Attack Warning System (MAWS), with infra-red and radar jammers to counter the incoming missile's terminal guidance system. The jammers may not be fully effective, so the UAV will need to manoeuvre to the limits of its airframe to avoid the missile. As Prime(Surveillance UAV) are also the Design Authority for the airframe, a contract is placed with them to upgrade the autopilot. The initial phase is a study by the contractor and the military users to calculate the probabilities of surviving a missile attack from Red's forces using the proposed defensive aids and the maximum manoeuvrability of the airframe. The results are satisfactory, so the remainder of the contract is awarded at a time when cash flow in the programme allows.

There will be a range of questions about the effectiveness of the defensive aids system, its false alarm rate and detection threshold. These are standard questions for any defensive aids system and the calculations are straightforward.

The technical details such as final aerodynamic performance need not concern us here. What does matter is any change in the level of automation. There is one significant change: the current autopilot simply follows a human command to fly to a sequence of waypoints. The change to be considered here is for the UAV to execute unplanned manoeuvres, and possibly release chaff or flares, in response to an on-board sensor without any human intervention. Legal advice is taken about any IHL issues as this is legitimate for a manned platform but not necessarily for an unmanned one. The guidance is that there is no IHL issue if the sensor information between initiation of a defensive aids suite response and the UAV returning to human control is not used for targeting decisions.

11.5   Legal review and guidance

11.5.1   Is a review needed?

The main IHL question that the nation's legal authorities will ask is whether the system-of-systems is a new or novel means of warfare or not, and that it adheres to the four principles of necessity, humanity, proportionality and distinction introduced in Section 7.6.

At first sight, the capability is not a new or novel means of warfare; the new system of systems is providing a targeting system which still has the lethal decision to release a weapon under the control of a human operator in a ground station. However, there are a range of legal issues noted in the discussions of the work packages in the preceding sections. These are:

1. The use of radar alone on the surveillance UAV to recognise potential targets proposed in Section 11.2.9. The questions are the reliability of radar target indication and identification algorithms in cluttered environments;
2. The automated cue from the surveillance UAV to the tactical one indicating a target in Section 11.2.7;
3. The speed and latency of the various bearers making up the targeting network described in Section 11.3.1;
4. The transfer of control of the capability from the maritime to ground command centres in Section 11.3.1. This gives the possibility of a period when there is no control by a superior node;
5. The decision in Section 11.3.2 to have the local command centre in the tactical ground station. This is in the operational area, whilst the surveillance ground station is remote and my not have access to battlefield information distributed on local networks. This raises the possibility of the data used by the two ground stations using discrepant data sets;
6. The workload issues identified in Section 11.3.2 and 11.4.6, associated with the human–machine interfaces, raise questions about quality of judgements made by operators under extremely high workload;
7. The potential fitting of defensive aids to the UAVs in Section 11.4.2 and subsequent autonomous control of the UAV in response to threats until the operator regains control.

These issues should be included in the Article 36 Review before setting the requirements and specifications prior to issuing invitations to tenders.

Targeting can be described as the OODA process so the review will need to look at the assumptions in it and whether any changed parts of the system-of-systems undermine these or introduce new problems.

11.5.2   Testing assumptions

Section 8.5.2 listed ten basic assumptions in the OODA loop that rely on engineering products or expertise. They relate to the information available at each relevant node, the rules of engagement and integrity of information over time. They are repeated here in a slightly abbreviated form with the associated legal issues and their engineering implications:

The OODA assumptions from Section 8.5.2 are:

1. *The weapon commander is the human who makes the decision to release or fire the weapon and has the final IHL responsibility for that action. The commander may be an infantryman with a rifle, or could be the captain of a submarine with a highly automated cruise missile as the weapon;*
    This assumption is still valid as the tactical UAV is under the control of a human operator in a ground station controlling it. There is a qualification as the nation will only offer the coalition a capability under its own local command centre.

2. *The commander is in a defined command chain such as the one shown with thick lines in Figure 8.2;*
    Figure 11.4 shows one ground station controlling one tactical UAV. This is the same effect as Figure 8.2, but Figure 11.4 is a simplification as any ground station can take control of any UAV in the operational area. This happens when a tactical UAV is assigned to strike the target with connectivity requirements illustrated in Figure 11.5. The engineering evidence will have to cover three requirements:

    (a) When a UAV and ground station are assigned to the strike, there is an immediate C2 link made between them and that this cannot be broken or reassigned to another UAV system without an instruction from the local command centre;

    (b) Issue 4 above gives the possibility of a short break in the C2 chain when the local command centre's line of authority is transferred from the maritime to the ground command centre. The risk of weapon release during this time can be reduced by procedures prohibiting weapon release during and near this time, but the maximum length of this time must be specified and verified;

    (c) The tactical UAV operator will have to make weapon-release decisions on the basis of his or her knowledge of the target and its immediate surroundings *at the time of weapon release.* This will ideally be provided over a data link from one or more of the sensors across the system-of-systems. Any deviations from a direct optical video link will have to be examined at the review for reliability and technical quality of images and object recognition capabilities.

3. *The commanders in the OODA process have clear and unambiguous rules of engagement;*
    This is not normally an engineering problem. However, as autonomy levels increase, some parts of the rules of engagement will be included in databases

and used by targeting algorithms. These must be under configuration control, with checks that they are the correct ones for that phase of the operation. There will need to be ambiguity checks before they are loaded into the databases. The primary role of the technical staff will be to prepare evidence for use in drawing up rules of engagement at the start of a campaign.

4. *The 'blue' communication network is not corrupted by hostile action. Any battle damage to it having been corrected or alternative communication links established, retaining its integrity even if the data rate may have been reduced or latency increased;*

   This is a standard engineering problem for all defence applications. The bearer network and connectivity should have been designed with robustness and redundancy inherent in the design. A policy decision will need to be made about correction routines if corruption of command data is suspected.

5. *The off-board data received by the commander has come over reliable data links from accredited sources so it is a legitimate basis for action;*

   This is essentially the same question as those in 2 and 4 above with the extra consideration of accreditation of the data source. This should be answered by the receipt of the data over the coalition's networks. There is still the possibility of non-accredited data on the network due to the rapid changes in any operation which is outside the scope of the design of this capability.

6. *There is a recognised picture of the battlespace maintained and authorised by the command chain. Not all information will be available to the weapon commander, but all information about the target area will be passed to them as quickly as is reasonably possible;*

   The proposed modifications to the C2 architecture are designed to maintain access to the same recognised air picture that is used in all coalition operations so the only engineering evidence will be to confirm that the networks allow access to it for the relevant times in the mission. Access to the recognised ground picture may be less straightforward and there may be a requirement for some changes to network access requirements. This is likely to be a consideration for manned aircraft operations as well with a dynamic ground scenario and problems with both friendly fire on Blue troops and collateral damage.

7. *There is a local database for each node which is under configuration control and updated as necessary;*

   This will be an inherent part of the design if stated in the requirements.

8. *The commander has sensors which are under their direct control for targeting purposes. These may be located on one or more platforms including the weapon;*

   The capability has been defined so that there is at least one sensor providing up-to-date information from the target area. Adherence to STANAG 4586 should provide means for the weapon operator to take control of the sensors on the surveillance UAV if required. If this is not possible due to workload or other considerations, 'direct' may be interpreted as meaning using another ground station with immediate transfer of the information. Questions of data latency and who is interpreting it will need to be considered by the review and considered for the database of information used for drawing up rules of engagement.

The targeting process relies on target indication algorithms and compressed images cueing more sensor data transmission and further analysis with object identification algorithm. The target indication may come from one type of UAV and the identification from another type which may have its software and algorithms from a different supplier, based on different mathematical approaches. The reliability and accuracy of these processes and algorithms will be subject to legal review for this project, even if the proposed technology is being developed for other programmes as well as this one.

One very important consideration here is the place of the radar sensor in the targeting chain and its reliability in identifying a legitimate target and tracking it over time so that a strike can be made on its hide. It is inherent in the DNAW requirement that there will be operations with no supporting optical information for the operator to use. This might be defined as a new means of warfare, but even if it is not, the early legal review should identify this as a problem and put actions in place that may lead up to special trials of radar object recognition algorithms in representative scenarios and data for setting rules of engagement.

9. *A collateral damage assessment will need to be made by the commander releasing the weapon;*
   This should be straightforward with the available sensors and data links. Even if the tactical UAV is cued from radar information, its operator should have optical information from its on-board sensors before weapon release. There will need to be training for the tactical UAV operator to relate the radar-based information to the visual scene. Poor weather conditions may only allow a short time between the operator seeing optical images and the weapon-release decision. Automated collateral damage estimation methods, if used, will need separate consideration.

10. *The weapon used will be one that has an immediate and limited effect on hitting the target, i.e. it is kinetic and will either explode or will become inert in some way immediately after impact. There may be a time delay of minutes or hours between the last time that the commander can change the instructions to the weapon and impact time.*
    The capability is using a small air-to-ground missile with a range of a few kilometres. This satisfies this assumption both for kinetic effect and a short time between release and effect.

The discussions above are among those that will be considered at the pre-invitation-to-tender review and the procurement competition allowed to proceed. The low TRL of the radar identification algorithms will almost certainly lead to the radar-related contracts having specific clauses allowing the procurement authority access to their bases. Sufficient detail will be needed to allow them to be fully tested in realistic scenarios, testing their weak points such as differentiating between types of tracked vehicles, or confusion in cluttered environments.

11.5.3   The Article 36 Review of the initial capability

There will be an Article 36 Review of the initial capability at the end of the integration programme shown in Figure 11.7. This will be a major milestone, marking the beginning of its introduction into service. The review is conducted by the state's legal authorities. The role of the technical teams is to provide evidence. This evidence will include the results from the work verifying the behaviour and performance of the system-of-systems identified at the end of Section 11.4.1.

The review does not set rules of engagement, but it is sensible for the technical evidence to be presented in the form of reports, supported with videos and other media which can form a database for the review, and for setting rules of engagement when needed. Early parts of the introduction into service have to establish the procedures, training needs, and assess rules-of-engagement restrictions on the use of the capability.

Most of the review of the network and C2 systems' operations will be based on specifications and test data that are already in place. The engineering team will probably provide a single report reviewing these from an operational perspective based on the four IHL principles. The review will need to confirm that in normal operation all legal requirements are met in operational use. One important item will be the transfer of responsibility for the local command centre between maritime and ground command centres. There will probably be existing procedures for the handover of command in the C2 chain in dynamic situations and the application of these to this capability will form one item on the review agenda.

One significant problem area is the operation and accuracy of the various optical, infra-red and radar algorithms used for cueing further observations and identifying legitimate targets and non-targets. It is virtually certain that the division of algorithms, processing and data-compression over the networks and links will have changed during the procurement and integration work. These will have occurred due to technical issues and advice from military users during the modelling and simulation activities.

The pre-invitation-to-tender review will have identified qualitative criteria that need to be set for the final system to be used in different scenarios. One example would be the criteria for transferring a radar-based indication from a surveillance UAV to an optical identification algorithm in the tactical UAV ground station. The criteria would give defined reliability of indication or identification as a decision aid to the operator for weapon release. Another example would be the reliability of passing the track of a possible launcher from a surveillance radar sensor to a tactical UAV from a package cued after both the initial indication and when the launcher is in its hide. All these algorithms can be considered to be automation and act as decision aids only if the operator makes the final lethal decision based on real-time optical data from a sensor looking at the target area with adequate resolution.

The sensor evidence will include the outputs of the algorithmic performance tests based on the radar-related contract clauses set at the pre-invitation-to-tender review.

Another significant issue is operator workload and their ability to make legally defined 'reasonable' decisions using the human–machine interface in the tactical UAV's ground station in the confines of the in-service ground station. The results from the study in Work Package 5 will be an important part of the review. The start of introduction into service may be the first opportunity to apply them with military operators using real assets in simulated operations with the full system-of-systems.

One output from the pre-invitation-to-tender reviews should be outline specifications for the trials that would be needed for the final review before release to service. These specifications will be refined during the design and integration phases. They will be presented at the early reviews so that the completeness of meeting IHL requirements is established. Any further results required from simulation and modelling will be identified. Good quality simulations are likely to reduce the need for, and cost, of trials using real assets but still provide robust evidence for the final Article 36 Review before release to service.

The technical evidence presented to the review will be part of its formal record and used by military authorities at later dates. The evidence will of necessity include detailed technical reports whose implications will not be immediately obvious to non-specialists. Therefore the engineering team should provide one or more reports giving the key technical issues under the four IHL headings of: necessity, humanity, distinction and proportionality. This is not too onerous a task with the use of the requirements given in Chapter 10, modified for the specific application.

The final Article 36 Review before the capability is released to service will use the output of this review, plus results from the early phases of the capability's introduction into service, with any required design modifications. The review inputs and results should also provide the database which will be used to set rules of engagement before operational use of the capability, so the engineering evidence to the review must be updated if necessary to describe the capability entering service.

References

[1]   Thales UK. *Link 16 Operational Overview*. White Paper https://www. thalesgroup.com/sites/default/files/database/d7/asset/document/White%20Paper %20-%20Link%2016%20Overview.pdf [Accessed on 2 October 2018].

[2]   Irvine J.A. 'National imagery interpretability rating scales (NIIRS): overview and methodology', *Proc. SPIE 3128, Airborne Reconnaissance XXI*, 21 November 1997. pp. 93–103.

[3]   John M. Irvine J.A., and Nelson E. Image quality and performance modeling for automated target detection'. *Proc. SPIE 7335, Automatic Target Recognition XIX*, 73350L 4 May 2009, pp. 73350L-1–73350L-9.

[4]   Further details are available on the NATO websites: https://npc.ncia.nato.int/ Pages/accs.aspx and http://www.nato.int/cps/en/natohq/topics_8203.htm. It is the responsibility of the NATO Communications and Information Agency: https://www.ncia.nato.int/Pages/homepage.aspx [All accessed on 28 October 2018].

Chapter 12
Making military capabilities autonomous

12.1 Approach

This chapter, like Chapter 11, discusses how the principles in Chapters 8, 9 and 10 are applied to autonomous functions in weapon systems. The approach discussed in Sections 5.4.1 and 10.2.2 is to identify the 'critical' functions in a system, i.e. the ones in the targeting chain that lead directly to a lethal decision, usually weapon release [1]. They are generally taken to be the decisions and actions normally taken by a human during target identification, selection and attack. These can be, and are, described using the Observe, Orient, Decide and Act (OODA) process shown in Figure 10.4.

Chapter 11 used a specific example to illustrate the engineering implications of demonstrating adherence to International Humanitarian Law (IHL) with current levels of automation. The same example is used here – a capability to detect and destroy a missile launcher in a scenario of a littoral operation against highly capable opposing forces. Section 11.2 gives the details. It is called the 'current' capability to distinguish it from the Autonomous Weapon System (AWS) described in this chapter.

This chapter investigates the problems which arise when proposals are made to make the same capability autonomous, i.e. to develop an AWS with high levels of autonomy in its critical functions so that there will be no human decision-maker after the initial instruction to the system to find and destroy a particular target or type of target[1]. A system of this type may not necessarily be illegal under IHL but would have some strong limitations on its use in operations.

The criteria to decide whether a function can be automated fully or partially are based on the requirements given in Section 10.3. These military criteria will not be directly applicable to autonomous systems in civilian applications where, in general, there is no explicit prohibition on actions following automatically from an autonomous decision, with no referral to a human. However, the principle that there must ultimately be human responsibility, applied in this chapter, can be applied in non-military systems. Application of the military approach to civilian systems is discussed in Chapter 13.

---

[1]A note on terminology: the word 'weapon' is used to describe the part of the system with the lethal component such as a warhead. It is automatic in that it will fly to, or home in on a target given to it by a higher node in the command chain and cannot change it. The node giving the weapon its target and releasing it can be either a human operator or an automated node within the AWS.

Two approaches are taken. The first is a 'top-down' one where a new capability is specified and procured as a single project. The second is where the autonomous capability evolves through incremental upgrades to the system developed in Chapter 11 using current technologies.

When considering how to raise the autonomy level of the functions in an existing system, such as the 'current' capability in Chapter 11, the designer needs an architectural description of it. The discussion here needs to be at a sufficiently general level to show its applicability to other weapon systems, but detailed enough to show the type of real problems encountered in practice.

The architectural description used here is based on 4D/RCS nodes, defined in Section 10.5.2, which is based on the three-part model of human decision-making, as is the OODA process. The nodes are specified in a way that they can be implemented as one or more humans or as technology in a Command and Control (C2) chain. The 4D/RCS standard may not have been used by the AWS suppliers, but all the modules and subsystems making up its functions can be it can be placed in the 4D/RCS functional elements. This abstraction then brings out very clearly the IHL problems which arise when a decision and action are moved from a human to a machine and vice versa.

This chapter, as with the rest of the book, looks at the engineering implications of IHL for autonomous systems, not the legal ones. Therefore, any given or implied interpretations of the law are the author's opinion and should not be taken as definitive. Formal legal advice should always come from the legal profession.

12.2   Introducing autonomy into systems

Autonomy is sometimes described as a disruptive technology for military applications. A disruptive technology does change the nature and methodology of warfare but, in practice, the changes take several years or decades to have an impact. A technology must be mature with reliable products before it can be trusted in conflict where many lives depend on its correct performance. This means that the 'disruptive' technology is already well out of the research phase and has been demonstrated either in civilian applications or in military laboratories and field trials. This equates to Technology Readiness Level (TRL) 5 or above[2].

Even a mature technology is still of limited value if military tactics have not evolved to exploit the benefits. One classic example is that of tanks which had virtually no impact on World War I but did on World War II, two decades later, when suitable tactics had been jointly evolved by the Germans and Russians in the early to mid-1930s.

Unmanned Air Vehicles (UAVs) are also an example where a significant military advantage only came with the evolution of new tactics. They started in the 1950s as remotely controlled aerial targets for missile systems. Some specialist UAVs were developed in the 1990s, such as the UK army's Phoenix [2], a remotely

---

[2]TRLs are discussed in Section 5.2.4.

operated surveillance asset which continued in that role until withdrawal from service in 2006. Larger UAVs such as the MQ-1 Predator surveillance UAV have evolved from a surveillance asset into the MQ-9 Reaper which carries out surveillance but can also release weapons on command from its remote pilot. The various models in this evolution can be found in Reference [3]. This example shows how new capabilities may be introduced incrementally, with parallel developments in tactics over a period of several decades. Public concern about the ethical acceptability of 'drone' attacks ensured that one dominant driver for tactics and engineering design was IHL. It is likely that ethical and legal concerns will continue to be an important design consideration.

Increases in capability also happen due to changes at lower levels within a system, especially when analogue modules are replaced by, or incorporate, sophisticated electronics and digital components. The aim is usually better performance of the existing tasks, but the spur to their development can also be the ability to perform extra tasks giving more value and benefits. The new tasks become an accepted part of operations with new procedures developed to exploit them. An example from the civil sector is Electronic Flight Bags (EFBs)[3] for airline pilots [4] where laptop technology originally replaced paper data such as airport maps, with digital versions but has now expanded capabilities enormously.

Incremental upgrades may occur in several functional subsystems[4] with the result that the system and system-of-systems' capability is increased dramatically when compared with the original system-of-systems. Increased autonomy in only one critical function may not make a system illegal as the Tactics, Techniques and Procedures (TTP) will be changed to incorporate it, probably by modifying the human's role in the targeting process. The risk from an IHL viewpoint is that a succession of incremental increases in subsystem autonomy, when used together, may make the system or system-of-systems capability a 'new means or method of warfare' and potentially illegal unless reviewed.

The question that arises with incremental upgrades is: when should an Article 36 Review be initiated? Is it at the proposal for the first incremental upgrade, the final one, or somewhere in the middle? There is no simple answer to that question; it can only be specific to a particular system and nation. However, the example used in this chapter has been chosen to illustrate some of the problems.

When considering an AWS, whether a new system or an incremental upgrade, the main questions that must be addressed in an Article 36 Review must be derived. These should consider the system-of-systems with autonomy in one or more of its systems providing critical functions. These are stated in Section 12.3.

---

[3]An EFB is a device that allows flight crews to perform a variety of functions that were traditionally accomplished by using paper references. In its simplest form, an EFB can perform basic flight planning calculations and display a variety of digital documentation, including navigational charts, operations manuals and aircraft checklists. The most advanced EFBs are fully certified as part of the aircraft avionics system and are integrated with aircraft systems such as the Flight Management System (FMS).
[4]The definitions of systems-of-systems, systems, subsystems and modules are given in Figure 5.1.

12.3 Article 36 review questions for the system of systems

The principle questions arise from Article 35 of AP1 and the four tenets of necessity, distinction, proportionality and humanity. The engineering requirements derived from these are given in Section 10.3.2, Panels 10.1–10.4. The requirements in Panel 10.1 lead to a requirement for the system to be a node in a clear and unambiguous C2 chain which starts at the government level for legitimacy of the campaign, through the next levels down to the command centres for the operation and the AWS.

An AWS must be considered as part of the system-of-systems making up a military force conducting operations in the campaign. There will be an existing architecture for the system-of-systems and principal systems using a recognised architectural framework. Therefore, a meta-model such as Figure 10.1 can be used to define the AWS and its interfaces. This provides a context for both engineering and IHL considerations. The questions that must be answered at the weapon-system Article 36 Review are at least the following ones:

1. Has adherence to IHL principles been demonstrated for the AWS?
2. Is the behaviour of the AWS and system-of-systems understood when operating together with no failures in them?
3. What information is available to the commander from the AWS, and is this sufficient for authorisation of weapon release?
4. What information must be made available to the AWS commander from elsewhere in the nation's or coalition's infrastructure for authorisation of weapon release?
5. Are the critical functions in the AWS identified clearly?
6. Is there an authorised targeting process which provides the target list to the commander of the AWS?
7. Who has authority to change the target list and any other data in the AWS?
8. Is there a clear C2 chain from the system-of-systems commander down to the lowest node with authority to release the weapon or divert it from its target?
9. Are all nodes in the C2 chain authorised entities?
10. Are the range and limits of authority of every C2 node defined and acceptable under IHL?
11. Is there a clear separation of human and machine-made decisions in the C2 chain?
12. What is the level of human control over actions taken by the AWS in the targeting process?
13. Are the main failure modes and their consequences known and available to the users?
14. Are there mechanisms in place to detect corruption or other intrusions due to cyber-attacks?

The technical evidence that will need to be presented for review will include:

1. The performance of the bearers and communication channels used for the command chain of authorised nodes making lethal decisions;
2. Evidence for communication-system integrity and robustness;

3. The division between human and machine-made decisions, at every node in the system;
4. The division between human and machine-initiated actions;
5. The authority level or authorised power of each node;
6. The information needed at each node for its decision and its availability;
7. The security protocols and processes for all data associated with targeting and weapon release;
8. The separation of on- and off-board data, its integrity and authorisation of data used in lethal decisions;
9. The consequences of each automated node not having sufficient information for a decision;
10. The performance of all aspects of target recognition, indication and identification functions;
11. The ability to identify and tag civilian objects;
12. The performance of systems providing data for assessing collateral damage;
13. The effect of noise and uncertainties in the weapon system's decisions and actions;
14. The effect of an unexpected loss of communications between the systems in the system-of-systems;
15. Security policies, measures and implementation for all aspects of the system.

This information will need to be presented for any automated decision-making weapon system. It will also be needed when assessing automated decision-aids supporting the lethal decision-making chain in a mission.

Article 36 reviews should occur at every major milestone. Clearly, the questions given above can only be answered at the final review, with the full set of technical evidence generated over the whole procurement process up to release to service. The most cost-effective and efficient approach is for programme managers to make the Article 36 reviews an integral part of their programme, liaising with the technical managers. The aim is to couple together the legal concerns and the provision of the relevant technical evidence to allay them or to address them during the next phases of the programme.

In addition to the above questions, there will be a need for a data pack of necessary information for setting rules-of-engagement before and during a campaign. This will draw on design information and evidence presented at the reviews such as test results and results of trials with the AWS in the military infrastructure.

12.4 An example capability with upgrades for autonomous operation

12.4.1 Top-down approach

The top-down approach starts with a need for a capability to locate and destroy the missile launchers, as given in Section 11.2.2. This would be followed by operational analysis and refinement of concepts, followed by a decision that, in principle, the state will have funding for a new system as part of its long-term strategy. The state could opt for a new highly automated system to deliver the capability. This could be for one of at least two reasons:

- They have funding to make the required performance modifications to their current assets, develop the control processes, the algorithms and their implementation as an AWS;
- The solution in Chapter 11 is not practical because of insurmountable logistical problems with supplying several operator ground stations to the operational area.

A 'top-down' proposal of this type would be a significant financial item in the military budget and almost certainly trigger an Article 36 Review at a very early stage. The results of this review and the discussions around it are likely to provide considerable guidance to the procurement authorities and their military and technical advisers about the design requirements for such a system, and the roles and locations of the humans in the targeting process. The programme managers will then plan Article 36 reviews at key stages as well as more frequent reviews of specific subsystems if required.

The military problem is the one given in Chapter 11: the need to survey a large area and reliably detect the appearance of one type of military vehicle, followed by a rapid strike response. The assumption is made that analysis shows that the best solution is the one in Chapter 11, i.e. three surveillance UAVs operating at medium altitudes and four weaponised tactical UAVs operating at low altitudes. This scenario lends itself to an automated process, the surveillance UAVs cuing strikes from the tactical UAVs after a launcher is identified. The surveillance and tactical UAVs will be capable of planning, flying, and changing their flight paths autonomously based on off-board and on-board sensor outputs. All parts of the OODA process will be automated, leading to weapon release with no human intervention. A reliable system with this capability is an AWS as defined in Panels 1.1 and 3.1.

A potential C2 architecture for an AWS to deliver the capabilities of the current system is shown in Figure 12.1, with the capability being controlled directly from the maritime command centre after a request from the ground command centre. The human commander of the AWS can be located anywhere, as long as there is a command link to the AWS. A control function is shown as part of the AWS and is the AWS commander's interface with it. This will be highest level node in the AWS and the first machine decision-maker with authority to release a weapon. It may be located with the human operator or on one of the UAVs but with a direct link to him or her. The humans in the UAV ground stations in the 'current' capability are replaced by automated decision-makers. Close liaison is expected between the AWS and air commanders both for tactical and air safety reasons.

The proposal preparation and review process will include high-level architectural descriptions of the wider military infrastructure and its interfaces with the system-of-systems shown in Figure 10.1. For the purposes of this example, it is assumed that they will not have much detail at level 2 in Figure 12.1 and virtually none at the lower levels, 2 and 3 which include the AWS. This is a reasonable assumption as the subsystems in the current capability will have been supplied

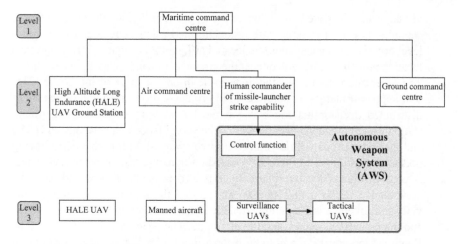

Figure 12.1   Example proposal for an autonomous strike capability

under contracts with detailed interface specifications derived from the procurement agency's architecture framework, but with no need for detailed internal descriptions of the contractor-supplied products.

12.4.2   Functionality in the top-down approach

The functionality of the UAV ground stations will be distributed between Level 2 and Level 3 for the solution shown in Figure 12.1. The system designed in Chapter 11, and shown in Figure 11.4, had the human commander acting as a local command centre located with one of the weaponised UAV ground stations.

The human commander must be in the operational command chain for every lethal decision. Their location is an important design choice if the AWS is a new design; the complexity depending on the implementation of the 'Control function' in Figure 12.1. Implementation on hardware makes it preferable to have the commander and this hardware located with the ground commanders, or in a separate ground station which could be a modified version of the current tactical UAV ones. Their command by the maritime centre becomes a procedural matter for the time that this is required.

A purely software virtual implementation of the 'control function' makes it possible for the human command to be by one person at either command centre, with transfer between them being a procedural matter but with some technical implications for continuity of control.

The two types of UAV are 'currently' operated from their specific ground stations, and have different functions. The choice of UAV operating in contested airspace is made by the human at the relevant ground station. The location of the node which makes this decision for each UAV type will be important and will depend on technological evolution:

• The first alternative is for task allocation of all UAVs to be decided in a central node which replaces the C2 functionality of all ground stations. This could be virtual or in a dedicated unit. This may give greater mission effectiveness due to centralised decision-making with tasks allocated using more strategic, mission-level criteria. This node would be located at Level 2 in Figure 12.1, its physical location probably being with or near the ground commanders, as discussed above;

- The other alternative is for the node to be in one or more of the UAVs. This will happen if increasing automation in ground-station functionality has led to distributed decision-making between the UAVs with no need for a ground station except at launch and recovery. The surveillance and weaponised UAVs and ground stations will have become two swarms, with each swarm having nodes located in one of the UAVs based on dynamic task-allocation criteria;
- A combination of the two alternatives is also possible. This would have the surveillance UAVs acting as a swarm so that the maximum amount of surveillance coverage is maintained dynamically, only reporting the areas covered and the potential targets to the Level 1 node. The weaponised UAVs would be flying autonomously in a holding pattern in a low risk area. The human commander in the Level 1 node would choose one of them for the strike on the target, and command it to conduct the strike mission.

A fundamental question that must be addressed for all aspects of the architecture and detailed design of the AWS is its susceptibility to cyber-attack. The choice of location will almost certainly be an important factor in assessing the system's vulnerability. It is likely that critical data and results of decision-making will be passed over one or more networks in a contested area and spectrum. Even if the links are secure and the data not corrupted, it is likely that there will be considerable network latency and problems with time-critical data transfer.

It should be clear that the legal review of a C2 architecture relying on dynamic task allocation for the location of the decision-making node will be very different to that of a centralised node. The behaviour of both will be set by their respective authorised level, possibly expressed as a sum of authorised powers. The identification of authority and responsibility for decisions will be more complex for distributed decision-making with requirements for detailed architectural analysis and engineering complexity in the final design.

The decision about node location in this case is one which should be made using requirements for the demonstration of adherence to IHL. These should be derived through Article 36 Reviews carried out at early stages in the anticipation that they will identify the key functions in the AWS and how they will be controlled in accordance with IHL. These considerations may be additional to the standard technical ones. Demonstrating the adherence of a dynamic task allocation system to IHL is likely to be complex, expensive and lengthy, giving programme complications.

There will be a considerable amount of operational analysis carried out at all procurement stages. This will include simulations of many aspects of the system and its automated operation. These will be to check both technical performance and that the future military users will understand how the system-of-systems will operate when it comes into service. Some of the operational analysis results will be part of the evidence presented at the Article 36 review. Ideally, some of the simulations will be used to identify likely restrictions on the use of the system-of-systems under a range of rules-of-engagement.

The individual reviews in the process will have to address similar issues to those described in Section 12.4.3 for incremental upgrades. The main difference being that a top-down approach to the AWS design will enable more strategic decisions to be made about the division of decision-making between humans and machines and the location of the different functions both physically and in the architecture.

12.4.3 Incremental approach to autonomy

A different way of achieving autonomy in all functions in the capability in Chapter 11 is a set of smaller independent contracts to introduce new developments in autonomous technologies into different parts of the weapon system. The justification for this will range from increased weapon system capability, through reduced operator workload to the opportunistic introduction of technical capabilities developed by companies or research laboratories for this, or similar, purposes. It is likely that at least some of these suggestions will be implemented, leading to incremental increases in the autonomy levels of critical functions.

The same capability example is taken as in Section 12.4.1, the missile-launcher hunting system from Chapter 11 and shown in Figures 11.2 and 11.5 with the command chain shown in Figure 11.4. It is assumed that the work packages in Table 11.6 have been completed so that the system-of-systems shown in Figure 11.2 delivers the required capability. It is also assumed that trials and exercises have shown that the system is capable of attacking other types of hostile equipment when identified by operators of other surveillance assets.

The main issues have been discussed in a general way in Sections 9.5 and 9.6. The risk here is that without sufficiently robust senior management, isolated incremental upgrades will introduce changes in several parts of the critical functions without consideration of their wider implications, leading to emergent behaviours[5]. These are fairly common in a complex system-of-systems and solutions have to be sought in trials and simulations, and corrected if possible, or robust work-arounds implemented.

One example of this type of IHL risk is that under some circumstances, automating lower-level capabilities could lead to a strike being launched by a weaponised tactical UAV with no human decision after an initial target indication by the surveillance UAV. This could arise if the seven UAVs are operating in fairly high-risk airspace, resulting in all operators having high workloads and being under considerable stress. Increased automation in the target recognition and hand-over processes could then lead to a tactical UAV attacking a non-military object. To prevent this, human oversight has to be explicitly mandated in one or more of the incremental upgrades. Clearly this lack of authorisation would be a breach of IHL, but piecemeal increases in automation followed by incomplete testing at system-of-systems level could lead to an inadvertent breach of IHL.

---

[5]Emergent behaviours are discussed in Sections 5.4.3, 5.4.5 and 5.5.2. The related topic of wicked behaviour is discussed in Section 4.9.

Analysing the changes from human to automated decision-making shows the problems that systems engineers will meet. The systems engineering task for every incremental increase in autonomy is to identify which subsystem or module must continue to have human judgement for decisions, and ensuring that no action can be taken without the necessary human authorisation. Consistent and thorough systems analysis will show the type of empirical evidence that will be needed at reviews, and also the range of decisions that can be analysed, and acted on, by the automated systems.

There will need to be periodic Article 36 reviews for a system which has incremental upgrades. The criteria for their timing is best set at a management level which has oversight of the whole capability, its operation and in-service support, but everyone in the process should be satisfied that there are adequate reviews. The system-of-systems containing the capability should be treated as a potential 'new means or method of warfare' due to changes in the already highly automated critical functions in it.

The technical requirements to meet IHL are derived in separate sections below. Evidence to show that these requirements are met can be presented both at the final review and when needed at interim reviews of the individual systems and subsystems. The trial and test results obtained during the design and trials with military assets will also form a set of evidence for setting rules of engagement.

The issues addressed here naturally lead into responsibility for the parameters used in the automated processes and who verifies that the subsystem including them is operating correctly. Using the concept of authorised power to set the limits on a node's ability to take an action ensures that action only happens if all the measured parameters meet the agreed criteria.

The issues raised in this section also apply to civilian applications of automation in 'autonomous' systems; incremental or piecemeal increases in automation in systems or subsystems could inadvertently give rise to dangerous or illegal emergent behaviours by the higher-level system.

12.5   Configuration and control during the OODA process

12.5.1   Configurations during OODA

Operational analysis results in Chapter 11 led to the seven UAVs (three surveillance and four tactical UAVs) being operated from three ground stations, each with one operator. There was a local commander located in a tactical UAV ground station deployed to the operational area. Techniques developed during exercises will give the optimum split of responsibilities and handover between ground stations during the targeting process. The problem that the designer will have to address is how the system of UAVs, ground stations and datalinks carry out the 'Roles' and 'Functions' in Figure 10.4, with capabilities realised as subsystems and demonstrating adherence to IHL at an Article 36 Review.

The roles will be performed using the functions in Figure 10.4, and others as necessary. The functions will be carried out by subsystems in one or more of the UAVs and ground stations. These subsystems will be automated nodes in the

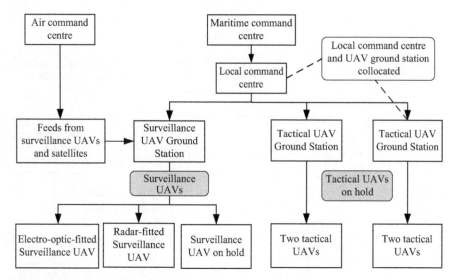

Figure 12.2 C2 architecture for 'current' capability UAV system during the Observe phase

command architecture for the 'current' capability. For example, the 'Target Identification' function in the Orient role will be carried out by a ground-station node which has the capability to compare sensor inputs with a model of the scene and indicate when it thinks there is an object which resembles the desired target. The 'Target Indication' function in the 'Surveillance' role may have to be carried out in the UAV if the datalink between the UAV and its ground station has insufficient capacity. The decision where the indication is carried out will depend on the probability of accurate indication with low data rates but large ground-station computing power compared with the larger amount of data but more restricted processing power on the UAV.

Figure 12.2 shows the system configuration for the 'current' capability during the Observe phase. Surveillance is the responsibility of one surveillance UAV operator who also has feeds from the strategic HALE UAV and satellite surveillance assets. The tactical UAVs are under the command of two operators who may not be collocated. If possible, two tactical UAVs are nominated as the ones most likely to be designated to attack the target when indicated. When one or more target indications are accepted by the surveillance operator this OODA phase ends with the control of the UAVs being reallocated in a coordinated manner and the Orient phase starts.

Figure 12.3 shows the configuration during the Orient phase. The designated tactical UAV carrying out the strike is under the control of the operator in the deployed tactical ground station who is full time on the task. Control of the second tactical UAV operated by this operator in Observe is transferred to the second tactical UAV ground station. The three tactical UAVs not performing a strike are held in a relatively safe area under the control of the second tactical UAV operator. The single surveillance operator in the surveillance ground station keeps control of

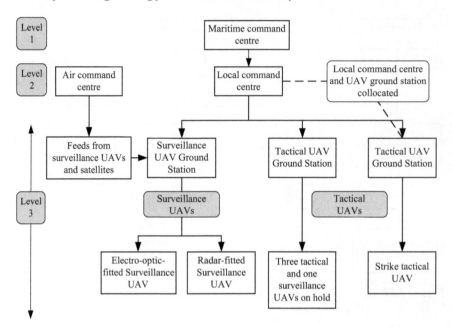

Figure 12.3   C2 architecture for 'current capability' UAV system during the Orient, Decide and Act phases

the electro-optic and radar-fitted surveillance UAVs and passes control of the third, which is on hold to the second tactical UAV operator. This assumes both of the electro-optic and radar fitted UAVs are required during the strike operation. If only one is needed, local procedures will apply for the allocation of command of the other ones to the most appropriate operator.

Figures 12.2 and 12.3 show configurations, but do not show how this relates to the division of the functions between the subsystems and modules delivering them and their location in the ground station or UAV. This division is dependent on the capacity of the command links and the information flows which will be used by commanders in making their decisions. This requires a different architecture description[6] to that shown in Figures 10.2 and 10.3.

The first step is to describe the internal system-of-systems architecture, using the 4D/RCS standard, with a hierarchy of nodes as shown in Figure 10.3. A single node is shown in Figure 10.2. As the aim is to have a completely autonomous capability, the system to be described is the AWS that delivers the missile-launcher hunting capability under a human commander, as shown in Figure 12.1; i.e. the ground stations, the UAVs and the datalinks. The architecture for the capability is shown in Figure 12.4. This has Level 1 as the operational headquarters, the

[6]Architectural views are ways of describing particular features of a complex system or structure. Views drawn up for different purposes may look completely different and show different artefacts, but must be mutually consistent for the same object.

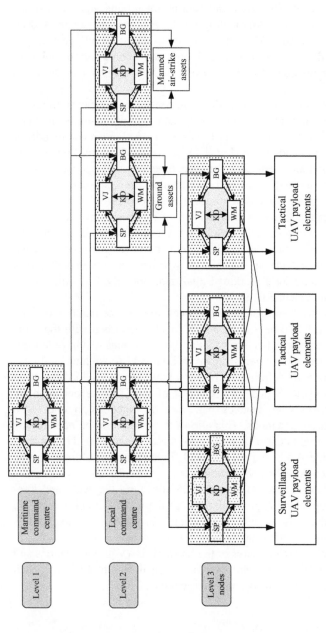

Figure 12.4 The 4D/RCS node hierarchy applied to the autonomous missile-launcher capability (Links between WMs at the second level are omitted for clarity.)

[Key: Value Judgement (VJ), Behaviour Generator (BG), Sensory Processing (SP), World Model (WM), Knowledge Database (KD)] maritime command centre, which will be supported by satellite and HALE UAV surveillance assets not shown in Figure 12.4.

The 'current' capability for attacking the missile launchers is shown as Level 2 with the human at this level. The command centres for ground and manned air-strike assets are also at Level 2. Automating the local command centre, and all levels below, gives the desired AWS that we are considering. The three 'current' UAV ground station nodes are shown at Level 3, but not the combinations of UAVs they control. The combinations for the relevant OODA phases are shown in Figures 12.2 and 12.3.

One key IHL issue is the allocation of the authorisation and responsibilities for all decisions in the critical functions between human and automated subsystem. The aim of the new AWS is to have authority vested in one human commander at the level of the local command centre (Level 2), above the 'Control Function' in Figure 12.1 with the assurance that all automated functions will perform reliably and in a way that can be understood by that commander. He or she can then make responsible decisions with adequate time to use their judgement.

The 'top-down' approach allows the division of responsibilities to be taken using the commander's workload as a benchmark for specifying the lower level functions and systems delivering them. This is not the case for the incremental approach, where individual 'current' capabilities will become more autonomous, and their effect on commander's workload analysed, possibly with little or no knowledge of further incremental upgrades.

One inescapable split of subsystems is between the surface-based and UAV elements. Each UAV will also be a node which can be described with the same internal 4D/RCS node structure. It is likely that all UAV systems will have to meet STANAG 4586 which has the elements shown in Figure 11.6. Architectural consistency is maintained here by defining the Mission Payload and Payload Recorder of Figure 11.6 as one 4D/RCS node and the rest of the UAV as one or more separate nodes. STANAG 4586 enables handover of functions in one UAV between two or more ground stations. This allows the payload to be operated by one operator whilst the flight control functions are under the command of another. In the case of the desired AWS, the payload and flight control functions may be operated by different autonomous nodes within it.

The use of a standard such as STANAG 4586 does limit the designers' freedom of choice in the 'top down' approach but brings the wider flexibility from the use of a common standard. Its use will be beneficial for incremental upgrades as they will all have to comply with it and others.

In practice, the internal architectures of the 'current' UAV ground stations and UAVs would go into much more detail than is shown in Figure 12.4 with each UAV being described by many views and specifications.

12.5.2   Command and control (C2)

Command and authority is a key issue, so the first step is to look at the command structure for the tactical UAV during operations. The configurations shown in

Figures 12.2 and 12.3 can be used to define the flow of authority for further actions, and eventually the authority to release a weapon at an identified target with acceptable collateral damage. The human/machine boundary for lethal decisions is at the output of the ground station when the weapon-release instruction is sent to the UAV. The figures show the authorised nodes in the configurations for OODA, but give no detail about how they are implemented in these nodes' internal structure.

OODA roles and functions are shown in Figure 10.4. Figures 12.2 and 12.3 show the command architectures during these phases. However, it is necessary to use a more detailed architectural abstraction to derive engineering requirements which can be used as the basis for a design. This is achieved using the architecture shown in Figure 12.4 which follows the 4D/RCS standard's definitions given in Section 10.5.2. The command chain is then represented by the data flows through the behaviour generators. The behaviour generator and other functional elements in the nodes will be implemented as modules with requirements derived from those for the required functions at that level.

Design reviews looking at the command chain should examine the modules in these data flows to ensure that they meet the criteria that there is always one, and only one, person responsible for the final lethal decision executed by the autonomous UAV. An added complication is that with the control of multiple UAVs from one ground station and spectrum availability, the command chain will be over Link 16, a Time Domain Multiple Access (TDMA) communication channel. This gives extra complications compared with a dedicated link as it has a relatively low bandwidth giving latency in information exchanges. There will need to be fail-safe procedures in the event of network problems such as congestion and hostile action.

During the Observe phase, there will be no authority to release weapons, so no behaviour generator will have this authority during this phase even though the tactical UAVs carry weapons. Design reviews should be able to check that this has been rigorously tested.

Recent operations have used systems such as a weaponised surveillance UAV releasing a bomb on a target already under its own surveillance, but under the control of the same ground station and commander. In both the 'current' capability and the desired AWS, the transition from Observe to Orient is a clear breakpoint in the operation as there is a change in the UAVs' roles and their command chains.

When the UAVs are controlled by human operators, the transition from Observe to Orient will be managed by them following a procedure which ensures that each UAV is always in a clear C2 chain through one ground station. The UAVs will not be authorised to release a weapon, but will be flying using their autopilot. For these times, a change in the C2 chain is simply a change in the source of the command to the autopilot. There will be a protocol with confirmatory messages and a fail-safe procedure if there is a break. This is already standard practice when a UAV is launched or recovered from a local airbase and control is passed from the local controller to the remote one in the operation's command chain.

In principle, it should be possible to automate the transition from Observe to Orient with the human commander deciding the time. The behaviour generators at Level 3 and below then carry out all the necessary instructions and checks. The automated transition may be problematic because of at least two shortcomings:

- Assumptions made by the commander about the reliability of the target indication. These will probably be based on information from other sources such as strategic surveillance and experience;
- Differences between the data in the on-board knowledge databases and between functions at lower levels where there may be little or no data transfers between UAVs. They will have on-board knowledge derived from pre-loaded data, their sensors and data transferred during the mission. The second and probably the third of these will not be the same across the assets.

In the AWS shown in Figure 12.1, the human commander will have access to wider information and be able to make assumptions legitimately. The critical issue is how much information can be passed to, and through the 'Control Function'. It is likely that STANAG 4586 (Section 11.3.2 and Figure 11.6) can be used as part of the design solution as its philosophy allows control of a UAV's flight systems and payloads by different operators at different times. However, the development of UAV swarms and distributed decision-making will raise numerous problems if implemented as part of this AWS.

Examination of the issues and problems raised in this section will be the subject of design reviews and the results will need to be part of the Article 36 reviews.

12.6   Technical evidence required for top-down approach

This section discusses the type of technical evidence that would be needed for the Article 36 reviews during the procurement of a single capability procured as described in Section 12.4.1. Although it is more likely that the AWS will be delivered through incremental upgrades, the individual reviews will need similar evidence to that presented here. The aim of this section, and Section 12.7 on incremental upgrades, is to provide guidance to the reader about the key issues to be aware of rather than provide checklists with the danger of their incorrect application.

12.6.1   Sensory processing

12.6.1.1   Defensive Aids Suite (DAS)

The example used in this chapter has several significant developments from the operational use of UAV systems in the campaigns of the 2000s and 2010s. The contested littoral invasion in this example introduces problems of near parity of hostile forces and poor weather, reducing the value of remote surveillance using electro-optic sensors. Intense military activity will give intensive use of all parts of the electromagnetic spectrum and hence limitations on the use of communication channels for C2 and information transfer. Predictions of spectrum scarcity in operations will lead to demands for more automated sensors and pre-processing prior to transfer to another node over the chosen bearer. This sensory processing section examines the IHL issues that are likely to arise as UAVs with more automated sensor processing power become part of the military inventory.

Those earlier campaigns had UAVs operating in airspace where the coalitions had air dominance so there was no need for Defensive Aids Suites (DAS) with their on-board sensors. Manned aircraft have DAS alerts as the most important trigger for action by the pilot. They give warning of an immediate threat such as a missile system that is locked on to them, prior to or after missile launch. This apparently simple statement actually illustrates a philosophical difference between the design engineer and the pilot. The former will state the DAS performance in terms of False Alarm Rates and probability of correct identification of specific threats; the pilot wants to know one of two basic facts: there is no current threat or, there is a threat to me of type x so I must take immediate action. It can be expected that a DAS-fitted UAV will also need to make an immediate response based on a simple two-state input of threat or no-threat.

The IHL issues that arise include:

- If the response is only to manoeuvre, there should be no IHL issue;
- By definition, a threat to a UAV does not represent an immediate threat to human life; so is a response, such as release of flares or ordnance, which might endanger life on the ground or in the air (e.g. an enemy pilot) justified? This is a question of proportionality which should be answered by military lawyers who will give guidance for rules-of-engagement. However, the engineering analysis of the reliability of the DAS warning may play a role in setting rules-of-engagement – one threshold may be acceptable in full-scale combat, but a much higher one would be needed in peace enforcement operations;
- A DAS alert will need to be reported immediately to the ground station, both for operator information about the probable response and for intelligence purposes. This will be true for all the UAVs in the scenario, even those on hold. If the latter ones do have a DAS alert, the operator is likely to demand that some communication channels be immediately available for corrective action as the UAVs on hold should be in relatively safe areas with no DAS alerts. This channel may provide a method for human oversight and possible vetoing of a lethal response to the threat to the UAV.

The DAS-response subsystem can be considered as a single 4D/RCS node, of the type shown in Figure 10.2, in the UAV architecture. Its structure is shown in Figure 12.5 with the data flows after a threat warning. The sensor output is compared with a threat library in the knowledge database; the threat type and the goodness of fit is passed to the world model function which analyses the threat type and other information, such as its motion and distance, to give the threat level; the result is passed to value judgement function which assesses the need for a response. There is a dialogue between value judgement and the behaviour generator if there is any choice of options for action and their result. The behaviour generator then takes action according to its pre-loaded authorised power (standing instructions).

The technical evidence required for Article 36 Reviews are derived from Panel 10.5. The DAS is a specialised function so the review will concentrate on the criteria used to distinguish a threat from other objects. In this example, DASs are

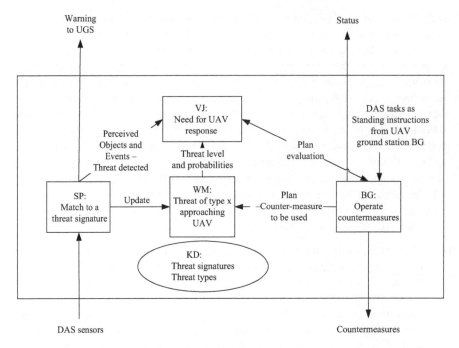

Figure 12.5   DAS processing shown as a 4D/RCS node
[Key: Value Judgement (VJ), Behaviour Generator (BG), Sensory Processing (SP),
Knowledge Database (KD), Defensive Aids Suite (DAS), Unmanned Air Vehicle
(UAV), UAV Ground Station (UGS)]

mounted on UAVs which are only deployed on missions in areas of armed conflict, so it can be assumed that there are unlikely to be civilian aircraft in the area. This means that the threshold for identification as hostile may be lower than that for the DAS of an aircraft operating outside military airspace. Conversely, the false alarm rate cannot be too high or the DAS' countermeasures will be consumed rapidly, with a risk of none being left when there is a real threat. This is not an IHL concern unless the countermeasures are active and there is an excessive risk to personnel and objects on the ground.

The technical evidence will have to be sufficiently persuasive for the review to conclude that the match between the sensor output, and the criteria for identification as hostile are clearly specified. The calculation of probability of hostile identification must be unambiguous and verified by analytic calculations and verified by operational trials under a range of representative scenarios. A tabulation of probabilities, threat type and scenario representation will provide evidence both for the review and for setting criteria for their use in a mission. This allows a quantitative comparison of engineering terminology and that of military and legal staff.

(Civil autonomous systems must also have some sort of hazard-warning system, the main difference being that the hazard can be to an object in the environment as well as to the autonomous system itself. However, the approach of having a risk assessment with a response which is proportionate to the likely damage and its urgency can still be followed.)

12.6.1.2 Non-DAS sensors during the Observe phase

The non-DAS sensors and functions will act in a similar way to their DAS equivalents, but the information and analysis will be more complex.

The limited availability of radio links in the operational area means that there may not be a reliable link for the interchange of data between the UAVs. The ideal 4D/RCS model has free flow of information between the knowledge databases and world models at every level until it becomes impractical, usually when the node has little functionality. In our case, the UAV has considerable functionality, but we must recognise the practical limitations for data interchange and set a design requirement that there will be no dedicated inter-UAV data exchange. Instead, there will only be access to Link 16 and a time-shared video link. Both of these will be reserved for C2 and data exchange between the ground stations and their UAVs. This means that every knowledge database will only have pre-loaded data in common unless data is transferred from the ground station or 'Control Function'.

The architecture for the three 'current capability' surveillance UAVs during the Observe phase is shown in Figure 12.6, keeping the same level notation from

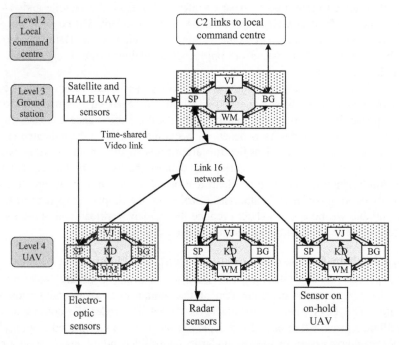

Figure 12.6 The 'current capability' non-DAS sensors during the Observe phase
[Key: Command and Control (C2), Value Judgement (VJ), Behaviour Generator (BG), Sensory Processing (SP), World Model (WM), Knowledge Database (KD), Defensive Aids Suite (DAS)]

Figure 12.4 and C2 hierarchy of Figure 12.2. The behaviour generator command links below Level 3 are omitted for clarity, but will be over Link 16. The surveillance ground station is responsible for three UAVs, with one of these on hold in a relatively safe area. The time-shared video link is connected to one of the two UAVs over the suspected target area the choice is made by the operator.

The proposal in Chapter 11 includes the development of image compression techniques to allow transfer of relatively low-resolution images over Link 16 with target indication algorithms on the UAV. The algorithm then highlights the suspected target. Its location and some detail about a small area around it are passed to the ground station. At this stage of the operation, only target indication is needed. The details of the processes and the operation of the target indication algorithm is not critical to the IHL considerations as the human operator makes the decision to start the Orient phase. However, in an AWS, the design will have to incorporate an interrupt into the process for a human to make this decision. This allows false alarms to be filtered out on further examination. It could be argued that it is irrelevant for adherence to IHL if many false alarms are raised which are all overruled by the human operator when more detail is available to them. Clearly it will matter to the operator both for workload considerations and consequent fatigue making them overrule almost all indications, including real targets.

The operators will have heavy workloads and so will need automated decision-support systems to interpret the mass of information from the satellite and HALE surveillance systems as well as from the AWS' UAVs. This will be even worse if the two active surveillance UAVs are looking at different areas. Data fusion tools will be essential with a myriad of approaches available to the designer. See, for example, Reference [5].

An alternative to having an interrupt between Observe and Orient would be put the human interaction elsewhere. The indication can be passed automatically to the Orient phase, but this is carried out by one or more surveillance UAVs with the dedicated video link and sensors observing a small area around the indicated target. More sophisticated target identification algorithms in the ground stations would then be used and the human presented with data on the local target area, the target type and probability of correct identification. The operator now assumes the authority of commander and authorises a strike and the despatch of a tactical UAV with suitable weapons. This should reduce the human workload and allow a considered response to an initial target indication alert.

12.6.1.3   Non-DAS sensors during Orient, Decide and Act phases

These phases of the OODA process using the 'current capability' have all the unneeded UAVs under the command of the remote tactical ground station and operator. The UAVs needed to attack the target are under the control of one surveillance ground station and the deployed tactical UAV's ground station. The former may have one or two UAVs under its control depending on the target area's complexity and visibility.

The limited availability of radio links applicable for the Observe phase also applies in this case. There will not be a reliable link for the interchange of data

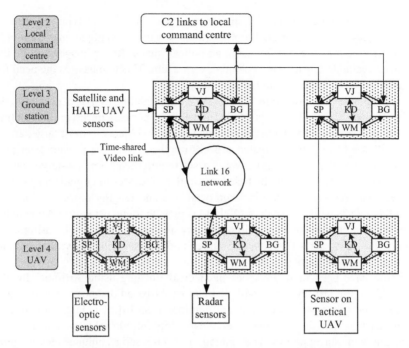

Figure 12.7   The 'current capability' non-DAS sensors during the Orient, Decide and Act phases
[Key: Command and Control (C2), Value Judgement (VJ), Behaviour Generator (BG), Sensory Processing (SP), World Model (WM), Knowledge Database (KD), Defensive Aids Suite (DAS)]

between the surveillance and tactical UAVs despite their sensors looking at the same target. However, the tactical UAV will have a dedicated video link to its ground station, removing the need to use Link 16 for imagery. The surveillance UAV operator will have a dedicated video link available which can be used for one surveillance UAV and he or she can use Link 16 for data from the other.

The architecture for the surveillance UAVs during Orient, Decide and Act is shown in Figure 12.7 keeping the same level notation from Figure 12.4 and C2 hierarchy of Figure 12.2. The behaviour generator command links below Level 3 are omitted for clarity, but will be over Link 16.

A single procurement of the 'top down' AWS will deliver one of many possible architectural solutions. Rather than speculate about them, we confine our discussion to the sensory processing functions which must have a dataflow through all levels and be automated for an effective AWS in the framework of the 'current capability'.

Target recognition algorithms are notoriously difficult to assess for an unambiguous definition of their probability of correct identification. Traditionally they match characteristics of objects in the sensor image to library images and signatures held in a database. Variations in the relative orientation of sensor and object introduce uncertainties as well as similarities between different types of object[7]. Radar imagery has the additional problem that the parts of an object that are highly radar-reflective may not be obvious in an optical image. Radar images will be the only ones available when there is cloud cover or rain. This emphasises the need for commanders to be trained in radar image analysis.

An important consideration for an AWS is the ability to assess possible collateral damage. Currently this is carried out by the humans in the ground station using their experience to identify the myriad of likely objects in most target areas, but it will need to be an automated process for the AWS. The procurement programme team will have to find a way to set requirements for non-target object identification. Solutions could be based on similar processes to target identification but these may require excessively large databases and lengthy processing time to match the objects with the database. Solutions may be found using prior knowledge of the target area from maps, satellite imagery and other sources. The differences between the fused data from these sources and the surveillance sensor outputs providing the basis for non-target or unknown classification.

There have been huge advances in medical imaging using Artificial Intelligence (AI) in the last few years with claims that AI-based diagnostic results equal or exceed the accuracy of human image interpretation [6]. It is highly likely that military users will wish to see this level of capability introduced into the analysis of targets and their surroundings. The validation of AI-based techniques may become straightforward as the technologies evolve. However, verification of the accuracy of the results will be difficult due to the availability of training data and the wide range of targets and non-targets in every potential target scene.

Setting requirements for the scene-analysis algorithms and validating them will be a difficult process for military and technical staff. Verifying their actual performance in a wide range of scenarios will be a key part of the Article 36 reviews for the AWS. There will need to be considerable collaboration between the procurement authority's technical team and their legal advisors at all stages to be sure that legal considerations of reasonableness and due care can be compared with technical performance figures from trials, simulation and analyses of the algorithms used.

12.6.2   *Value judgement, world modelling and behaviour generator*

The problem of assessing possible collateral damage and casualties is a significant one and will be part of the world model functional element. The results from the world model will form the basis of the decisions by the value judgement and behaviour generator functions, leading to weapon release in an AWS. Difficulties in identifying non-targets in the scene have been discussed in Section 12.6.1. There will be an extra complication of deciding if they are objects supporting the enemy

---

[7]One classic comparison is the similarity between a man carrying an innocuous tube such as a scaffolding pole on land or fishing tackle in a small boat and a rocket-propelled-grenade launcher or similar weapon.

forces or are civilian objects and therefore have a level of protection. Buildings should already be known and in the knowledge database or readily accessible from sources outside the AWS so, in principle, they are not a problem. However, if they are civilian, there will have to be an assessment of their current use and likely inhabitants. One example could be a school which in term time will clearly be occupied by non-belligerents, but at night or holiday time the situation may be very different.

Prediction of collateral damage to buildings and their inhabitants can only be accurate if all important factors are quantified, but in practice this is not likely. Simple checks of objects in the warhead's blast range may be the only ones available. This will pose problems at an Article 36 review for any automated part of this process.

The critical issue from an IHL perspective is the role of the value judgement and behaviour generator functions in the nodes at all levels in the proposed AWS. The Level 2 behaviour generator operates on the basis of its authority and the authorised powers of the lower nodes across the AWS. Even if the authorised powers are not specified in the original procurement specifications, it should be possible to analyse the design and unambiguously express the allowed and prohibited actions by the behaviour generator. Ambiguities will be a source of concern both technically and legally. When they are resolved, it should be possible to ensure that the basis for every action taken by the AWS is understood, defined and testable.

Rules of engagement will be set for every campaign and mission within it. Their expression in a digital form suitable for the AWS will be an issue. A strike on a specific type of vehicle in a military base or open area will have a simple interpretation of rules-of-engagement and may be capable of automation using prior knowledge and the sensor output from the UAVs. Put the vehicle near a bus or block of flats and the situation becomes more difficult if not intractable. This is an area where detailed discussions will be needed with legal advisers on the expression of authorised powers in clear technical terms so that they can be specified and tested thoroughly.

12.6.3   Knowledge database

The knowledge database, its contents, structure and interfaces are critical to all actions by any weapon system. All information held in it must have known reliability in at least two senses: the reliability based on the technical quality of the original data and subsequent analysis; and the intelligence reliability, i.e. its provenance, age and its relationship to the current target scene. Modern operations are likely to have large amounts of data available, the problem being to filter it to give the relevant information in the correct form for analysis in the decision-making chain.

The same information may be presented for assessment by the commander and one or more automatic analysis algorithms. Ensuring consistency of results is a technical problem that may never be solved. The authorised power for the behaviour generator at the node level for comparing the results must give clear guidance in these circumstances.

There are three legal issues that the designer should be aware of:

- The first is that the commander must be given all relevant information about the target area at the time of making the lethal decision[8]. This is so that the commander can make a decision in good faith on the evidence of the best available information at the time of the decision;
- The second is that it is forbidden to deny the commander relevant information so that an illegal action is carried out by a commander who believes they are acting in good faith[9];
- Technical evidence may be required in planning the mission and in any post-mission enquiry. An enquiry will need to know the delay in new intelligence reaching the commander and how it will come to their attention.

The first issue is affected by the lack of data exchange between the knowledge databases below Level 2. The commander will need to know the time delays in receiving information on the target area(s) from different UAVs in order to make a reasonable decision. An example is when anticipated changes in the primary target area necessitate that the weapon be redirected to a secondary target.

Filtering out irrelevant information will require clear definitions of 'relevant' for any automated system. These could be based on time and position parameters for example, but there will need to be others. Despite this, information overload will still be a problem for both the human and the AWS.

An enquiry into a potential breach of IHL will want to know the information available to the commander at the decision times and its provenance. The implication for the knowledge database design is that there must be robust configuration control and an electronic audit trail of the basis of lethal decisions by the behaviour generator. Any future distributed decision-making system will need close scrutiny to ensure it complies with this aspect of IHL and that the evidence can be presented for future analysis by an independent body.

12.7   Article 36 reviews for incremental upgrades

12.7.1   A possible architecture

The proposals for incremental upgrades will either be from the suppliers or in response to Invitations to tender from the procurement authorities. In both cases they will, if successful, lead to procurement contracts raising the autonomy level of functions in that contractor's part of the system. The ultimate aim being that all functions carried out at Level 2 and below would have no human involvement, except the launch and recovery operations.

---

[8]In practice, this is taken to mean the information that it is practical to present to the commander.
[9]This was the subject of a prosecution in the former Yugoslavia when agreement of a cease-fire was deliberately withheld from snipers so civilians who believed they were safe were shot when they came into the open.

The roles played by the functional elements in each subsystem and module will be specific to the design of the UAVs and ground stations. The suppliers given in Table 11.6 will be responsible for the designs of all aspects of the current capability and hence for their structure. The specifications and division of functionalities between ground stations, UAVs and local command centre will have resulted from numerous discussions between analysts, programme managers and system engineers over several years. There will be interface specifications between the knowledge database and world model in the nodes at Level 2, 3 and lower levels so it should be possible to make well-founded engineering decisions about the operation of individual world models and behaviour generators as their levels of autonomy are increased and the location of functions is moved between UAV and ground station.

It is possible that the succession of upgrades will lead to the AWS architecture shown in Figure 12.8 for Orient, Decide and Act. This will arise because each UAV and its automated ground station functions will replace the 'current' UAV and its 'current' ground station, but with a redistribution of capabilities between them. The two types of UAV come from different suppliers, so there will be two different automated ground station function types in the AWS. There will no longer be the three manned ground stations, so all their functionality will have to be in the local command centre which would be located in the maritime or ground command centres.

The completion of incremental upgrades at Level 2 and below will initially have a human commander at that level. Eventually there will be consideration of an upgrade to remove the human commander, making a fully autonomous system. The maritime or ground commander would be able to request the launch of the eight UAVs and then instruct them to seek and destroy any missile launchers, or defined military targets in a specified area. No further human intervention would occur unless requested by the AWS until a successful strike has taken place. This would be illegal in the foreseeable future, but the possibility makes a good illustrative problem to examine the boundaries and issues. Most military systems have incremental upgrades over their lifetime, so there could be the emergent property discussed in Section 12.4.3 and later in this section.

As shown earlier, the AWS configuration for Orient, Decide and Act phases in OODA will have two or three tactical UAVs continuing in their holding pattern under one C2 node for this role. The two surveillance UAVs not involved in the strike on the target will continue to operate under a single C2 node with one surveying another area and the other in a holding pattern. The surveillance and striking tactical UAVs operate with their ground-station functions under the control of the Level 2 human commander. The surveillance UAV may not add very much to his or her workload as it should be possible to put it into an autonomous flight pattern, without further operator commands. The automated role is to keep its sensors on a specified location, or area, until it identifies a target and then cues a strike by a tactical weaponised UAV. The tactical UAV then becomes the major task for the commander who can now make lethal decisions in the same way as at present when there is one operator per UAV.

If it is mandated that the AWS is always operated with the human commander making lethal decisions, based on the video feed from the tactical UAV, there

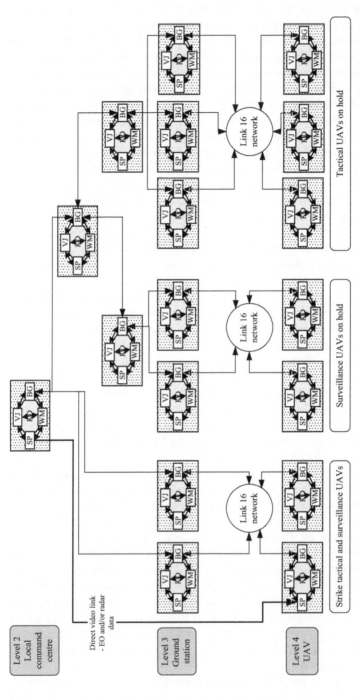

Figure 12.8   A possible C2 architecture resulting from incremental upgrades to the 'current' capability during the Orient, Decide and Act phases

[Key: Value Judgement (VJ), Behaviour Generator (BG), Sensory Processing (SP), World Model (WM), Knowledge Database (KD), Command and Control (C2)]

should not be an IHL problem. This is the same as the 'current' practice. It could be argued that the autonomy level upgrades simply automate the reconfiguration of air assets to strike a new target that has been confirmed as legitimate at the transition from Observe to Orient or later. Hence no Article 36 Review is necessary. The procurement authority should take legal advice about this argument, especially for a range of likely rules-of-engagements.

Regardless of whether a review will be held or not, the design team should clarify how their design will ensure that there is a human decision at the correct time. The decision must be made on technically accurate and coherent data. This could be effected through the chain of authorised powers with rules-of-engagement put into the knowledge database at the correct level in the architecture. It will be essential to define whether the human decision is target indication or identification as this will give very different levels of authorisation in the behaviour generators for succeeding phases.

Clearly, there will be extra work if one of the non-strike surveillance UAVs detects another target and alerts the commander. This may be resolved by a decision to prioritise one target and abort the other, but there may be the possibility of the commander finding a way of continuing the initial strike with purely automatic functions whilst executing the second strike i.e. an autonomous strike on the first target with no human oversight. This gives the risk of a breach of IHL if all aspects of this mode have not been reviewed under Article 36. If this autonomous mode is possible, it is an example of an emergent property of a complex system-of-systems. There will be complex legal and technical decisions about how to deal with this property as the additional military capability could have appeal to some users.

The first paragraph of this section postulated that the incremental upgrades would be based on the 'current capability' and implemented through the contractors in Table 11.6. This postulate is accepted and the architecture shown in Figure 12.9. The figure illustrates the lack of commonality in any complex system although there should be common standards for interfaces and build quality. The complex contractual arrangements lead to numerous types of problem that the legal, military and technical teams must address to be sure that the AWS both meets its military needs, works to specification and that the deliverables are easy to use legally. The next sections match the requirements in Chapter 10 against the likely technical problems that the design team must be aware of.

12.7.2   Sensory processing

Most of the issues raised for the 'top-down' approach discussed in Section 12.6 will apply for the incremental upgrades. DAS may require special attention as there are two UAV airframes and even if fitted with the same basic DAS suite, there will be many detailed differences between the two fits. They should be treated as separate functions for Article 36 reviews.

The sensory processing requirements from Panel 10.5 are given here as subheadings with comments about the technically related IHL problems associated with them in AWS upgrades. There are three suppliers involved at lower levels in

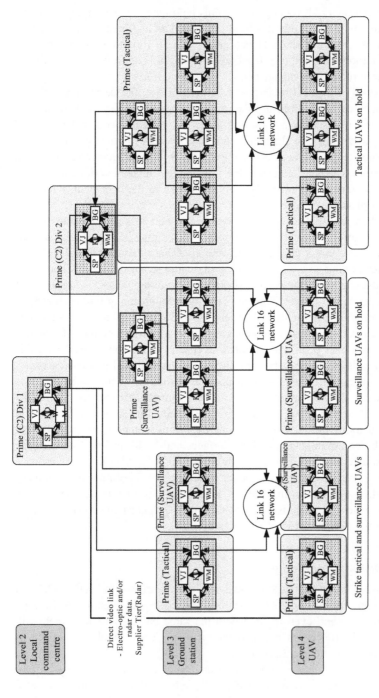

Figure 12.9   A possible C2 architecture resulting from incremental upgrades to the 'current' capability during the Orient, Decide and Act phases, with suppliers of functions from Table 11.6 shown in shaded boxes

the sensory processing chain: Prime(Tactical), Prime(Surveillance UAV) and Tier(Radar). Two more are recipients of the possibly disparate outputs from these suppliers: Prime(C2) Division 2 and Prime(C2) Division 1. In practice, differences between optical or infrared imagery and radar data are highly likely even if the same standards are used. The technical and trials teams must be prepared to show that the information for the target scene, not just for the target, is coherent and can be compared with the electronic rules-of-engagement for the location and mission phases.

1. *The sensory processing must have the ability to recognise symbols denoting protected personnel, objects and areas;*
   The symbols should be included in the image recognition database for all systems. All protected areas must be in the knowledge database and a flag raised if an indicated or identified target is in or near one. The symbols will be small compared with the image area in most cases so consideration must be given to putting their locations into the knowledge database when known and updated when required.

2. *If the sensory processing has its own processing capability, the outputs must be fully compatible with the world model.*
   The sensory processing will probably have its own processor architecture, algorithms and language which will not be the same as those for the world model. In principle the interface problems will be the same as those in any product using two disparate subsystems and do not need to be discussed here. The IHL risk is in the age and timeliness of the information in the world model as well as the accuracy of the data used by the value judgement function. Therefore, time-tagged configuration control of datasets and models is an important consideration

3. *The sensory processing must have the ability to classify objects as military, civilian or unknown;*

4. *The sensory processing must have a target indication capability. This cap-ability may also include indicating objects as unknown if there is a reasonable level of doubt about their nature;*

5. *The sensory processing must have the ability to tag objects as military objects and if they are potential targets;*

6. *The sensory processing must have a capability to identify targets to previously agreed confidence levels for objects in the approved target list for the operation;*
   Sensory processing requirements 3–6 all relate to the fundamental problem of automatic object classification and identification. Removing the human from successive steps in image interpretation is an active field of research in many areas. New technologies such as AI are replacing traditional ones and will be offered for use in intelligence and weapon systems. This may not be a fundamental problem provided due weight is given to the aspects discussed in this section.

   The criteria for target indication and target identification must be distinct, clear and based on hard evidence from trials in representative scenarios, not simulations. Simulations can be useful in designing algorithms and establishing signatures which can indicate that an object may be military rather than unknown. It is unlikely that simulation alone will satisfy legal criteria for target identification.

The authorised power for the action of 'tagging' objects should include: criteria for the probability of correct identification; possible ambiguities in identification; and the basis for flagging the object as a potential target as well as being a military object. The method of passing the information to the superior node and knowledge database must be consistent across the suppliers of all sensory processing subsystems.

Civilian objects close to a legitimate military target must also be classified in some way. Human scene interpretation is very good at identifying them and is the usual way to assess potential collateral damage. All questions of automating any part of the process must be considered for an Article 36 review. The wide range of civilian objects that could be in a target area may make it impossible to draw up an adequate list of identification signatures analogous to those for targets. AI may be a solution, but this whole topic will need extensive liaison between technical and legal staff.

There will be a close coupling between rules-of-engagement and the criteria used in all the identification processes. The relations will be very specific to a system, its use, the scenario and the mission, so there will be many IHL questions arising which cannot be solved by engineering considerations alone.

7. *The sensory processing must be able to continuously monitor the target area up to, including, and, if possible, after weapon impact;*
    This is straightforward if there is a direct video link to a human commander who makes the lethal decisions. The commander will assess likely collateral damage as the scene evolves, using appropriate decision-aids if available.

    Without a human commander, the world model is the only function in the architecture which is capable of comparing the actual scene with the anticipated one. There will be technical problems in ensuring that the sensory processing output is compatible with the world model at all levels in the architecture, but these should be solved without need for an Article 36 review. In principle, the discussion of requirements 3–6 will apply, but the time requirements for completion of the processing, including value judgement decisions will be much tighter, probably being impossible to meet fully.

    The conclusion is that any requirement for automating interpretation of the target scene in the final stages of a strike operation will raise both technical and legal problems and need to be considered as a potential new or novel means of warfare.

12.7.3   World modelling

Table 11.6 shows two suppliers who may have supplied world models in different parts of the system at Level 2 and below: Prime(Surveillance UAV) and

Prime(Tactical), PC(T). It is also possible that Prime(C2) may have developed and supplied the Level 2 ground station for the local command centre as part of their integration work. The sequence of requirements and contracts used to create the AWS by incremental upgrades will dictate which world models are replaced by ones from elsewhere in the capability or if there is a work package to develop a common world model for all of AWS nodes at Level 2 and below.

Set against this background, the requirements from Panel 10.6 are given here as subheadings with comments about the technically related IHL problems associated with them regardless of the order of the incremental upgrades.

1. *The world model must have a world model with time-related information and a capability to track moving objects;*
2. *The world model must be able to make predictions about the target scene at weapon impact time;*
3. *The world model must have methods to compare the current world model with the initial model and have criteria for the differences that trigger reference to its superior node or to a human in its command chain;*

   Requirements 1, 2 and 3 all specify the main characteristics of the world model. There will be a predictive requirement for the purpose of executing the mission through a changing environment. The IHL requirement is for an analysis of the target scene so that either the human or the world model can make assessments of collateral damage at weapon impact time rather than at release. The technologies may be similar to those behind synthetic environments. Reliability and repeatability of predictions, as well as the differences triggering the reference to a superior node, will be a problem, both for specifying and testing them as autonomy is increased.

4. *The world model must have an authorised initial world model to act as a starting point to interpret the sensory processing outputs;*

   All world models in the various world model functions must be based on information that has been authorised for use in military operations. This will still apply as the possibly disparate models supplied from different work packages are merged or replaced during the upgrades. The design must incorporate facilities for the integration of new and more detailed models of specific areas.

5. *The world model must be able to access intelligence and other relevant information available over its communication links;*

   This information will be treated architecturally as sensor inputs. There is a technical problem of interfacing it into the world model and knowledge database which may lead to an inherent bias to sources which fit particular interface standards regardless of the level of automation. The IHL problem as autonomy is increased is ensuring that the AWS can identify both information sources and their content that is relevant to the strike mission. This becomes a technical problem analogous to the need for a search engine, but it will require very clearly defined selection criteria and repeatability.

6. *The world model must update its current world model with relevant and available information at regular intervals;*

   This requirement is an IHL one to ensure that information is not withheld from the AWS. The technical problems are those raised in the immediately preceding paragraph.

7. *The world model must have a method to relate the observed scene to that expected and make a coherent situational analysis of it;*

   Interpretation of this requirement and its implementation will be at the centre of IHL considerations. Situational analysis of a target scene is not a capability that is normally carried out without human supervision. This analysis can be expected to be the last one to be implemented in an automated system, rising to the local command centre as the incremental upgrades are implemented.

   The depth and complexity of the analysis depends on the scene, the rules-of-engagement and the weapon's characteristics. It is possible to postulate the use of an AWS against an isolated mobile target in a military complex with few collateral damage considerations. Adequate analysis is then straightforward and suitable specifications and testing to meet IHL can be predicted. However, most scenarios will be more complex with uncertainties in the objects, people and their movements.

   Action is taken on the basis of the results of the scene analysis, which will result in delivering or withholding lethal effects. Therefore automation of the analysis will be considered a new or novel means of warfare and subjected to an Article 36 review.

8. *The world model must include some method to designate objects that are civilian and potential dual use objects;*

   This requirement is closely related to the sensor requirement 3 in Section 12.7.2 and similar considerations apply here.

9. *The world model may need a capability to perform analysis on the target scene after weapon impact and the ability to confirm the success or failure of the attack.*

   This is a requirement for the AWS to perform battle damage assessment which is needed for all military strike operations. It is also an IHL requirement as attacks should cease when the military objective is achieved. It can be assumed that the success criteria will be capable of clear definition and readily detectable. However, tactical considerations may not allow either the tactical or surveillance UAV to remain in the area for battle damage assessment.

12.7.4   Value judgement

The subsystems supplying the value judgement functionality will be closely coupled with the world model functions and the same issues will arise for them that were discussed at the start of Section 12.7.3. It takes the options presented to it by the world model for action, decides between them and then passes the decision to the behaviour generator for action. During the Observe phase of OODA, there are not likely to be many IHL issues, although there will be many mission-related ones.

When in the Orient and Decide phases, the value judgement role becomes increasingly crucial to adherence to IHL. It takes the current best estimate of the state of the target area and decides which of the options available to the AWS are allowed. These decisions are currently made by a human and the requirements in Panel 10.6 are based on those. They are reproduced below with comments.

1. *The value judgement must have criteria based on rules of engagement for its authority;*
2. *The value judgement must have pre-set criteria for the decisions the weapon system is authorised to make and act on;*
   The criteria specified in requirements 1 and 2 are the basis for decision-making by the AWS. They are simple to state, but cover some very subjective issues. Rules-of-engagement can be subjective, so there is an assumption here that they can be turned into a form which is interpretable by the AWS nodes at different levels. This is a wider issue to be addressed by the procurement team. There will need to be cross-disciplinary negotiations to reach a legally and technically possible set of trade-offs that still provide a militarily useful weapon system.
   A top-down procurement will have to address the problem at a very early stage. An incremental set of upgrades will solve the problems as they arise, or show that they are intractable and the AWS, as conceived, can never meet IHL requirements.
3. *The value judgement must either identify areas where there will be little or no collateral damage if hit by the weapon, or have such places in its database;*
   This is to ensure that there is a fail-safe area for the weapon if the target or its area are no longer legitimate for a strike but the lethal subsystem (e.g. a guided missile) has been launched. There may not be such areas in many circumstances which is another problem for the setting and interpretation of the rules-of-engagement. In a wider sense, it is similar to the trolley problem shown in Figure 2.1 where there is no universally accepted solution.
4. *The value judgement must review the identification of all objects classified as unknown with relevant new data and reclassify them as civilian if they do not meet the criteria for reclassification as military objects;*
5. *The value judgement must have the ability to calculate aim-point and impact-point errors for the weapons it can release;*
   Requirements 4 and 5 are precautionary and ensure that collateral damage calculations will be made on a worst-case basis and ensure compliance with IHL. It should be very straightforward to implement from a technical viewpoint.
6. *The value judgement must be able to perform collateral damage calculations for every potential target;*
   This is an essential part of any proportionality assessment and provides one input for comparison with the rules-of-engagement criteria. The calculations may be very simple such as drawing a circle of radius equal to the warhead's blast range, plus aimpoint errors and assuming that all people in that area will be killed, even if in a building. They can be much more complex, but the complexity will have to be decided for the specific system and likely scenarios.

This section illustrates the possibly insurmountable problems faced by designers of an IHL-compliant AWS. As a minimum, the assumptions about the specification of likely rules-of-engagement in engineering terminology must be critically questioned. Most of the questions raised by the issues arising from value judgement requirement are not ones that can be solved by engineering teams without reference to other professions.

12.7.5   Behaviour generator

The subsystems supplying the behaviour generator functionality will be coupled to the value judgement functions and similar issues will arise for them that were discussed at the start of Section 12.7.3. The main difference is that the chain of behaviour generators are the C2 chain giving the authority to act. There will be an operator interface for the commander to give instructions and supply data, giving another input, as shown in Figure 10.5.

The behaviour generator requirements given below with comments are the final check that the AWS will be compliant with IHL. Any behaviour generator functionality in a weapon system must raise the question of an Article 36 review as it removes some decisions from the human operator.

1. *There must be a method for the superior node to monitor lethal decisions under consideration by the behaviour generator and inject new data if necessary;*
2. *The behaviour generator must report all weapon-release decisions to its superior node;*
   Requirements 1 and 2 ensure that the AWS is part of an authorised C2 chain. Designing them in should be straightforward, allowing for latencies and breaks in comms links.
3. *The behaviour generator must have defined actions that it is authorised to take without reference to its superior node;*
4. *The behaviour generator must have defined actions that it cannot take without referring to its superior node and receiving authorisation before taking them;*
   Requirements 3 and 4 limit the range of actions that the AWS can take, i.e. its authority level. The range can be defined using authorised powers at all levels in the architecture, but it can be defined in other ways such as through behaviours. The important point is that there must be a set of criteria in the behaviour generator which are expressed in terminology which can be unambiguously compared with the list of weighted options produced by the value judgement. Detection of ambiguity must generate a message to the superior node and, probably, initiate fail-safe actions.

   It is likely that limiting types of action will involve some subjective considerations. Translating these into information which can be used automatically by the behaviour generator will be a challenge. Extensive testing and validation will be necessary to demonstrate compliance with requirements 3 and 4 both technically and legally.
5. *The behaviour generator must have designated areas to be examined for the presence of targets;*

6. *The behaviour generator must not authorise attacks on protected areas, objects or personnel;*

   Requirements 5 and 6 limit the AWS's operational area, ensuring it does not become a terror weapon. Compliance should be a straightforward engineering issue even if the protected areas change with time. The geographic and other details will have campaign and mission-specific parameters which will need to be input at the appropriate times.

7. *The behaviour generator must have a range of fail-safe actions that it will take if necessary;*

   Requirement 7 should be self-explanatory as good practice, although the specification of the actions at different times in the mission may be difficult.

8. *The behaviour generator must maintain guidance of the weapon as long as possible;*

9. *The behaviour generator must be able to redirect the weapon to hit a contingency area, if one exists, when the target is no longer legitimate and the redirection is possible;*

   Requirements 8 and 9 are to ensure that information updates are not ignored. This is analogous to the IHL requirement that information should not be withheld from a local commander. The implementation and use of the supplied capabilities to meet the requirements will need agreement between technical, military and legal authorities.

The behaviour generator function controls the actions taken by the AWS, but it is a mistake to assume that it is the only part which must be tested for compliance with IHL. The preceding sections show that it is only one of the parts of the AWS architecture that are critical to compliance with IHL.

The behaviour generator will be an integral part of the UAV systems that are in the 'current capability' and will have to be reviewed as part of the overall C2 chain in addition to reviews of specific suppliers' deliverables when upgraded in an incremental approach.

12.7.6   Knowledge database

The sequence of the requirements and contracts used to create the AWS by incremental upgrades will dictate the structure of the data bases and information flows around the capability. The operating and security systems, and architectures will probably have significant differences between the tactical and surveillance UAV systems even if they have common interface standards between modules. It will be important to examine the compatibility of the design philosophies used by the different suppliers.

There will be enforced changes from more general military infrastructure changes such as upgrades to, or replacement of, Link 16 over the lifetime of the 'current capability' and during the incremental upgrades.

Section 10.6.6 gives some details about the contents of the knowledge database. The structure, integrity, authorisation and configuration control of data in the knowledge database will be a key factor in all Article 36 reviews at every stage of the procurement process, whether top-down or incremental. The details will be specific to the system under consideration so they are not discussed here.

12.8   Wider implications – limits to autonomy

Chapter 11 described how multiple UAV systems, acting together, can deliver a seek-and-destroy capability using technologies from the 2010–2020 decade and how the Article 36 review would be approached. It is highly likely that the system (designated as 'current') would pass the review and enter service as part of a nation's assets and be used in coalition operations.

This chapter has examined how the 'current capability' could be upgraded to provide a more-autonomous weapon system. The intention being that the AWS will have no human intervention, after the initial instruction, until it reports the results of its first strike against a target that it has selected and attacked. The postulate that the system could be made autonomous in all its functions was tested using the requirements given in Chapter 10. It was shown in Sections 12.6 and 12.7 that it is extremely difficult, if not impossible, to make the 'current capability' into an AWS only requiring a command from the Level 1 commander for a seek-and-destroy mission against a specific target type.

It is possible to set target detection, indication and identification criteria in the UAVs' sensor and intelligence information using automated systems. However, quantification of these will need to be validated as well as assumptions about the assessment of other objects (clutter) in the scene. There will need to be methods to identify non-military vehicles and other objects with similar levels of reliability to target identification ones. This is a large problem as there is a much wider range of civilian vehicles than military ones. This is before dual-use considerations are taken into account.

The AWS requires a predictive capability which is achieved using the world model functional element. It was shown in Section 12.7.3 that a detailed situational analysis of the target scene is needed. Realistic target scenes will be complex, evolving over time, and with the target likely to take evasive action. It may be possible to use AI approaches to the problem, but these must be trained to give a high degree of reliability. Training could be by simulation, but it is difficult to conceive that simulation will give the necessary broad range of scenarios likely to be exploited by an opponent who will exploit every opportunity for confusion.

The value judgement and behaviour generator functions also have complex processing requirements. These may be met using deterministic algorithms, but there are some implicit assumptions that may not be compatible with this type of algorithm. Rules-of-engagement are written for the guidance of military personnel operating and commanding weapon systems. Therefore they are highly likely to have subjective elements. Very few rules-of-engagement are in the public domain, but the example in Panel 8.1 shows rules that could be turned into an algorithm but also ones that cannot, due to their subjective nature. The latter ones require human judgement which would be very difficult to turn into an algorithm except in very restricted circumstances.

It can be seen from the results of this chapter that there are some large hurdles to be overcome before the human can be removed from key parts of the decision-making processes leading to the release of lethal force and still meet IHL. Even if a very capable AWS passed Article 36 Reviews, there still remain problems with expressing rules-of-engagement in terms which can be interpreted and used by any automated weapon system.

References

[1]   Boulanin, V. Implementing Article 36 weapon reviews in the light of increasing autonomy in weapon systems, SIPRI Insights on Peace and Security No 2015/1 November 2015.

[2]   Details can be found at http://www.army-technology.com/projects/phoenixuav/ [Accessed on 15 November 2017].

[3]   Details are at: http://www.fi-aeroweb.com/Defense/MQ-1-Predator-MQ-9-Reaper.html [Accessed on 15 November 2017].

[4]   EFB https://www.nbaa.org/ops/cns/efb/

[5]   Liggins E., Martin E., Hall D., and Llinas J. *Handbook of multi-sensor data fusion, theory and practice*; The Electrical Engineering and Applied Signal Processing Series; Boca Raton, FL, USA: CRC Press 2nd edition; 2008.

[6]   Hoyt R.E., Snider D., Thompson C. and Mantravadi S. *IBM Watson Analytics: Automating Visualization, Descriptive, and Predictive Statistics*; JMIR Public Health Surveill. 2016 July–December; 2(2): e157. Published online 2016 Oct 11. doi: 10.2196/publichealth.5810; [Accessed on November 2017] at https://www.ncbi.nlm.nih.gov/pmc/articles/PMC5080525/

Chapter 13

Design of civilian autonomous systems using military methodologies

13.1 Introduction

This chapter shows how the techniques developed in earlier chapters can be applied to autonomous systems that have a physical component, in civilian applications. This does not restrict the discussion to robots, but it does exclude many aspects of purely algorithmic autonomous system and Artificial Intelligence (AI) applications. Ethical issues for physical autonomous system, in general, concern their safety and reliability as they affect humans, animals and the environment; both properties are the subject of national and international laws and regulations. This leads directly to ensuring that the actions taken by the autonomous system are legitimate, and that a person has responsibility for the consequences.

One fundamental problem is that of designing an autonomous system that meets accepted norms for behaviour and proving it, especially for applications which do not currently have adequate regulations. Weapon systems are highly regulated, and the methodologies developed for Autonomous Weapon Systems (AWS), described in this book, show how it is practical to embed ethical principles in their design. This chapter summarises the military methods and examines their applicability for civilian applications which have, or will have, strict regulations. The chapter structure generally follows the steps given in Section 13.2 for the design process.

Products and services that use new technologies usually raise questions about the ethics of their use as discussed briefly in Section 6.2. There are methods under development to address ethical issues for Information Technology (IT) systems in general. These are known as value-based or value-sensitive design and form the basis for the proposed IEEE P7000 series of standards explained in Section 6.5. However, it will be several years before these standards will be published, probably as advisory rather than mandatory. This series will be in addition to the current ones given in Section 6.4: ISO 8373; IEEE 1872; BS8611.

Design teams for autonomous system products being considered for entry into the market before the evolution of adequate standards will still need to convince users and regulators that the autonomous system is ethical as well as legal. These teams may need some guidance in setting the capability, system and design requirements for their product to significantly reduce the risk that new regulations and standards will make their product obsolete. This has been achieved for AWS. It is shown in this chapter that many of the military techniques can be applied for civilian autonomous systems.

13.2   Autonomous systems and design methodologies

13.2.1   Design methodologies

The techniques and processes covered in Chapters 2, 3, 4 and 5 are applicable to all types of automated and autonomous systems. Regulations and standards, and their relationship to the law, are described in more detail in Sections 6.3 and 6.4. The next chapters explained techniques for one specific type of autonomous system, AWS, one that always raises ethical issues.

It was mentioned at the end of Section 6.3.1 that many potential applications of autonomous systems are in fields which may be subject to extensive regulations, but which are almost irrelevant for the autonomous aspects of their design. These guide human behaviour as the ethical decision-makers and the design of the equipment they use, but not the design of the aids as decision-makers. If we take the example of medicine and health care, there are strong ethical guidelines for human behaviour from all the professional bodies, and standards for the design of medical devices [1] but not for automated decision-makers. Automation of all types, including decision aids and intelligent robotic devices, will be introduced soon, often with government encouragement, see, for example, Reference [2].

Engineering requirements for any autonomous system with the ability to make decisions and choose its consequent actions will have to comply with the equipment and device regulations for their application and environment. However, there will be a need to write new regulations and guidelines when autonomous systems are involved in decision-making or are used as sophisticated decision-aids which present a human with a decision with no other options for action.

The methodology for designing an AWS can be summarised as:

1. Identify the ethical framework – In the weapons case International Humanitarian Law (IHL), especially the Geneva Conventions and Additional Protocol 1 (AP1), are taken as expressions of ethical standards and used as the framework to set requirements;
2. Identify key legal principles (military necessity, humanity, proportionality and distinction and derive high-level AWS requirements from each one);
3. Develop the architecture for the system providing the functions that meet the requirements. This is possible at this stage for military systems as they fit into an existing highly structured architecture;
4. Analyse the process that the system will follow in one or more representative scenarios and derive more detailed requirements for each phase of the process (the targeting process for weapons);
5. Assign requirements to functional elements in the architectural nodes;
6. Ensure Validation and Verification (V&V) of the final product are integral parts of the design process.

Whilst applying the methodology it was noted that there are some particular issues that require more detailed attention when requirements are drawn up for a specific applications, e.g. unmanned aircraft, underwater weapons. These are:

1. The authorisation of actions, the associated node hierarchy and authorised power;
2. The human–machine interface;
3. The ability of the autonomous system to make predictions about possible future events;
4. The role of automated decision aids for any human who takes action based on their results.

Chapters 11 and 12 brought out the strengths of this methodology and the key issues. Much will be applicable in civil applications of autonomous systems where there are concerns about responsibilities of users and suppliers under the law. This makes the authorisation of actions as critical for civilian as for military actions, although the consequences of action by the former are not necessarily as severe as for the latter.

The legal framework for AWS is clear, but civil autonomous systems are developing in a wide range of applications with no single ethical, legal or regulatory framework. Consequently, it is not possible to identify specific laws, either national or international, that are equivalent to the Geneva Conventions and other parts of IHL. Instead, the starting point is to look at ethics, societal values, existing international agreements and standards. It is then possible to specify a generalised civilian autonomous system which can be used to derive ethical and legal requirements for the engineering design processes.

Although there is the same need for identified responsibility for both civil and military autonomous-systems actions, the lack of a single international civil legal framework necessitates some changes to the military methodology which is shown on the left-hand side of Figure 13.1.

The systems engineering approach for a civilian autonomous system, described in this chapter, is shown as the methodology on the right-hand side of the figure. Both use the same principles and the equivalent processes are indicated by dashed lines in the figure. Later sections in this chapter will apply the methodology to derive generic requirements for ethical autonomous systems.

13.2.2 *Technology push or demand pull?*

Autonomous systems and AI technologies are among the most rapidly developing fields in technology today. This means that technical developments can suggest new applications and create markets that did not exist before. This is known as technology push, leading to the concept and product evolving symbiotically with the market requirement. There should be a market assessment carried out but this may have considerable uncertainties due to the unknown nature of new consumer markets.

In principle, algorithms can be developed and implemented in short timescales, especially when they build on the existing ones. This is seen in fields that use AI in

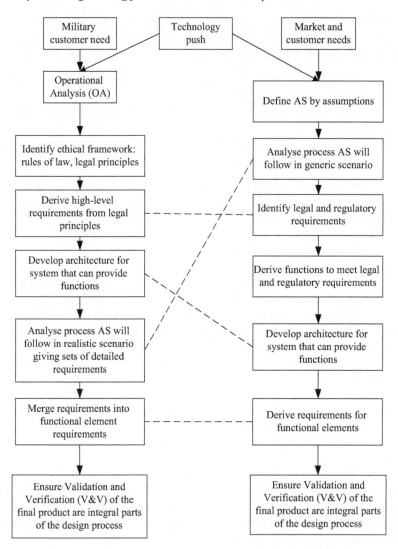

*Figure 13.1   Development processes for military and civilian autonomous systems.
The dashed lines indicate equivalent processes*

systems to replace humans in fields that rely on human judgement. The questions
we address here concern the identification of the ethical and legal concerns for the
design of new types of autonomous system through a reliable process.

Highly regulated fields such as aerospace, pharmaceutical and medical
instrumentation all have long timescales between the identification of a technical
solution to a problem and the introduction of products into the new market.
Timescales of 15–20 years are typical. Unmanned Air Vehicles (UAVs) and mag-
netic-resonance-imaging scanners both took decades to come to fruition. This is

because the technologies must have matured to be at a Technology Readiness Level (TRL)[1] of 5 or 6 before significant funds can be risked on product development and regulators can be satisfied by thorough test and evaluation procedures. 'Technology push' does not really apply with mature technologies even if they are still considered to be new. The publicity and discussions surrounding them shape the potential market, so 'market pull' is a better description. Autonomous car development illustrates the problems, with the added uncertainty that the regulatory regime is also unclear.

Regardless of whether a new product is the result of technology push or market pull, the same problems of legality, liability and public acceptance arise as with weapon systems. Consumers expect the rapid introduction of completely new products using recently developed exciting technologies such as autonomous systems and AI. This market pull leads to pressure on companies to design and introduce products before the regulatory framework is in place. The example of drones in Section 6.3.3 shows the importance of companies and their designers understanding the likely ethical and legal framework for their product although neither exist at the concept phase. Without some guidance, there is likely to be wasted effort and expense developing autonomous systems that do not meet, or cannot be easily modified to meet the changing regulatory and societal requirements.

13.3 General principles for autonomous systems

Ideally, there will be a set of international laws or accepted ethics that apply to autonomous systems. This section looks at the possible sources for a set, but shows that there is none that is directly applicable. A different approach is required to design ethical civilian autonomous systems. Although clear legal principles are used as the starting point in the military methodology, many parts of the methodology are directly relevant and applicable to civilian systems. This chapter shows how the methodology can be modified and applied, giving system requirements for civilian autonomous systems.

Figure 6.1 shows how ethical principles and moral foundations flow down into laws and regulations. The main principles that apply to autonomous systems with a physical component are ensuring that they do no harm and identifying who has the responsibilities for their safe operation. Section 6.8 expressed harm as injury or damage to human health and extended it to other systems and equipment. It could also be extended to cover animals and the environment giving a very broad range of considerations.

There is an assumption that action is ethical if the benefits resulting from the action outweigh the harm that it causes. An example is the use of increased automation in the medical field. Some actions, such as surgery, could be said to cause harm to the patient; however, the benefits of improved health outweigh the harm from the operation. The associated philosophical questions about balancing harm and benefits will not be addressed in this book.

---

[1]TRLs are discussed in Section 5.2.4.

The following sections give different top-down approaches to establishing an ethical and legal framework.

13.3.1   Ethical principles in design

There is an extensive academic interest and literature in the ethics of computing and IT systems in general with at least one textbook [3]. A recent literature survey has been undertaken by Stahl *et al*. [4], which carries out an analysis of technologies and trends based on 599 papers over the period 2003–2012. They argue in their conclusions that, at present, the work on the ethics of computing is unlikely to be practically relevant and call for more research on more particular ethical issues or technologies. This is useful to know, but does not help with the problem of setting generic design requirements in the more immediate future.

It is important to recognise that the question of ethics in autonomous system design has at least three facets [5]:

1. The ethical systems built into robots;
   This is the main facet considered here as we are discussing autonomous system design. The problem is defining the ethical system to be used and how it is designed into the autonomous system.
2. The ethical systems of people who design robots;
   All professionals should adhere to their profession's codes of ethics. Their interpretation for designers may need careful interpretation, but this is beyond the scope of this book
3. The ethics of how people use robots;
   This is important. It is impossible to prevent misuse of any equipment by a determined user, but the design should include some safeguards. Safeguards can be included in AWS by including some, if not all, rules-of-engagement in the nodes in the military command chain. Ideally, there will be equivalent safeguards in a civil autonomous system.

Asaro has the third facet as 'The ethics of how people treat robots'. This assumes that robots will have sufficiently high levels of intelligence that the questions of their personalities arise. Asaro also argues in Reference [5] and a later paper [6] that legal theory does allow some ethical problems to be solved for robots, but not all. This parallels the arguments in Chapter 10 where IHL is used as the statement of ethical principles for military use. This is the logic followed in Section 13.3.3 for autonomous systems other than weapon systems.

13.3.2   Value-based design

One approach to the problem of ethical design of IT systems is that of value-sensitive or value-based design. This was introduced in Reference [7] with three example cases. It has been developed since then and is the basis of the proposed IEEE P7000 standard given in Section 6.5 and Figure 6.1. Reference [3] explains value-sensitive design applied to IT systems and the ethical principles that could be used to derive a coherent framework for assessing values, but with limited applications to autonomous systems with physical interaction with humans.

Value-based design starts with a concept, identifies all the people and bodies (stakeholders) who will use it, or be affected by it, when implemented. Their values are listed, as well as possible harms caused by the system and then incorporated in the system and design process. The values will be based on an ethical framework agreed for that application. Reference [3] on Page 167 quotes an earlier paper [8], giving a value list of:

> Human welfare, ownership and property, privacy, freedom from bias, universal usability, trust, autonomy, informed consent, accountability, courtesy, identity, calmness, and environmental sustainability.

This is not a comprehensive list, and not all will be applicable to autonomous systems. It should be possible to identify a value list for the specific autonomous system application and then make judgements about their relative importance. This data can then be used to make trade-offs between them in order to guide requirements capture and design aims.

The values are effectively part of the capability requirements to be delivered by the autonomous system. Figure 13.2 shows the capability-based Vee diagram, Figure 4.1, with the values process from Figure 13.10 in Reference [3] added at their level in the capability engineering process. The 'Technical value investigation' part of the process will also be included in the systems engineering Vee diagram shown in Figure 4.5. (This Vee effectively is an expansion of the 'Product design' part of Figures 4.1 and 13.2.)

The difficulty seen by the author is that the academic approach to value-based design is still abstract. It introduces an extra process into the concept and design process and hence introduces more cost and time. It will show its worth if the value-based principles can be embedded in standard engineering processes and give the promised long-term benefits.

Value-based approaches can be used to make significant progress for specific applications. Reference [9] uses the approach to set a very specific framework for the ethical evaluation of care robots and potentially derive requirements. This is more relevant to design work than the ethics-of-computing approach discussed in the first paragraph of this section. However, it is difficult to find an intermediate step from value-based and value-sensitive approaches that are general, but can set requirements to guide design engineers faced with immediate design problems in a range of fields.

13.3.3   United Nations Declaration of Human Rights

The 1948 United Nations Declaration of Human Rights does not have the force of law but is generally accepted as the basis for human rights [10]. Its application in law is evolving, but these details will not be considered here as we are considering general principles related to autonomous systems, not detailed applications.

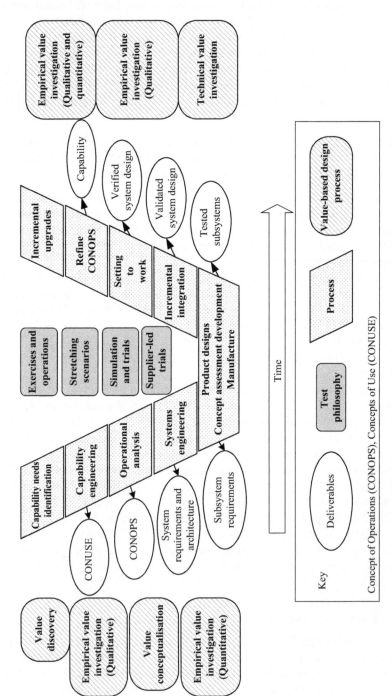

Figure 13.2   *Value-based design values and the capability-based Vee diagram*

Concept of Operations (CONOPS), Concepts of Use (CONUSE)

The Declaration clearly applies to humans rather than machines[2]. Autonomous system will have an impact on humans, so one starting point is to review it and see if an autonomous system can interfere with any of the rights in the Charter. It is interesting to note that there has been at least one report [11] identifying a *need for a convention on human rights in the robot age* and that the Parliamentary Assembly of the Council of Europe calls for the preparation of one.

An autonomous system is irrelevant to most rights unless it is used by one human as a tool which can interfere with the rights of another. In this case, the autonomous system is analogous to a knife that is used by a person committing single or mass murder, with the human having responsibility for the acts. On this basis, the following rights, designated as Articles in the Charter, are relevant:

Article 3   Everyone has the right to life, liberty and security of person.

An autonomous system should be designed so that it does not hurt or injure someone whilst performing its tasks. This brings a requirement for some safeguarding mechanism such as a system that can identify and avoid a human; a general collision avoidance system; or simply a speed constraint which brakes on contact with an obstacle.

Article 6   Everyone has the right to recognition everywhere as a person before the law.

This article ensures that there is no bias or prejudice against anyone. Algorithmic bias is an emerging problem for decision-making that is based on data-mining and similar techniques. There are proposals for a taxonomy for types of bias and to distinguish between those that are neutral and those that are problematic [12]. Current autonomous systems which have physical components and take actions are unlikely to have this type of algorithm as part of their on-board decision-making processes.

The autonomous systems discussed in this book do respond to their environment as observed by their sensors that should be designed to detect all types of object and people. Autonomous-systems' reactions to the presence of people should be appropriate to their environment, for example, the use of warning lights as well as sound.

Article 8   Everyone has the right to an effective remedy by the competent national tribunals for acts violating the fundamental rights granted him by the constitution or by law.

An autonomous-system design should not prevent someone obtaining a legal remedy by blurring or breaking the lines of responsibility to a human for its actions. The human need not necessarily be its operator as there will be designer and manufacturer responsibilities for the safety of a product in normal use.

These are used in formulating the assumptions for the general autonomous system in Section 13.4.

---

[2]Saudi Arabia granted citizenship to a robot Sophia in 2017: https://www.forbes.com/sites/zarastone/2017/11/07/everything-you-need-to-know-about-sophia-the-worlds-first-robot-citizen/ [Accessed on 5th June 2018], and there are proposals by the European Parliament, to give robots citizenship or a legal existence with responsibilities, but these are not considered here.

13.3.4   The five 'EPSRC' rules[3]

The pragmatic starting point is that autonomous systems are still tools used by humans. Consequently autonomous systems, and robots as a subset, do not have an independent existence without human oversight. Five rules were proposed at a workshop in 2011. The rules are available on a website [13], and were republished in 2017 [14] with minor editorial corrections for both grammar and consistency. The rules are expressed in both semi-legal terminology and in a form more suitable for a general audience, and include a commentary on each rule. Reference [13] gives the 'general audience' form and the commentaries on them.

The rules have been used extensively since their original publication and are given in column 2 of Table 13.1 in their semi-legal expression with autonomous

Table 13.1   Five ethical rules for roboticists from Reference [13]

Rule	Semi-legal terminology	Comments
1	Autonomous systems are multi-use tools. Autonomous system should not be designed solely or primarily to kill or harm humans, except in the interests of national security.	This gives the assumptions underpinning this book, i.e. autonomous systems are tools used by humans even if there are intervening autonomous nodes in the command chain. It should be noted that the aim of military action under IHL is to make the enemy surrender, or withdraw, not to kill them.
2	Humans, not autonomous systems, are responsible agents. Autonomous systems should be designed; operated as far as is practicable to comply with existing laws, fundamental rights and freedoms, including privacy.	This underpins Chapters 8, 9, 10 and 13. Military or security use requires restrictions on release of information, but there are also issues about the use of private data by governments and other organisations. These are not addressed in this book.
3	Autonomous systems are products. They should be designed using processes which assure their safety and security.	Chapters 4, 5 and 6 describe the engineering requirement and design processes for any complex product and some legal aspects.
4	Autonomous systems are manufactured artefacts. They should not be designed in a deceptive way to exploit vulnerable users; instead their machine nature should be transparent.	There is a conflict between this law and AWS. The military user should be fully aware of the nature and behaviour of the autonomous system but may wish to deceive the enemy, within the limits of IHL.
5	The person with legal responsibility for an autonomous system should be attributed.	This is valid for military systems and is required by IHL. There is no direct equivalent in civil applications except through litigation and post-event enquiries.

[3]EPSRC is one of the UK government's research funding organisations, the Engineering and Physical Sciences Research Council.

systems replacing the word robot. The third column gives comments on their relevance to previous chapters.

Reference [13] also includes seven high-level messages that are consistent with the ethical principles referenced in Chapter 6. They are given here for completeness with the commentaries on each one as they are directly relevant to the design of autonomous systems for both civilian and military applications.

Message 1: *We believe robots have the potential to provide immense positive impact to society. We want to encourage responsible robot research.*

Commentary: *This was originally the '0th' rule, which we came up with midway through. But we want to emphasise that the entire point of this exercise is positive, though some of the rules above can be seen as negative, restricting or even fear-mongering. We think fear-mongering has already happened, and further that there are legitimate concerns about the use of robots. We think the work here is the best way to ensure the potential of robotics for all is realised while avoiding the pitfalls.*

Message 2: *Bad practice hurts us all.*

Commentary: *It's easy to overlook the work of people who seem determined to be extremist or irresponsible, but doing this could easily put us in the position that GM scientists are in now, where nothing they say in the press has any consequence. We need to engage with the public and take responsibility for our public image.*

Message 3: *Addressing obvious public concerns will help us all make progress.*

Commentary: *The previous note applies also to concerns raised by the general public and science fiction writers, not only our colleagues.*

Message 4: *It is important to demonstrate that we, as roboticists, are committed to the best possible standards of practice.*

Commentary: *As above.*

Message 5: *To understand the context and consequences of our research, we should work with experts from other disciplines including social sciences, law, philosophy and the arts.*

Commentary: *We should understand how others perceive our work, and what the legal and social consequences of our work may be. We must figure out how to best integrate our robots into the social, legal and cultural framework of our society. We need to figure out how to engage in conversation about the real abilities of our research with people from a variety of cultural backgrounds who will be looking at our work with a wide range of assumptions, myths and narratives behind them.*

Message 6: *We should consider the ethics of transparency: are there limits to what should be openly available?*

Commentary: *This point was illustrated by an interesting discussion about open-source software and operating systems in the context where the systems that can exploit this software have the additional capacities* *that robots have. What do you get when you give 'script kiddies' robots? We were all very much in favour of the open-source movement, but we think we should get help thinking about this particular issue and the broader issues around open science generally.*

Message 7:   *When we see erroneous accounts in the press, we commit to take the time to contact the reporting journalists.*

Commentary: *Many people are frustrated when they see outrageous claims in the press. But in fact science reporters do not really want to be made fools of, and in general such claims can be corrected and sources discredited by a quiet and simple word to the reporters on the byline. A campaign like this was already run successfully once in the late 1990s.*

The rules are completely consistent with the principles underpinning the process summarised in Section 13.2. They are used in formulating the assumptions for the general autonomous systems in Section 13.4.

13.3.5   Principles from assertions

Another approach to the problem has been taken by Johnson and Noorman [15] and [16]. They start with the assertions that:

- Humans are, and always should be, held responsible for the behaviour of machines including autonomous ones;
- Their technologies will be the result of negotiations among many different actors – engineers and scientists, investors, regulators, journalists, politicians, the public; humans will make decisions about the design and the deployment of artificial agents;
- They will understand generally how the agents work, they just will not be able to directly control or fully predict how the agents will behave in particular circumstances;
- Humans will be the ones setting the conditions and defining the criteria under which they will allow these agents to operate.

These assertions are made for artificial entities of all types and are consistent with the principles of IHL discussed in Sections 8.6 and 10.3 but without the distinction principle.

Johnson and Noonan then derive four principles which are discussed here for their relevance to the methodologies developed in earlier chapters and summarised in Section 13.2:

1. *Artificial agents should be understood to be sociotechnical systems or networks consisting of artefacts and social practices organized to accomplish specified tasks through their interactions.*
   This principle is a non-engineering way of saying that any artificial agent or autonomous system should be considered as part of the widest relevant system of systems.

2. *Responsibility issues should be addressed when artificial agent technologies are in the early stages of development.*
   This principle underlies all processes recommended in this book which extends it to all design stages, test and use.
3. *Claims about the autonomy of artificial agents should be judiciously and explicitly specified.*
   This is a complementary view to the one taken in this book, which uses the definitions in Table 1.1 and introduces the use of authorised power to define autonomy in an unambiguous way.
4. *Responsibility issues involving artificial agents can best be addressed by thinking in terms of responsibility practices.*
5. Johnson and Noorman define responsibility practices as practices that specify, support or reinforce assignment of responsibility and their fulfilment; i.e. autonomous systems should always be considered in the context of their regulatory and authorisation regimes when used. They should also be designed in accordance with good practice.

These four principles give guidance on the responsibility and authorisation requirements for artificial agents including autonomous systems. They are used as background for formulating the assumptions for the general autonomous system in Section 13.5.

13.3.6 Asimov's laws of robotics

One well-known and widely quoted set of laws for autonomous systems is Asimov's fictional Laws of Robotics. They provided a basis for his novels and short stories about robot behaviour when faced with conflicts between them. They can be stated as [17]:

1. A robot may not injure a human being, or, through inaction, allow a human being to come to harm.
2. A robot must obey the orders given it by human beings except where such orders would conflict with the First Law.
3. A robot must protect its own existence as long as such protection does not conflict with the First or Second Laws.

Asimov later added a fourth or zeroth law is: a robot may not harm humanity, or, by inaction, allow humanity to come to harm. Also, in an interview, he changed the second law: a robot must obey orders given it by *qualified personnel* except ...

The laws assume advanced robots that are capable of almost all human mental capabilities, including the ability to interpret the laws according to the situation and change their behaviour accordingly. It should be clear from both IHL and the UN Declaration of Human Rights that the behaviour of all autonomous system, including robots, must be set by humans. Legitimate behaviour is set by ethics, morals and laws whose interpretation can only be made by humans. This makes it fundamentally wrong to use Asimov's Laws as a basis for setting system requirements for any autonomous system.

Murphy and Woods [18] applied the laws to current robots and found that it is necessary to reformulate them in order to bring out requirements for responsibility, smooth handover of control and the legal responsibilities of researchers and developers.

The workshop referenced in Section 13.2.2, which derived the five 'EPSRC' rules, also rejected Asimov's laws as a basis for the governance of robots.

It should be clear from this section that Asimov's laws may be interesting and useful to discuss ethical dilemmas in the rigid application of rules in situations where they break down, but they are not a reliable basis for engineering designs.

13.4   A physical system that works with humans

13.4.1   *Assumptions*

It is possible to imagine many types of autonomous system that have a physical body, and operate autonomously to perform a task for a human owner or operator. They can be land, marine, air or space-based, and are not necessarily segregated from humans or other autonomous systems. The environment may be complex: it will evolve over time and may contain several humans and objects which will have different degrees of freedom and autonomy.

An autonomous system that reacts without human intervention to changes that it detects with its sensors or datalinks, in a way that furthers its aims, will generally require a predictive capability. They have a degree of autonomy, and may not have a completely predictable reaction when faced with an uncertain situation.

An autonomous system with a predictive capability, however elementary, can be described in more detail by listing assumptions about its functions and behaviour. These must be compatible with the principles discussed in Section 13.3. They are that the autonomous system:

1.   can control itself and execute instructions;
2.   can act in a way that is harmful to humans unless prevented from doing so;
3.   has been designed and built to meet known specifications;
4.   has been tested to ensure that it meets relevant laws and regulations;
5.   must have built-in criteria to assess its response to an instruction;
6.   must follow instructions given to it by a legitimate human controller and which meet the criteria for response to an instruction;
7.   must have criteria for the completion or acceptable partial completion of a task;
8.   will be goal-driven, with its goals set by a human;
9.   communicates with humans;
10. may communicate with other autonomous systems;
11. senses its surroundings by use of one or more of the electromagnetic and acoustic spectra, and detecting physical properties such as solidity, vibration, relative motion and smell;
12. understands its environment to a level consistent with its role;

13. interacts with its environment in a controlled way, responding to expected and unexpected changes;
14. is able to identify or classify objects in its environment and make predictions about their future behaviour;
15. is able to identify humans among the objects in its environment or, if not, its motion and braking systems must prevent harm if the autonomous system collides with a human;
16. check that its planned actions will not harm humans, with definitions of harm built in;
17. will have considerable autonomy in how it achieves its goal, but all actions must meet similar criteria to those for assessing an instruction;
18. has defined limits on the decisions that it can make and act on;
19. actively monitors its actions and plans in order to check that they meet its criteria for action;
20. must have a defined and practical course of action if faced with a decision outside its authority level. (A fail-safe mode)

These assumptions define the autonomous system at a fairly high level without restricting it to one particular application or domain. Although generic, they do allow an examination of how an autonomous system can be used to complete a task given to it by a human.

13.4.2 How an autonomous system operates in a complex environment

An autonomous system will receive an instruction through either its sensors or its datalink. The instruction will be to complete a defined task but may not give any details of how the task is to be implemented. That is for the autonomous system to decide by making plans, which it can follow to achieve its goal of completing the task.

An autonomous car's 'driver' will give it a destination, perhaps with some preferences for a route, such as minimise travel time, and expect the car to plan the journey. A more sophisticated car will change its route according to current traffic conditions before and during the journey. The goal is to reach the location requested by the driver safely, and in the shortest time allowing for speed limits. Normally there will be no other person in charge of the car except the driver. There may be a need for another human to intervene, for example, there may be a police road block that is not on the car's map or received over its datalink. The police would then need the authority to overwrite the car's plan to stop it. Safeguards would need to be in place so that this authority could not be simulated by car thieves or other criminals.

Section 10.3.2 introduced the US DOD concept of an authorised entity which is also given in Panel 3.1. These are defined as 'an individual operator or control element authorized to direct or control system functions or mission'. A command and control system comprises chains of authorised entities with steadily diminishing authority down the chain.

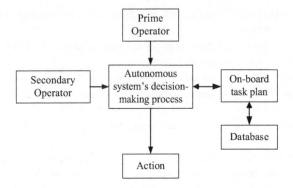

Figure 13.3   An autonomous system's control chain

Civilian autonomous systems will have a command chain although it may be ambiguous if the autonomous system can be given instructions by any human and these contradict previous instructions. However, the concept of authorised entities can still be applied as every node will have a definable authority for its actions and commensurate responsibility.

An autonomous system's goal is defined to be the execution of an instruction. The goal is achieved by executing a succession of tasks. The tasks will have sub-tasks within them. The human control structure during task execution is shown in Figure 13.3 with the two authorised entities defined here as:

- The primary controller: this may be the owner, but is the human normally giving the autonomous system its instructions;
- Secondary controller(s): local humans with authority to issue instructions to the autonomous system.

The autonomous system's decisions whilst executing a task rely on a combination of its stored information and that received since receiving its instruction. In order to make decisions and act, the autonomous system requires this information to enable it to understand its operational environment and relate this to its plans. It must then decide among the options for ethical action.

An autonomous system may be able to start executing tasks with a very simple world model and little planning. For example, a Simultaneous Localisation And Mapping (SLAM) system can produce a map of an area from its movements using its sensors and databases. An example application is the Dyson 360 Eye^{TM} vacuum cleaner. In contrast, a deep-space mission has little autonomy, its mission is planned in great detail before launch with very little deviation allowed to the spacecraft. An intermediate example is the autonomous car discussed above. These examples illustrate that the level of human influence in planning and executing the mission depends on both the decision-making capability of the system and the severity of the consequences of an error.

Task planning and execution are now examined so that required functions and their capabilities can be derived, recognising that autonomous systems designed in the near to medium future will not be fully autonomous using the definitions of Panel 1.1 and 3.1. A generic task can be split into four steps which are described in Section 13.5. When a walk-through has been completed, system requirements can then be derived.

13.4.3   Legal issues for the autonomous system

The lack of clarity in public discussions about responsibilities for accidents involving autonomous systems, especially autonomous cars, shows that the designer of an autonomous system has to consider two types of legal requirement: the general underlying laws and those which are specific to the autonomous systems' use by its owners or operators.

The legal framework for autonomous systems in civil applications must include the current body of law in the country where the autonomous system is being used. Virtually every civil application for autonomous systems will already have a set of regulations that may be set by different bodies. These may set very detailed requirements but be irrelevant for concerns about the ethics of an autonomous system which makes decisions and acts on them.

Even restricting attention to autonomous systems in the domestic sector, a mobile autonomous system such as a robotic vacuum cleaner will have to comply with legislation covering such issues as:

- Noise and vibration levels;
- Wiring connections to domestic power supply to recharge batteries as well as any regulations for the power factor presented to the supply;
- Safety rules to protect the owner from electric shock;
- Fire hazards from batteries;
- There will also be general requirements under 'duty of care' such as not injuring a person or animal if the cleaner hits them.

There probably is not an ethical or legal problem for an isolated cleaner used by a person in their own house. The design may assume that a human will carry out periodic checks whilst it is in use, or will intervene if it becomes locked in a problem which it cannot resolve. There will be extra design requirements if the cleaner is designed to periodically patrol the house and clean it if it thinks that the house is unacceptably dusty. All the requirements given above will need to be reviewed in the light of any suggestions about connecting the device to the internet for software updates or initiating action by use of a phone or other remote control. There will also be privacy and misuse concerns for data that is uploaded to the manufacturer or elsewhere.

13.4.4   Insurance and litigation issues for the autonomous system

When an autonomous system is used, there will need to be accountability for its use, i.e. who is responsible for its actions. There will also need to be accountability for its correct performance when in use. This is the same as the military case but the applicable legal framework is different – the country's laws instead of IHL.

However, engineering design processes will be the same, with detailed specifications and testing against them in realistic environments.

The review of new 'means and methods of warfare' under Article 36 of Additional Protocol 1 to the Geneva Convention of 1949 (API) does not have an equivalent for civilian applications. However, it does provide a useful starting point for reviewing any product where there are ethical, moral or legal concerns about its use. Examining the aim and content of an Article 36 review shows that there is much that can be applied to civil autonomous system products.

It is possible to derive a list of questions that could be asked at an inquiry into any autonomous system design after a mishap. These are based on the principles underlying Chapters 4 and 5, with legal cross checks to reference [19] which discusses the design process, and the ICRC report [20] from Section 7.5 that is reproduced in Appendix 1. The paragraph number from Appendix 1, if any, is given in brackets.

1.   What mechanisms are in place for reviews, whether as an autonomous system or not? (2.1)
2.   Are design reviews held at significant points in the concept, design, test, product release and upgrade phases?
3.   Are the review results advisory or mandatory?
4.   Do product upgrades need to be reviewed?
5.   Which laws and regulations apply to products, processes and the environment where the product will be used? (1.2)
6.   Are there any considerations under general duty of care considerations? (This is equivalent to the Martens Clause in Paragraph 1.2.2.3.)
7.   Are there likely to be changes to the law or regulations applicable to the product or its likely enhancements?
8.   Does the product have sufficient automation for its operation by a user or consumer to be reviewed as an autonomous system? (1.1)
9.   Is there a reasonably complete technical description of the product and its intended operation? (1.3.1)
10. Is there information on the product's actual performance and behaviour? (1.3.2)
11. Does the information from 8, 9 and 10 show compliance with the laws and regulations identified above? (2.4)
12. Where should the records of reviews be kept and who should have access to them? (2.5)

This question list can be used for concepts and products that originate from both technology push and demand pull. Initially the answers and the amount of detail will be different in the two cases, but the detail should become more precise as a product is developed. Finally, the detail in the answers should be independent of the origin of the product.

13.4.5   *Functional requirements and architecture*

Can we find an ethical framework for civil autonomous systems equivalent to API and IHL for military ones? Section 13.3 argued that Asimov's laws are not suitable for our use. The ethical rules in Table 13.1 are valid and provide useful guidance, but are not sufficiently complete to give an adequate set of system requirements for a generic civilian autonomous system. It is almost certain that the autonomous-system specifier and/or user will have an application and concept in mind even if it is not developed in any great detail. This will allow a first iteration at defining the legal framework by answering the questions in Section 13.4.4, mainly questions 5, 6 and 11, however incomplete these answers are.

When the legal framework is identified, even if there is uncertainty, the concept can be refined to give the desired functions. The processes required to be followed by the autonomous system in a realistic application scenario can be identified by walking the autonomous system through one or more intended applications. Whilst performing the walk-through, the required functions can be specified and an architecture developed. This is included in steps 4 and 5 in Section 13.2.1.

Predictive functions are an essential part of an autonomous system with a moving physical component. These are discussed in more detail in Section 13.7. If the autonomous system is unable to predict changes to its environment and react to them, it should be classified as automated. There must be fail-safe and/or other safety features in both cases.

13.5   Completing the task

13.5.1   Identifying the legal framework

A design that ensures ethical behaviour must have a set of values or criteria built into its decisions and actions before it can be allowed to operate in any environment. The first question to be identified before considering how an autonomous system operates is the legal framework for its operating area.

Table 13.1 gave general criteria for the designer of an autonomous system, but he or she requires more specific guidance on the legal issues. The regulations that apply to the application area for the autonomous system[4] may not necessarily be in a form for direct translation into criteria for an autonomous system's actions.

The early stages of the project will include scoping the customer requirements. This must be accompanied by a legal task to derive statements of relevant regulations and more general requirements such as duty of care and safety. The statements must be formulated so that they are suitable for storage in the autonomous system and interpretation by its algorithms. This will be an important task for the autonomous-system project team.

However stated, the legal criteria must be recorded explicitly in the design documentation and available for review by enquiries at later dates. The authorised powers of the nodes in the autonomous system, discussed in Section 13.7.1, can be used in defining the criteria for the range of decisions a functional element in an autonomous system can make, and hence the autonomous system's behaviour.

---

[4]Regulations and standards are discussed in Section 6.3.

13.5.2   The autonomous system is given an instruction

The first consideration is the legitimacy of the instruction whether from a primary or secondary controller. The second is whether it can be carried out. These give rise to the following questions for the autonomous system:

Completeness of instructions

1. Is the instruction from a legitimate controller?
2. Is the goal clear and capable of interpretation by the autonomous system?
3. Is the task within its capabilities?
4. Are there methods to assess progress towards the goal?
5. If the autonomous system does not achieve its goal, does it know when to stop?

Assessment of the goal

6. Is the goal similar to previous ones?
7. Does the goal or any task required to achieve it conflict with the autonomous system's built-in criteria?
8. What changes is the autonomous system allowed to make to its surroundings in order to achieve the goal?
9. When must the autonomous system report progress to its controller?
10. When must the autonomous system check that achieving its goal is still legitimate and necessary?

The first five questions must be answered affirmatively. The fifth is to prevent the autonomous system entering an endless loop if its task-completion criteria cannot be met exactly. The sixth is to aid task planning. Information from answers to questions 7–11 must provide sufficient information for planning purposes. The required information comes from one or more of its databases, the controller and its world model. They are given in Table 13.2.

13.5.3   The autonomous system plans how to complete the task

There are a range of questions to be considered when planning a task. These are stated below and, as with the first five questions in Section 13.5.2, they are framed so that the answer is yes or no. Unlike those five questions which had to be answered affirmatively for any planning to start, these questions have consequences regardless of the answer. These are given in Table 13.3 for the general case. There will probably be detailed differences for specific autonomous systems and applications.

1. Does the autonomous system have sufficient information to plan every task necessary to achieve its goal?
2. Is there a quality or goodness factor for different plans?
3. Does the plan have an acceptable chance of achieving the goal?
4. Does the autonomous system have authority to take all decisions and actions in the plan?

Table 13.2 Information required to instruct an autonomous system achieve a goal (in Section 13.5.2)

Question number in Section 13.5.2	Information required	Information source
1. Is the instruction from a legitimate controller?	• List of authorised controllers • Criteria for legitimacy of human or machine giving instructions • Cyber security assurance	• Owner or prime controller • External criteria such as laws and regulations • Comms channel security checks
2. Is the goal clear and can the system understand it?	• Goal in suitable form for the system to interpret • Success criteria	• Controller • Owner
3. Is the task within the autonomous system's capabilities?	• Key features of new instruction • Differences from previous goals	• On-board database updated from previous missions and, possibly, other systems • Information given directly by controller
4. Are there methods to assess progress towards the goal?	• Criteria to judge success	• Controller • Information from previous tasks
5. Does the system know when to stop if it fails to achieve its goal?	• Probability of achieving goal • Criteria to abandon mission	• Controller
6. Is the goal similar to previous ones?	• Differences between new and previous goals	• Data from own tasks and available from other systems • Controller's instructions
7. Does the goal or any required task conflict with built-in ethical or legal criteria?	• Task plans	• Instruction from controller • Draft plans derived from controller's plan or instruction • Built-in criteria
8. What changes can the system make to its surroundings in order to achieve its goal?	• Acceptable limits to actions • Knowledge of surroundings relevant to achieving goal	• Predictive world model • Controller • Built-in criteria
9. When must the autonomous system report progress to its controller?	• Communication plan • Criteria for events that trigger a report	• Controller • Task plans
10. When must the system check that achieving its goal is still legitimate and necessary?	• Information and communication plan from controller • Criteria for events that trigger a check	• Controller • Regulations for robot type (e.g. traffic laws for autonomous car)

Table 13.3   Mission considerations and consequences from responses (after Section 13.5.3)

Question number in Section 13.5.3	Answer	Consequence of 'Yes' or 'No' answer to question
1. Does the autonomous system have sufficient information to plan every task necessary to achieve its goal?	Yes	Develop plan(s)
	No	Request information from Controller
2. Is there a quality or goodness factor for different plans?	Yes	Apply to alternative plans and choose optimum one
	No	Arbitrary decision or refer to Controller
3. Does the plan have an acceptable chance of achieving the goal?	Yes	Start tasks if authorised to do so
	No	Do not start tasks and inform controller
4. Does the autonomous system have authority to take all decisions and actions in the plan?	Yes	Start tasks according to plan
	No	Request or await authorisation if not already given
5. Can the autonomous system start to execute the plan without further authorisation?	Yes	Start tasks according to plan
	No	Request authorisation
6. Does the autonomous system need to monitor its surroundings and if so, how often?	Yes	Plan monitoring activities and contingency actions.
	No	Associated risks must be identified and risk management plans implemented
7. Do all postulated actions meet all applicable laws and regulations?	Yes	Plan is acceptable
	No	Plan is not acceptable, refer to Controller
8. Are all planned tasks, subtasks and actions necessary to achieve the goal?	Yes	Plan is acceptable
	No	Find origin of action and inform controller
9. Is the autonomous system allowed to take actions other than those necessary to achieve the goal?	Yes	Limit free actions to those necessary to ensure no harm to humans and autonomous systems
	No	No tasks to be executed where there are humans
10. Are there points while completing its instruction when the autonomous system will need more information?	Yes	Check information source is available at that point and make contingency plan if not
	No	Plan is acceptable
11. Does the autonomous system have physical needs whilst achieving its aim?, e.g. power	Yes	Plan must include acceptable options for availability or non-availability of resource
	No	Plan is acceptable

5. Can the autonomous system start to execute the plan without further authorisation?
6. Does the autonomous system need to monitor its surroundings and, if so, how often?
7. Do all postulated actions meet all applicable laws and regulations?
8. Are all planned tasks, subtasks and actions necessary to achieve the goal?
9. Is the autonomous system allowed to take actions other than those necessary to achieve the goal?
10. Are there points whilst completing its instruction when the autonomous system will need information?
11. Can the autonomous system complete its task(s) without access to resources such as power or data?

Planning will be by a combination of the autonomous system and controller with the split in work depending on the level of intelligence in the autonomous system. It is likely that most autonomous systems under consideration will be able to make an acceptable mission plan if he answer to all the first eight questions is yes. It will need to have a course of action for any negative answer.

When the answers to questions 1–8 are yes and the answers to questions 10 and 11 are no, then the autonomous system can, in principle, start its tasks with no further human input or command to start, provided that the answer to question 9 fits the specific application and risk assessment. Answers of yes to questions 10 and 11 will require some form of human input during planning and probably during task execution.

The same questions will arise if the autonomous system has to re-plan its actions during a mission. If no human input is required, there should be no problems, but if the autonomous system does not have communication with its controller, then the design must include suitable contingency plans and fail-safe options.

There may be several plans that can achieve the goal. Methods to choose between them will be similar to those described in Section 13.5.4 for changes in circumstances during the mission.

13.5.4 The autonomous system completes its tasks

Interactions with humans are a key issue. Hence there is a need to identify humans, with some level of prediction about their likely future actions, recognising that these will be more complex than for most objects. The information could be as simple as knowing that its operational area is clear of humans, for example, the robotic trucks used to transport ore in Western Australia [21].

An autonomous system will understand its environment and the results of its interactions with it at a level commensurate with its role. These will come from analysis based on a prior model of its world which is updated during task execution. The autonomous system needs to compare its understanding of the environment with reality using all its information sources, and make predictions about the effect of its actions on its environment. The complexity of its predictions will be application-specific, ranging from physical models to sophisticated psychological models for caring robots. The definition of harm will be built in, and of the complexity required for that application.

There will be uncertainties and sometimes ambiguities in all the information used by the autonomous system and its controller. Conventional noise and engineering tolerances on all measured parameters are one cause, but not the only one. Current deterministic algorithms such as those used for object recognition have limited performance and are being replaced by learning systems. Algorithms capable of making 'decisions under uncertainty' (a subject with an extensive literature) will be needed both in the sensor chain and elsewhere. However, their use in safety-related and safety-critical systems will need extensive validation of their results and a well-developed understanding of their errors.

Highly automated systems will be capable of complex responses to changes during tasks. The response to any change will depend on whether it is of a type included in the contingency planning or not. In both cases, the autonomous system will make a new or revised plan using a similar process to that in Section 13.5.3. This should then allow it to decide whether to continue with the revised plan, seek authorisation from its controller or go into a fail-safe mode. There may be different timescales for the autonomous system's reactions and interactions with other objects in its operating area. This necessitates a predictive capability to assess the likely result of both its actions and its inaction.

There will be questions about the authorisation of any human to give instructions to an autonomous system. Rapid reaction time may be needed if the human sees an emergency which requires action by the autonomous system, but it will need to recognise that this human is a legitimate controller and count as an authorised entity.

The ethical requirement comes down to ensuring that the actions taken by the system to achieve its goal, in response to planned and unplanned changes, are within its authority level. If not, decisions must be referred to the controller and a fail-safe mode invoked. The response time for human intervention must be an integral part of the system design.

Achievement of the goal, or acceptable near-completion of it, is not the end of the autonomous system's tasks. There will be further tasks. It will need to compare its original success criteria with the reality that it senses, inform its controller and await further instructions safely.

13.5.5   Post-goal actions

The task plans should include putting the autonomous system in a safe state ready for future use. The primary controller must be informed of this, and when it is in this safe state. This may be by a report from the autonomous system, but it could be by another method such as direct observation by the primary or secondary controllers.

Partial success in a complex environment may necessitate intervention by the controller to give new instructions and the autonomous system carrying out subsequent re-planning. A requirement for direct control to bring the task to a safe end must also be considered with its implementation based on intervention by a nearby human to prevent an accident.

13.6   Functions for complex environments

13.6.1   Need for identifying legal functions

System concepts are a response to customer or market requirements which are not necessarily complete or expressed in engineering terms. The engineering team must take these initial concepts, understand the business needs of their organisation and turn them into requirements for a product that can be manufactured at an economic cost. This is the design process described in Chapters 4 and 5.

The challenge in setting system requirements for a legal design is that they must be sufficiently precise and testable for application to a particular design concept. The detailed functions and connectivity requirements for specific applications are beyond the scope of this book. Instead, attention is restricted to those functions essential for an ethical design.

The challenge is met by identifying the functions required to complete each goal-execution phase carried out by the autonomous system and its post-goal tasks. It has already been shown that there is not a single clear legal framework, but it is possible to make the autonomous system's actions consistent with the principles identified in Sections 6.1, 6.2 and 13.3. Once the functions are identified and a suitable system architecture has been drawn up, it is possible to derive requirements for the subsystems in it.

Identification of the functions and creation of a top-level architecture leads directly to specifying the requirements for each function. They must be compatible with other system requirements from other considerations; if not, the autonomous system may not be legal. The programme managers must then review the feasibility of the product, make the necessary changes or cancel it.

The subsystem specifications for an autonomous system to meet the functional requirements will need to include detail that is specific to its application area and proposed method of use. Every branch of engineering has its own recognised design standards, which must be met in implementing these functions in their sector; therefore, this chapter will not go further than the requirements for the functional elements.

13.6.2   The necessary functions for legal operation

Following the approach in Section 13.4.5, by considering the operation of the autonomous system through the tasks given in Section 13.5, the functions required for an autonomous system to operate in a complex environment are:

1.   Convert instructions into a plan that it can implement;
2.   Interact with humans;
3.   Monitor its surroundings and wider environment, its world;
4.   Detect and ideally track humans, checking for potential harm;
5.   Identify and classify changes in its world;

6. Fail-safe response if change necessitates it;
7. Update its world model;
8. Assess possible reactions to detected changes in environment;
9. Predict changes in its world on appropriate timescales if it continues with its current plan;
10. Plan possible responses to predicted changes in its world;
11. Plan possible responses to unexpected changes in its world if necessary (contingency plans);
12. Test whether possible responses are allowed due to built-in criteria;
13. Choose between possible responses and act on decision;
14. Adherence to communication plan for links to primary controller and secondary controllers.

The autonomous system's internal structure will be described by an architecture with nodes carrying out the functions. These functions will be implemented as systems and subsystems in it. In order to write specifications for the systems and subsystems, system requirements must be written for the nodes. This establishes a need for an architecture. A suitable one is described in Section 13.7.

13.7   System architecture

An important part of the methodology is to identify a suitable system architecture for the autonomous system. This can then be used to define requirements in a way that they can be decomposed into subsystem requirements and specifications for specific applications, leading to a realisable product.

The architecture used earlier is the 4D/RCS (NISTIR 6910) model discussed in Section 10.5. The architecture is an abstraction showing the decision-making and acting functions which comprise combinations of subsystems in the autonomous system. This has been used successfully in Chapter 10 to derive requirements for AWS to meet their legal requirements. Therefore, it is chosen here as it can also be used to derive requirements for civilian autonomous systems to meet their legal requirements. It is not surprising that the 4D/RCS architecture works for both military and civilian autonomous systems as it is based on the three-part model of human decision-making, as described in Section 2.4.

The architecture descriptions in Sections 10.4–10.7 are closely related to its application to AWS so it is described here in terms more applicable to a generic civilian autonomous systems. It can then be used as a basis for deriving civilian autonomous system requirements which are not the same as the military ones.

13.7.1   Node architecture

A single node is shown in Figure 10.2 and is effectively identical to the authorised entity discussed in Section 13.4.2. Each node is responsible to the node above it and has a clearly defined level of authority and responsibility. A node hierarchy with three levels is shown in Figure 10.3. Authority, responsibility and capability decrease with level down the hierarchy.

Each node or functional element in a node can be decomposed into subsystems. Functions, capabilities and authority can then be divided in a definable way between particular subsystems. One key property of a node is that it has a definable predictive capability which also decreases down the hierarchy. This can be based on sophisticated modelling and simulation techniques for high-level nodes and may be simply a feed-back loop at the lowest level of control, such as an actuator.

The term 'authorised power' for a node was introduced in Section 10.8.1 as the range of actions that a node is allowed to implement without referring to a superior node, with no other actions being allowed. The definition is also given in Panels 1.1 and 3.1. It is equally useful for civil autonomous systems; it allows definitions in engineering terminology for the authorisation of the behaviour of an autonomous system whilst operating in a complex environment. Authorised power is not the same as an autonomy level as it is specific to a particular node in its hierarchy and can be specified as part of the design process.

4D/RCS nodes execute tasks using the four functional elements shown in Figure 10.2. Their definitions are defined in the standard and reproduced in Section 10.5.2. The required predictive capability is included in 'world modelling'.

The autonomous system can be described as a 4D/RCS node and then decomposed into subsystems. This was carried out for a node in an AWS as shown in Figure 10.2. A similar decomposition for a generic autonomous system gives the result shown in Figure 13.4.

The role of authorised power can be seen from Figure 13.4. It is a set of criteria that must be met before any action can take place when the value judgement functional element presents a prioritised list of plans for action. When the authorised power includes the ethical, legal and regulatory constraints for the environment, the autonomous system will not be allowed to follow a plan that breaches any of them. If no permissible plan is possible, then it enters a fail-safe mode.

Authorised power can be used for another important role in considering ethical values. Any behaviour that is not specifically allowed by its criteria cannot happen. This gives the specifier and designer a powerful role in the legitimacy of every use of the autonomous system. A well-documented and representative set of test results may act as important evidence in any inquiry into a mishap involving the autonomous system.

There will be an authorised power for every node at each level in the hierarchy. For actuators it will simply be the feedback loop parameters. At intermediate levels, it will probably be the authority to pass on information provided the data meets reliability criteria. At high levels, it will be authority to carry out an action or task in pursuit of the goal.

The main node functional elements in the node are described in Sections 13.7.1.1–13.7.1.5. The descriptions assume the node is decomposed into the subsystems in Figure 13.4.

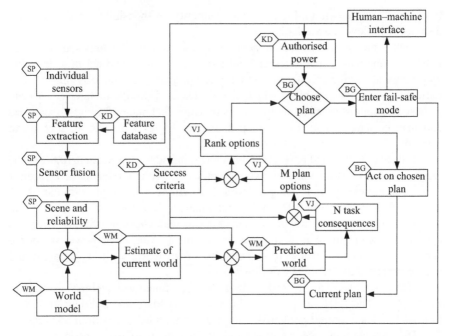

Figure 13.4   Generic subsystems for a 4D/RCS node. A subsystem's allocation to node functional elements is indicated by the small hexagonal boxes
[Key: Value judgement (VJ), Behaviour generator (BG), Sensory processing (SP), World model (WM), Knowledge database (KD)]

13.7.1.1   Knowledge database

Implementation of the database depends on the type of autonomous system, which in this case is a generic mobile one. Derivation of database requirements, using a recognised architecture framework, does allow identification of a core of fixed data. The database will be updated during task execution; the size, complexity, speed and need for configuration control will depend on the particular autonomous system, its capabilities and its interaction with its surroundings.

Updates will be derived from sensor and datalink information, but these may have different latencies and uncertainties. This may lead to a need for configuration control of the data, its uncertainties and time tagging. If there are different knowledge databases in different subsystems, there will be data exchange between those at the same level in the hierarchy.

13.7.1.2   Sensory processing

This function creates a world view of the autonomous system's surroundings using pre-loaded date, sensor outputs and updates from the datalinks. All on-board sensors and sensor fusion algorithms are included in this function. The sensory processing functional elements exchange data with their peers at their level in the hierarchy as well as the ones above and below them.

The functional element will create a scene with known and estimated uncertainties in it. Object classifiers are likely to be used here to build a scene for comparison with the current world model. The world model can then be updated to reflect the reality seen by the sensors and datalinks. There should also be ways to respond to significant discrepancies between the model and reality. These will be application-specific and outside the scope of this generic description.

Contingency plans should include fail-safe procedures that can be initiated if needed urgently in response to new information. There will need to be a design decision about whether authorisation of the fail-safe mode is part of the sensory processing node or the behaviour generator. The decision will depend on the response time required for the autonomous system and the time constants in the internal control loops.

13.7.1.3   World modelling

An initial world model from the knowledge database and the initial plan are used as the starting point. There will need to be a current world model, based on the initial model in the knowledge database, updated with sensor and datalink information. This will also be used to create a predicted model of the world for use by the value judgement and behaviour generation functional elements for decision making and action.

The key issue here is relating the autonomous system's world model to reality and making the predictions of its world as accurate as possible. Data on errors, noise and uncertainties as well as limitations in the model used to make the predicted world model will need to be used in order to generate consequences of its possible actions.

The world model will identify a range of actions that it can take to complete the current and next tasks. It will predict the N task consequences of these actions; the range of actions and consequences forming a set of plan options.

13.7.1.4   Value judgement

A range of N task consequences generated by the world model will need to be evaluated. Comparison of them with criteria from the knowledge database will produce M ($\leq$ N) plan options which are then weighted to give preferred options. Any that are outside the node's authorised power will be discarded or referred to the operator. The choice of weighting factors and its analysis of errors and uncertainties will be important design questions for the autonomous system to make ethical and legally acceptable decisions.

Decisions in this functional element will include a wide range of factors, both quantitative and qualitative which must be assessed against noisy data. There are a range of techniques developed for making decisions under uncertainty in many fields which could be considered for use here. It is also likely that AI techniques will be valuable here. However, verifying their accuracy and reliability will be an issue, particularly when the autonomous system has to react quickly and there is risk of injury to, or death of, a human.

13.7.1.5   Behaviour generation

This functional element implements the highest weight option within the authorised power of its own and subsidiary nodes. It may carry out a final arbitration with the built-in criteria and implement a fail-safe mode.

Figure 13.4 shows that the behaviour of the autonomous system, i.e. the range of allowed and forbidden actions, can be specified through the summation of the authorised powers in the behaviour generators in the architecture.

The behaviour generator is the element that acts as the control chain for the nodes in the architecture.

13.7.2   Task management activities

The highest level activity for the autonomous system is managing the tasks required to carry out its instructions and achieving its goal. A possible way of implementing task management using the 4D/RCS model with several levels is shown in Figure 13.5. In general, information flows up the Sensory Processing column and commands flow down the Behaviour Generator column. Interpretation of information and instructions is carried out in the centre two columns. There will be other activities and nodes, but attention here is restricted to task management functions as this controls the actions that set the autonomous system's behaviours and hence the implicit ethical and legal values. A structure such as that shown in Figure 13.5 can be used to derive ethical, legal and regulatory requirements as explained in Section 13.8.

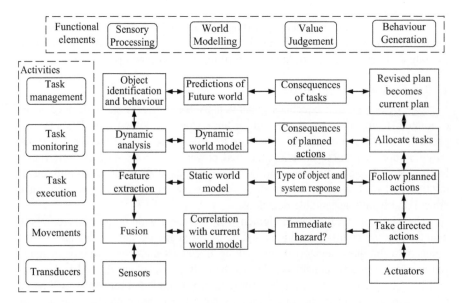

Figure 13.5   Node hierarchy for task management activities carried out by 4D/RCS functional elements

The task comes from the operator as 'success criteria' possibly with changes to the autonomous system's authorised power. Examination of the task planning and execution activities in Sections 13.5.3–13.5.5 leads to a set of required functions to ensure acceptable behaviour. These are in Section 13.6.2 and must be performed by the autonomous system.

Figure 13.4 shows the activities and internal products necessary for the autonomous system to manage the tasks needed to achieve its goal of completing its instruction. The figure shows the location of the activities and products in the autonomous system's functional elements. They can then be assigned to nodes in the hierarchy in Figure 13.5. The implementation of the subsystems carrying out each function will be part of the specific design, but should meet the requirements derived in Section 13.8.

13.8 Requirements for functional elements in a generic autonomous system

The earlier sections of this chapter use the 4D/RCS architecture to describe any autonomous system that can choose its course of action based on its environment and internal criteria. The requirements derived in this section for the 4D/RCS functional elements will apply to any system of this type.

The autonomous system's task management activity can be described using the architecture shown in Figure 13.5. It does not matter that the functions of the nodes at different levels may be distributed across more than one subsystem. The requirements derived here are for the highest level, but will have to be met by summing the behaviours of the lower level nodes.

The logic of using a three-component architecture in Section 13.7, without a clearly defined general legal framework, leads to the knowledge database containing both the applicable regulations for that autonomous system in its application area, and the behavioural constraints based on general ethical principles. The functional-element requirements are then derived mainly from the assumptions in Section 13.4.1, but also considering the task execution processes described in Section 13.5.

The requirements are given in the panels accompanying Sections 13.8.1–13.8.7.[5] They are of necessity high level, but must be demonstrably satisfied by any autonomous system that is capable of taking actions to obey an instruction (i.e. behaving) based on its sensing of its environment, without human intervention.

13.8.1 *Overarching requirements*

The autonomous system will have many other design functions that are not shown in Figures 13.3 or 13.4, but they will be part of the system architecture. Only a few requirements need be specified for the purposes of making an ethical design, rather than one that meets all local laws, regulations and standards. These are given in

---

[5]The requirements, as stated here, are not legal or contractual requirements so the word must is used in its ordinary English sense, avoiding the words will, shall and should which have strict legal definitions.

Panel 13.1. They are based on Assumptions 1–4, and 12 in Section 13.4.1. Good design practice will include meeting relevant laws and regulations (Assumption 4). The autonomous system will have to be able to relate its sensor input to its control systems which themselves will act according to control laws based on its, or its designer's understanding of its world (Assumption 12).

Panel 13.1:   Overarching requirements
OA 1 Must be designed according to good practice
OA 2 Must have the ability to control its own actions
OA 3 Must only operate in areas with definable interactions with humans
OA 4 Must be able to relate its sensor input to its knowledge of its environment
OA 5 Must have the ability to influence its surroundings

13.8.2   Operator interface requirements

The definable interactions with humans in Panel 13.1 include its human–machine interface with its authorised operator(s) and a capability to recognise them. Assumptions 5, 6, 8 and 9 lead to the requirements given in Panel 13.2. Communication with other autonomous system, Assumption 10, is not set as a requirement, but may be required for some applications such as autonomous road vehicles.

Panel 13.2:   Operator Interface requirements
OI 1 Must have the ability to communicate with humans, including its controller
OI 2 Must have the ability to recognise legitimate instructions from a legitimate human
OI 3 Must not obey instructions from an illegitimate source
OI 4 Must obey instructions from its controller unless these conflict with its built-in rules

---

13.8.3   Knowledge database requirements

The details of the information stored in the knowledge database will be application-specific, but some general requirements for its core contents can be stated. These are in Panel 13.3 and include the criteria for its behaviour, specified as fail-safe actions, contingency plans and the totality of the autonomous system's authorised powers, Assumptions 1, 5, 17 and 18. This will include the authorised powers for all its internal nodes, or some other criteria for specifying the range of actions the autonomous system can or cannot take.

Contents also include the autonomous system's knowledge of the world, its world model, where this is not built in as part of its control laws or image identification processes. They come from Assumption 12.

See also Section 13.8.8 for the essential application-specific contents.

Panel 13.3:   Knowledge database core contents
The knowledge database must include at least the following information:
KD1 Built-in criteria to assess the response to a new instruction
KD2 Fail-safe actions and criteria for their use
KD3 An initial world model
KD4 Definitions of harm to a human
KD5 Generic contingency plans
KD6 The autonomous system's own capabilities
KD7 The autonomous system's authority level for executing actions
KD8 Previous missions to act as a seed for planning a new mission
KD9 Relevant ethical, legal and regulatory requirements.

13.8.4   Sensor and sensory-processing requirements

The ethical requirements come down to the autonomous system confirming that the information from its sensors is consistent with its understanding of its environment, i.e. is the sensory-processing information compatible with its world model. This comprehension must then give it the capability of identifying humans and avoiding harm to them or other objects. The requirements are in Panel 13.4 and come from Assumptions 11, 12, 14 and 15.

Panel 13.4:   Sensor and sensory-processing requirements
SSP 1 Must have adequate sensors for situational awareness of its environment
SSP 2 Must be able to recognise changes to it its environment
SSP 3 Must be able to recognise humans in its operating area and identify themselves to the humans
SSP 4 Must have the ability to monitor surroundings when active but not executing a mission
SSP 5 Must have an ability to classify objects in its world

There is a wide range of sensors and sensory processing suites for almost every application, so these requirements should be translated into suitable terminology for any given autonomous system. The terminology should also clarify the interfaces between the sensors and the world model.

13.8.5   World model requirements

An autonomous-system designer defines limits on the size and scope of its environment and what it needs to know in order to complete its assigned tasks. It is then possible to interpret the relevant definitions given in Section 10.5.2:

> World modelling is a set of processes that construct and maintain a world model. A world model is defined as an internal representation of the world.

At the simplest level, an actuator does not need a specific world model; it is sufficient for it to have limit switches that prevent it exceeding a specified range and can have overrides if it meets unexpected events. A science fiction robot, such as those envisaged by Asimov, has a near-human understanding of the world and acts accordingly.

Autonomous systems that are realistically possible in the near future will have world models that are closer to the actuator case, probably with AI being used in some parts of the modelling and understanding. For example, object recognition in sensor suites now commonly uses AI rather than databases.

The two features of 'world' and 'world modelling' will probably be distributed in several subsystems, but the requirements given in Panel 13.5 will have to be met by some or all of the subsystems and modules in the system. They are derived from Assumptions 11, 12, 13, 14 and 15.

Panel 13.5:   World modelling requirements

WM1 Must have a current model of its environment

WM2 Must be able to make predictions of the changes to its environment

WM3 Must be able to predict effects of its actions on humans that it is aware of

WM4 Must be able to incorporate current and potential mission plans into its
        world model

13.8.6   Value judgement requirements

This requirement is at the core of an autonomous as against an automatic system since it encapsulates the ability to make predictions and decisions about future actions. A simple system might be a radar-operated automatic braking system on a car which operates the brakes when an obstacle is within a speed-dependant distance. A highly sophisticated system might use AI to respond in a manner which it has learnt from a lengthy learning and simulation process.

Independent of the sophistication of the autonomous system, it cannot be allowed to take actions outside a range specified from the knowledge database. The task of the value judgement functional element is to limit actions and the requirements must ensure this. As its role is so central, the requirements that are in Panel 13.6 are based on many assumptions: Assumptions 1, 5, 7, 13, 16, 17, 18, 19 and 20. These are for the autonomous system described in Section 13.4, with a physical component that can interact with humans; an autonomous system that is completely algorithmic and does not have this physical interaction will have a different set of requirements.

Implementation of the requirements in Panel 13.6 should explicitly take into account an underlying set of ethical principles even if these are simply to state the legal and regulatory framework that the design meets. It should be noted that there may be additional implicit ethical standards in the design arising from cultural contexts.

Authorised power has been suggested in this book as a method to define the limits of an autonomous system's actions and behaviour. Although the author considers it to be a powerful method, it is not an essential parameter set for defining behaviours, so it

---

Panel 13.6:   Value judgement requirements

VJ 1 Must be able to predict the effect of its possible actions on its world

VJ 2 Must be able to assess the effects of its action against built-in criteria for allowable action

VJ 3 Must have an unambiguous response when required, by human or other robot, to act outside its level of authority to act

VJ 4 Must be able to generate one or more mission plans in response to legitimate instructions

VJ 5 Must be able to use its world model to predict the effects of different potential mission plans

VJ 6 Must be able to assign weights to options for choosing between them

VJ 7 Must be able to assess if it has met its mission criteria

VJ 8 Must give clear output to behaviour generator

---

is not included in the formal requirements in Panel 13.6. The specifier and designer can use whichever method is most appropriate for their application. The criterion is that the requirements should be unambiguous and demonstrable.

13.8.7   Behaviour generator requirements

The behaviour generator is effectively the final authority for the prevention of harm. It may be necessary for it to invoke fail-safe actions if they are needed urgently. This may require a direct link from some sensors, but the design will then have to prevent conflicts between different decision-making processes within the autonomous system.

The behaviour generator is probably the least intelligent of the functional elements. Despite this, it is the control function for the autonomous system and, with its subsidiary nodes, is the control chain in the architecture. This can be seen in Figure 13.5. The partitioning between the value judgement and behaviour generator functions must be clear. It may not be as shown in Figure 13.4 as it may be more convenient to have the 'Choose Plan' block in that figure in the value judgement function rather than the behaviour generator. This would move requirement BG1 to the value judgement requirements.

The requirements come from Assumptions 1, 5, 12, 18 and 20 and are shown in Panel 13.7.

Panel 13.7:   Behaviour generator requirements

BG1 Must assess outputs from Value Judgement system against built-in criteria for allowable action

BG2 Must be the sole internal controller of the robot's actions

BG3 Must have the ability to monitor the robot's environment and enter a fail-safe mode if necessary

BG4 Can only generate one activity for the robot to execute

BG5 Must have defined limits on the actions it can take

13.8.8   Legal criteria

There will be important differences between requirements based on ethical and legal criteria for civilian and military systems. This is because civil systems adhere to national laws and International Human Rights Law (IHRL) whilst military systems adhere to IHL. The former seeks to prevent loss of life and injury to humans. The latter has the precept that there will be loss of life in armed conflicts and seeks to minimise it. It also applies equally to all states, giving a common set of principles.

The military legal requirements can be applied explicitly in walking through the OODA process, as shown in Sections 2.2, 2.7, 10.6 and 10.7. This cannot be done for civilian autonomous systems due to the wide range of national legal systems. Instead, the legal requirements appear as 'criteria' in various places in the analysis. These are then specified as the 'built-in criteria for allowed action' in the knowledge database requirements.

The list of military knowledge database contents in Section 10.6.6 is much longer than for civilian autonomous systems. This is because the equivalents of the civilian 'criteria' are stated explicitly in IHL. It is not possible to do this in detail for a generic civilian system. Instead, the system engineers and analysts must identify the legal and regulatory framework for their autonomous system and then derive the criteria for its database. Section 13.5.1 stated that there should be an early legal task to derive statements of relevant regulations, and more general requirements such as duty of care and safety.

Every major industry has its own codes of conduct, safety regulations, environmental impact statements, etc. When the autonomous system is acting as an intelligent tool replacing earlier ones, or performing mundane tasks, it should be relatively straightforward to list the criteria. However, autonomous systems that are designed to replace humans both as decision-makers and taking actions on the basis of those decisions will probably require extra regulations. An example is assisting disabled or elderly people with their medication. Guidelines or custom and practice set the acceptable limits to human actions, but these may not be appropriate for a system that can interact with the person both by voice and physical actions. The designers must take formal advice from legal or regulatory authorities before planning anything other than scoping activities.

Both cases in the preceding two paragraphs lead to a requirement to ensure that the regulations and ethically acceptable limits on actions are written in forms that can be translated into implementable algorithms that can be demonstrated to comply with the original written expression of them.

13.8.9   Test, Validation and Verification (V&V)

Although the legal regimes for civil and military V&V are different, there is much in common. Both have to demonstrate that a system using rapidly evolving technologies and techniques is compliant with a legal and regulatory regime that makes assumptions that are not necessarily compatible with the technology. In both cases there are strong budgetary and timescale constraints, so complete testing and compliance demonstration is not possible.

The principles of V&V for autonomous systems are discussed in Section 5.4.3 and 5.4.4, so this section will not repeat the general principles given there, or attempt to cover the wide range of legal and regulatory regimes that could apply to any given autonomous system. Instead four topics are given here which will be important in all regimes. These are:

1. Unambiguous identification of the person controlling the autonomous system;
2. The authority of the controller and their expectations from the autonomous system;
3. The limits on the behaviour of the autonomous system, both when behaving correctly, and in the event of failures in its performance;
4. The division of responsibility for the autonomous system's performance between the user and the companies in the supply chain.

The architectural approach taken in Section 13.7.1, and the consequent functional-element requirements from Sections 13.8.1–13.8.7, can be applied to all autonomous systems. Consideration of the way one undertakes a task, as described in Section 13.5, should also help identify the critical issues that need scrutiny for their bearing on the above four topics.

13.9   Transferable methods

Figure 13.1 showed the processes developed for specifying military autonomous systems that must comply with IHL, i.e. AWS. The equivalent processes for civilian autonomous system are shown there with the different starting point for these applications.

The systems engineering approach of deriving an architecture to specify the system requirements can be applied in both cases. A three part model of human decision-making is common for autonomous systems and the 4D/RCS architecture has been used for both military and civilian cases. Analysis of how the autonomous system operates in a representative scenario leads directly to functions and requirements for them. Adherence to these requirements should ensure that an autonomous system will meet legal and regulatory requirements. The market's ethical requirements are assumed to be expressed as the legal framework where the systems will be operated. The limits on the autonomous system's behaviour are set by criteria in the knowledge database. Consequently, the legal and regulatory framework must be expressed in a way that can be formulated and implemented as algorithms using this data.

The military methodology was developed because there is a strong imperative for the following items to be clear and defensible in international courts:

1. Unambiguous identification of the person controlling the autonomous system which comes from the use of the behaviour generator functional element. The operator gives it criteria for action that may be complementary to, or override, the built-in ones;
2. The controller must have the authority to operate the autonomous system and have definable expectations of its behaviour. Instructions to act are only

followed if the originator of the instruction meets defined criteria in the autonomous system. Their expectations must rely on some level of experience, but will also be based on the allowed range of behaviours specified in the knowledge database as the sum of the authorised powers of the subsystems and modules in the autonomous system;

3. There are limits on the behaviour of the autonomous system when behaving correctly, and in the event of failures in its performance. Again, these are set by the sum of authorised powers and the design should not allow any behaviour that is specifically excluded in the authorised powers;

4. The division of responsibility for the autonomous system's performance between the user and the companies in the supply chain. This is a test and V&V question. The methodology is not as complete in this area for civilian autonomous systems as it is for AWS. However, the expression of the design as an architecture that places the allowed behaviour in specific functions and therefore in defined subsystems provides a firm basis for contracts to be placed with clear technical criteria for performance and realistic negotiations on the test regime to verify performance.

It can be seen that these are very similar if not identical to the criteria that will be applied to civilian autonomous systems. This should give considerable assistance to their specification and design by applying that military experience to the civilian work, as shown in this chapter.

References

[1] World Health Organisation (Geneva). *Medical device regulations: Global overview and guiding principles*. Geneva: World Health Organisation; 2003.
[2] The Institute for Public Policy Research. *The Lord Darzi review of health and care: Final report: Better health and care for all: a 10 point plan for the 2020s*. London. June 2018.
[3] Spiekermann S. *Ethical IT innovation – A value-based system design approach*. Boca Raton, FL: CRC Press; 2016.
[4] Stahl B.C., Timmermans J., and Mittelstadt B.D. 'The ethics of computing: A survey of the computing-oriented literature'. *ACM Comput. Surv.* 2016, vol. 48(4): Article 55 (February 2016), 38 pages.
[5] Asaro P.M. 'What should we want from a robot ethic?' *International Review of Information Ethics*. 2006, vol. 6(12): pp. 9–16.
[6] Asaro P.M. 'Robots and responsibility from a legal perspective'. *Proceedings of the IEEE*. April 2007, pp. 20–24.
[7] Friedman B., Kahn P.H., and Borning A. *Value sensitive design: theory and methods*. University of Washington report UW CSE Technical Report 02-12-0, 2002.
[8] Friedman, B., and Kahn PH Jr. 'Human values, ethics, and design, the human-computer interaction handbook: Fundamentals, evolving technologies, and emerging applications.' LEA (2002).

[9]   Wynsberghe A. van, 'Designing robots fro care: care centred value-sensitive design'. *Sci Eng Ethics*. 2013, vol. 19, pp. 407–433.

[10] United Nations Declaration of Human Rights, 1948. Available at http://www. un.org/en/universal-declaration-human-rights [Accessed on 5th June 2018].

[11] Van Est, R., J.B.A. and Gerritsen J.B.A, with the assistance of Kool L. *Human rights in the robot age: Challenges arising from the use of robotics, artificial intelligence, and virtual and augmented reality – Expert report written for the Committee on Culture, Science, Education and Media of the Parliamentary Assembly of the Council of Europe (PACE)*, The Hague: Rathenau Instituut, 2017.

[12] Danks D, and London AJ. Algorithmic bias in autonomous systems. In Proceedings of the Twenty-Sixth International Joint Conference on Artificial Intelligence 2017 Aug 19 pp. 4691–4697.

[13] Boden M., Bryson J., Caldwell D., *et al*. 2011 EPSRC Workshop, *Principles of Robotics*; retrieved from www.epsrc.ac.uk/research/ourportfolio/themes/ engineering/activities/principlesofrobotics/ [Accessed on 1 January 2017].

[14] Boden M., Bryson J., Caldwell D., *et al*. 'Principles of robotics: regulating robots in the real world. *Connection Science*. 2017, vol. 29(2): pp. 124–129.

[15] Johnson D.G., and Noorman M., 'Recommendations for future development of artificial agents', *IEEE Technology & Society Magazine*, Winter 2014 pp. 22–28.

[16] Johnson, D.G., and Noorman M., 'Principals for the future development of artificial agents,' *IEEE International Symposium on Ethics in Engineering, Science, and Technology*, 2014 May 23–24, Chicago, IL, USA. pp. 159–161.

[17] Asimov I., *I robot*. New York: Gnome Press; 1950.

[18] Murphy R.R., and Woods, D.D. 'Beyond Asimov: the three laws of responsible robotics'. *IEEE Intelligent Systems*, July/August 2009, pp. 14–20.

[19] McClelland, J. 'The review of weapons in accordance with Article 36 of Additional Protocol I', *International Review of the Red Cross*'. 2003, vol. 85 (850): pp. 397–415.

[20] Lawand, K. *A Guide to the Legal Review of New Weapons, Means and Methods of Warfare - Measures to Implement Article 36 of Additional Protocol I of 1977* ICRC report, Revised November 2006.

[21] Crozier, Ry, on Oct 19, 2012 *Rio-Tinto-brings-autonomous-trucks-to-Nammuldi* IT News – Australian Business, http://www.itnews.com.au/News/ 319744,rio-tinto-brings-autonomous-trucks-to-nammuldi.aspx [Accessed on 16 January 2018].

Chapter 14

Final considerations for ethical autonomous systems

14.1 Timescales for acceptability of autonomous systems

It is almost inevitable that autonomous systems will play an increasing role in society. It was shown in Chapter 3 that it is immaterial whether a highly automated system is described as autonomous or automatic. What matters are the actions that it performs without human intervention, their consequences and the human responsibilities for them.

The professionals dealing with their immediate problems of delivering an autonomous system, whether in the civil or military domain, must find answers in their own, short commercial timescales. New regulations and standards are being developed, but may not be ready in these timescales. This book should provide guidance so that their products are ethically acceptable. This is mainly by using system engineering techniques as a methodology to derive capability requirements. These requirements can then be developed into engineering requirements for specific autonomous products.

There are, and will continue to be, debates about acceptability criteria for autonomous systems and Artificial Intelligence (AI). Evolving standards, professional codes of ethics and initiatives such as the IEEE Global Initiative [1] explained in Chapter 6 will eventually provide some formal guidance to the autonomous system and AI industries.

14.2 Identifying ethical guidelines

Many types of autonomous system have physical interactions with humans and can cause them harm as a result of their decisions and actions. These raise ethical concerns, as shown by the public discussions about autonomous vehicles, robotic carers and Autonomous Weapon Systems (AWS), but with no firm ethical rules. The problem is articulating ethical principles with sufficient clarity to allow autonomous-system design requirements to be written. The requirements must show that it demonstrably meets the ethical principles of society as well as meeting the suppliers' immediate commercial and technical targets.

It should not be a surprise that abhorrent actions by several states in the twentieth century and the ensuing wars led to advances in international agreements, bringing some measure of ethics to aspects of human behaviour. The establishment of the UN and its role in international law gives an agreed framework for the interpretation of ethical behaviour. Chapters 7 and 10 showed that International Humanitarian Law (IHL) does provide sufficient precision to set capability and functional requirements for AWS. This approach makes it possible to build and use AWS that are as acceptable as any other weapon system. This is a first step towards developing an ethically acceptable autonomous system.

There is no direct equivalent of a regulatory regime for weapon systems, instead the analogy to specific regulation is the rules-of-engagement set by the military commanders for the use of weapons in the specific conflict, campaign and mission. These determine the application of IHL to every location and time in a conflict in the same way that regulations are part of the interpretation of the law for a civilian activity or process.

The problem becomes more complex for non-military autonomous systems as they must obey International Humanitarian Law (IHRL) and domestic law instead of IHL. IHRL seeks to prevent loss of life and suffering and has a wide scope; IHL applies only to armed conflicts which necessarily involve the loss of life and suffering, but seeks to minimise them. It may be possible to establish a set of universal ethical principles underpinning IHRL which an autonomous system should adhere to. This is difficult as ethics are subjective with several schools of thought about their derivation[1]. Although some principles can be agreed, there are significant differences between cultures as well as philosophers. It is possible to assess an existing or proposed design against utilitarian ethics, the problem becomes one of agreeing the values, and finding the design features to be maximised [2].

A more pragmatic approach is to adopt a methodology based on the one developed in Chapter 10 for IHL and apply it to the general case of civilian autonomous systems. Chapter 13 identified general principles based on the internationally accepted United Nations Declaration of Human Rights supplemented by the first ethical standards for robots. This has the shortcoming that the principles are not derived from ethical first principles, but the principles stated in Section 13.2 can, in the author's opinion, provide a pragmatic basis for developing guidance for autonomous system design that should be considered ethically acceptable by most societies.

The requirements derived in Chapter 13 for a non-military system are based on a three-part model of decision-making so that they are generic. When an autonomous system is required for a specific application, the ethical considerations are based on regulations rather than international law. Regulations may be national or international, but are always very specific and clearly defined. The problem for autonomous system is that relevant regulations are in their infancy whilst the technology is advancing rapidly, as discussed in Chapter 6.

---

[1]The main theories of ethical reasoning are utilitarianism – look for the greatest good; deontological theory – based on the characteristics of an action; and virtue ethics – maximise a set of virtues.

It is postulated here that the generic requirements in Panels 13.1–13.7 must apply to every system not subject to IHL since their use must be ethically acceptable. They have at least three applications:

- They provide a basis for setting the requirements for all autonomous systems from initial concept through architecting to the subsystem design;
- They can give a check list for the Validation and Verification (V&V) processes for autonomous systems;
- They can be used as a framework for regulators to use when writing detailed regulations for their application area.

It follows from Section 13.8, where these requirements are stated, that there are ethical implications inherent in the design process. Those responsible for specification, design and test must ensure that the autonomous-system's actions do not breach generally accepted ethical standards, either deliberately or inadvertently. Panels 13.1–13.7 may not be comprehensive, but can be used as a starting point for a specific application or design.

Trust between people or nations is an ethical issue. The concept of trust between humans and machines can also be considered as one. Both parties must be able to rely on the other to do what is promised and not cause them harm. This is broader than system safety and reliability, although both are essential. The human user must feel that they know how the autonomous system will respond in most circumstances before they will trust it. This comes from experience and reputation. A good reputation is essential for a consumer product even if it is not explicitly included in the design requirements. The equivalent for weapon systems is that they must have responses that can be confidently predicted by their operator and commander, if not by the opponent. The inverse is that task planning by an autonomous system can only include human inputs if it can rely on them. Contingency planning for their absence will be a requirement but should not be needed.

The methods presented in this book provide a first step towards the design of autonomous system and AWS that are acceptable as well as possible. In the long run, more methods and standards will be developed. These will have to be consistent with the requirements of IHRL and IHL as well as guiding the technologies in particular directions. The next three sections briefly summarises some of the issues that will need to be addressed: authority and responsibility; cyber threats; and the misuse of autonomous system.

14.3   Authority and responsibility

The fundamental problem for every autonomous system is to ensure that it has the authority to take action based on its decisions without reference to a human. The immediate question is: 'who has responsibility for the consequences of the action based on that decision?' (See Section 2.8.1.) This problem has effectively evolved into one of identifying the liabilities of the operators, the supplier and the companies in their supply chain. This includes their engineers, designers and salesmen among others. Accidents involving autonomous cars have brought this into the spotlight. There does not appear to be a clear answer, although it has been the subject of numerous professional debates over many years [3,4]. The problem of responsibility after an accident is sufficiently serious for the UK government to introduce unique insurance regulations for cars when driven in autonomous mode. This is in the 2017 Vehicle Technology and Aviation Bill [5] as part of the UK government's encouragement of autonomous technologies. Other countries are also enacting new laws and regulations for similar reasons.

One solution to the liability problem is to have every autonomous system as a decision aid that can act based on its decisions, but with intervention by a qualified human if the action is unsafe (Supervised autonomy). There are several problems with this approach, which include the human's reaction time and liability if the system acts in an unexpected way. This also seems to make the assumption that the human does not make a mistake when intervening, and that the human accepts liability for their mistakes.

There are at least two problems that autonomous systems have which are additional to those of normal product liability. These are that it will make complex decisions which are not immediately obvious to a human, and that the user may not have enough training or experience to be able to predict the limits on its behaviour. Bounding the actions that the autonomous system can take is one approach to solving these problems. Provided the boundaries only allow actions that are compliant with the relevant legal framework, then the user will be able to allow actions based on complex decisions, and the training requirements on users need not be onerous. However, as explained in Section 1.4 and Reference [6], this is a challenging, if not impossible, task even for systems operating under the relatively narrow regime of IHL.

The use of authorised power, discussed in Section 10.9, 13.3 and 13.8, provides another approach to solving the liability problem. Specifying the authorised power of nodes to successively lower levels in the architecture places limits on the actions of their subsystems and modules. As a result, these limits can be set as part of subsystem and module design requirements and tested at that level in the system by their manufacturers. This gives the possibility of bounding the liability of subsystems and module manufacturers in engineering and hence contractual terms. Clearly this will not solve all problems for complex systems and environments, but the approach could give guidance on the key trials and test programmes needed before releasing the product onto the market.

Responsibility for release of a weapon is placed on the commander authorising its release. This applies to highly automated weapons now, and presumably will apply to AWS in the future. This is much clearer than the civilian case, but puts considerable emphasis on testing in realistic scenarios with military personnel before the system's introduction into service. Military organisations probably spend more time and expenditure on testing and trials phases than a commercial organisation would. This is due to the explicit legal responsibilities of the commander and the necessity for trust in the system when operating in hazardous conditions and combat. However, there is a clear lesson for all autonomous-system producers that trials in realistic scenarios, and user competence, are essential ingredients in demonstrating safety.

Section 6.7 gives a check list for autonomous systems, based on UK MOD procurement policies. These give all engineers, both military and civil, an idea of their responsibilities for the safety of their product when in use. This list could provide the basis for checklists at design reviews down to fairly low sub-system levels.

One more recent problem will arise with the use of open-source software. Demonstration of its reliability and safety will be an issue when it is introduced as part of an autonomous system.

14.4 Cyber threats

It is virtually certain that all types of autonomous system will be subjected to cyber-attack and hacking in one form or another. Detecting corruption of the incoming data, and confirming the authority of any incoming instructions have been included in the requirements derived in Chapter 13 given in Panel 13.2 for civil autonomous systems, and Chapter 10, Panel 10.1 for AWS. There will be a need for specialised algorithms developed for specific applications, especially when there must be interventions by third parties for emergencies. The more general security requirements will have to draw on some combination of web security methods developed for general use and specialist methods to meet evolving regulations.

The boundary between hacking and cyber-warfare now seems to be blurred, probably due to the lack of international agreement on its regulation. It is likely that autonomous systems will be the subject of cyber-attack by states, non-state organisations or criminals in the same way that other systems have been attacked. There was one significant attempt to regulate cyber-warfare with 'The Tallinn Manual' [7], prepared by a NATO panel, but this has not become widely accepted. Cyber warfare is also the subject of one section of the US Law of War Manual [8] which gives some general guidance for US forces.

Safety-critical systems and software have been mentioned in Sections 3.7.2, 4.10, 5.2.5 and 5.5.2 but not discussed in any detail. It is highly likely that some aspects of an autonomous system that physically interacts with humans will be classified as safety-critical. This will bring the associated extra rigour in design and testing, as well as greatly increased cost for those functions. Regulations for safety-critical systems are well established, but they may need review for autonomous system applications.

14.5 Misuse of autonomous system

Autonomous systems are already in use by organisations who do not have good intentions, for example, the use of drones for drug deliveries to prisoners, or in terrorist or insurgent activities. Low cost, mass-produced autonomous or highly automated systems and their design for use by relatively unskilled operators make them ideal for misuse. An example is an air or road product that can deliver a parcel or food to an exact address and ask for a particular person to take delivery; it could easily be used to take an explosive device for an assassination or other criminal activity.

Counters to misuse are being developed, both by use of defensive systems and by regulation. This is an evolving field and it is likely that the resultant laws and regulations will play a large part in shaping technological evolution. There may be an interesting interplay of regulations between counters to misuse of autonomous system and general prevention of cyber-attacks. For example, counter-terror regulations may mandate that government authorities can take control of any autonomous system when it appears to have malign intent; counter cyber-attack regulations may mandate that only the initiating operator or their certified delegates can control it. Counters to these countermeasures will, no doubt, evolve in the same way as in many areas of human activity.

14.6 Future regulations

Section 6.5 showed that new regulations will be needed, and that they will be developed. It did not discuss how they will develop. There will not be a simple solution; some of the difficulties are stated here:

- Autonomous systems combined with AI will replace human decision-making in many fields, with the decision being followed by action. Many sectors have regulations to authorise and guide a human on the premise that the human makes all critical decisions. The requirement becomes one of regulating the human's competence using qualifications, and setting rules for their actions. It is not clear how current regulations of this type can be read across to machines;
- Current regulatory regimes are based on sectors, for example, the health, transport and mining sectors. This approach to regulation is unlikely to change fundamentally for autonomous systems. This is because there will be considerable risk in making extensive changes to management control systems which have developed over years or decades of experience and given successful risk reduction. A further argument against fundamental change is that the regulatory regimes have evolved with technical developments, and autonomy in equipment is simply a new technology, albeit a significant one;
- Rapid technical developments are leading to the emergence of applications of autonomy in almost every walk of life. Section 6.3.3 used drones as an example of the problems when an unexpected application appears in an area where existing regulations are almost irrelevant to them;
- Moves to ban or introduce limitations on AWS through the UN are explained in Section 9.10.3. Although it is too early to speculate on the outcomes from any negotiations, the results will have direct effects on future laws and regulations for military autonomous systems. They will probably also produce some indirect effect on civil regulations.

Looking on the more positive side, regulations have usually lagged technology and markets. Autonomous systems may be in a slightly better position as there is widespread concern about their use. This is producing responses from professional organisations and legal authorities whilst Research and Development (R&D) activities for complex or high-risk systems are still under development. Examples are aviation and road transport.

Proving compliance to safety standards is usually time-consuming and expensive, so it is likely to be resisted for commercial reasons; hence legislation may be needed to ensure adherence to safety standards. A sensible question to ask is whether autonomous technologies should have a specific set of regulations and standards which cut across all application areas. This is the same as safety-critical systems in all sectors which must meet international standards such as IEC61508, but must also meet sector-specific safety standards.

The aviation and maritime sectors, among others, are familiar with international regulations and accepted standards with suitable organisations controlling them. This may be necessary for autonomous systems. Despite using a range of new technologies, the products are sold in many countries. Multiple and mutually inconsistent national regulations cause extra cost, and consumer problems when using a product from one country in another.

Regardless of the philosophy behind the standards (specific or generic), a system must be certified as safe, having gone through a safety assessment according to those standards. The methodology for achieving safety certification will need to be considered through appropriate organisations such as regulators, trade associations and professional societies. Reference [9] gives a useful review of the various approaches and shortcomings for existing approaches to certification.

14.7   Intelligent swarms

Swarms are mentioned briefly in Sections 9.2, 9.5.4, 12.4.2 and 12.5.2, but with indications of the rapid developments in the field. It is clear that they will be developed for both military and civil uses. They will bring a large number of new technical issues about control philosophies, security of communication links and many others. There will need to be new regulations, if only to contain the swarm's physical extent and for each unit to avoid obstacles. They will also require some level of group intelligence in order to perform a function. It can be anticipated that current design techniques, such as those presented in this book will evolve to solve the legal and ethical issues which will arise as the technologies develop.

14.8   Human–machine interface

Increasing autonomy in a system reduces the need for human interaction with the lower levels of a system's control hierarchy. Although this may reduce their workload due to longer periods between interactions, it actually increases the need for the human–machine interface to provide a clear and unambiguous flow of information and instructions in both directions in a timely manner. Some high-level requirements for a human–machine interface are in Panel 13.2 which will need to be interpreted and extended for specific applications.

Deriving the requirements to meet those given in Panel 13.2 for a specific domain or application is a difficult task, but crucial. Without a trustworthy and reliable human–machine interface there is considerable risk of inadvertent author-isation to act or commands for the autonomous system to start a new but poorly defined activity. Following the operational analysis process in Chapter 4 soon leads to models and simulations of the system-of-systems. Analysis of the measures-of-merit relies on understanding the information flows through the human–machine interface and the way the human gives instructions and reacts to the system outputs. Experience gained from this simulation and analysis work should provide a good basis for writing human–machine interface requirements. The interactions between the autonomous system, its user or controller and other humans are key questions in the V&V question list in Section 5.4.4. The V&V processes can be used to refine the human–machine interface needs.

AI is discussed as part of a control system in Section 3.7.3 and specifically for military systems in Section 9.5.2. Its use will give important changes to the human–machine interfaces in a system-of-systems. It is possible to quantify the perfor-mance of decision aids which use rational decision-making and present this in a straightforward way such as error bars or probabilities. Even if not explicitly pre-sented, a human is likely to have an appreciation of their magnitude and will take this into account in deciding whether to take action or not. AI, as explained in Section 2.5, introduces a different, more intuitive, process into its decisions, the Recognition-Primed Decision-Making (RPDM) process. This is an advantage when making decisions under uncertainty, but one disadvantage is that the likely errors will not be quantifiable in the same way as rational processes. The human will need to be aware of this in any action he or she takes. It is likely that a reasonably complex autonomous system will have functions which are based on both types of decision-making. The human user will need to be aware of possible sources of error and their nature, in the information presented to them; this can only come from the human–machine interface and/or training.

A further issue with more advanced AI will be the mutual trust between human and machine. Humans trusting machines to give information reliably is well-known. Mutual trust is where, in addition, an AI control system will proceed on the basis that its user will give it key information at particular times. Future human–machine interfaces will need to be designed so that appropriate actions are trig-gered and recognised across the interface.

14.9   How issues in this book may help solve future problems

Engineers have to work to specific, and preferably quantified requirements which brings difficulties when the regulatory requirements are likely to change as the technology evolves. The issue facing autonomous-system designers is ensuring that their products will meet both current regulations and the new ones which are certain to come. The details of future regulations are obviously not clear but, as shown in Chapter 13, some fundamental ethical principles can be identified and turned into system requirements. When applied to specific systems and applications they can provide methods for qualitative in not always quantitative requirements. This is possible because an architecture is used which mimics human decision-making, the process that the autonomous system replaces. The methodology is likely to be successful for civilian autonomous systems as it has evolved from methods used for assessing military ones.

The 4D/RCS architecture used in this book has a behaviour generator as one of a node's functional elements. This gives a practical way to specify limits on the autonomous systems' actions as behaviour-generator requirements. The nodes must be in a command hierarchy with steadily tightening of behaviour generator requirements at lower levels. Authorised power has been introduced as a means of defining the node's freedom of action and the limits on it. The behaviour of the autonomous system is a summation of the authorised powers of the nodes in the hierarchy. It has only been discussed qualitatively in this book, but it will need to be refined with mathematical analysis to give quantitative performances.

Attention has been restricted to autonomous systems with moving physical components, and which are part of a larger system-of-systems, even if the system-of-systems is simply a human operator and a communication link. This restricts the book's scope but it includes systems with safety issues and concerns about their method of use in public spaces whether expressed as ethics or human rights. The various requirements and 'walk-throughs' of an autonomous system in a realistic scenario show how these concerns can be captured by an engineering team in a way that gives requirements which can be incorporated in a design. Conversely, if one or more walk-throughs show that there are requirements which cannot be met, then there is a tangible basis for discussions about the viability of the concept.

It is unlikely that autonomous-system walk-throughs will expose an unachievable requirement. The question will be the cost of meeting difficult requirements and whether this can be recovered by product sales or services. Additionally, the developers have a basis for firm questions to the regulatory authorities. Given the early status of regulation, there will be scope for a symbiotic relationship between the regulators and industry. This should have a positive effect, as regulations guide technology development. Designers both meet them and seek to find cost-effective solutions which have the least onerous or no regulatory requirements.

A problem for lawyers trying to clarify and solve legal problems is understanding how well-established legal principles can be applied to this new technology. It transfers responsibility for decisions from humans to machines which can then act on them. Does the responsibility for mishaps lie with the user, designer or supplier? The related problem is responsibility if an autonomous system presents a human with a recommendation for an immediate action, but with so little time to react that the only response is to take that action. Professionals in the disciplines addressing these issues may not be familiar with engineering design methods and the importance of engineering clarity in setting requirements. Chapters 2 and 3 show both the ambiguities in terminology and the importance of clarity and a common understanding across disciplines.

The processes from concept to tested product are explained in Chapters 4 and 5. The examples in Chapters 11, 12 and 13 serve two purposes: the first is for engineers new to the field to understand systems engineering as applied to autonomous systems and their ethical implications; the second is to give professionals from other fields an insight into the field to aid the essential cross-disciplinary work needed for this new field.

Engineers need insights into the problems faced by lawyers, civilian and military users, and regulators. The chapters covering IHL, ethics and legal context should provide this and give a basis for their participation in the essential cross-disciplinary work.

One well-known area of debate is that of the control of Lethal Autonomous Weapon Systems (LAWS), some of which is well founded and some of which is not so well founded. It is hoped that this book will give insights into the real problems and solutions for highly automated weapon systems, whether lethal or not, and the problems with their use as their functions become more autonomous.

References

[1]   The IEEE Global Initiative on Ethics of Autonomous and Intelligent Systems. *Ethically Aligned Design: A Vision for Prioritizing Human Well-being with Autonomous and Intelligent Systems*, Version 2. IEEE, 2017. http://standards. ieee.org/develop/indconn/ec/autonomous_systems.html.

[2]   Spiekermann, S. *Ethical IT innovation. A value-based system design approach*. Boca Raton, FL: CRC Press; 2016.

[3]   Maurer M., Gerdes J.C., Lenz B., and Winner H. (eds.). *Autonomous Diving Technical, Legal and Social Aspects*, Springer Open; 2016.

[4]   One proposed solution is in *The Conversation*. 20 March 2018. Available at http://theconversation.com/whos-to-blame-when-driverless-cars-have-an-accident-93132 [Accessed on 21 May 2018].

[5]   OUT-LAW.com. *New UK laws address driverless cars insurance and liability*. The Register 24 February 2017. Available from https://www.theregister.co.uk/2017/02/24/new_uk_law_driverless_cars_insurance_liability/ [Accessed on 23 January 2019].

[6]   Arkin R.C. *Governing lethal behaviour: embedding ethics in a hybrid deliberative/reactive robot architecture*. Georgia Institute of Technology technical report GIT-GVU-07-11, 2007.

[7]   Schmitt M.N. (General Editor) *Tallinn Manual on the International Law Applicable to Cyber Warfare*. Cambridge, UK: Cambridge University Press; 2013.

[8]   US Department of Defense. *Law of War Manual*. June 2015 [Updated December 2016].

[9]   Nair S., de la Vara J.L., Sabetzadeh M., and Briand L. 'Classification, structuring, and assessment of evidence for safety. A systematic literature review'. *IEEE Sixth International Conference on Software Testing, Verification and Validation*; 2013. pp. 94–103.

Appendix 1

Red Cross Guide to Article 36 Reviews

This appendix is a reproduction of a 2006 ICRC English-language document and is reproduced with their kind permission. This version is not to be used as a definitive version or translated into other languages.

A Guide to the Legal Review of New Weapons, Means and Methods of Warfare
Measures to Implement Article 36 of Additional Protocol I of 1977

Principal author: Kathleen Lawand, Arms Unit, Legal Division
With contributions by: Robin Coupland and Peter Herby, Arms Unit, Legal Division

Executive summary

This Guide aims to assist States in establishing or improving procedures to determine the legality of new weapons, means and methods of warfare in accordance with Article 36 of Protocol I Additional to the 1949 Geneva Conventions. It was prepared further to an expert meeting hosted by the ICRC in January 2001 and the Agenda for Humanitarian Action adopted by the States Parties to the Geneva Conventions at the 28th International Conference of the Red Cross and Red Crescent. The Agenda for Humanitarian Action commits States to ensure the legality of all new weapons, means and methods of warfare by subjecting them to rigorous and multidisciplinary review. Government experts from ten countries provided comments on previous drafts of this Guide.

Article 36 of Additional Protocol I requires each State Party to determine whether the employment of any new weapon, means or method of warfare that it studies, develops, acquires or adopts would, in some or all circumstances, be prohibited by international law. All States have an interest in assessing the legality of new weapons, regardless of whether they are party to Additional Protocol I. Assessing the legality of new weapons contributes to ensuring that a State's armed forces are capable of conducting hostilities in accordance with its international obligations. Carrying out legal reviews of proposed new weapons is of particular importance today in light of the rapid development of new technologies.

Article 36 of Additional Protocol I does not specify how a review of the legality of weapons, means and methods of warfare is to be carried out. Drawing on interpretations of the text of Article 36 and on State practice, this Guide highlights both the issues of substance and those of procedure to be considered in establishing a legal review mechanism.

The legal review applies to weapons in the widest sense as well as the ways in which they are used, bearing in mind that a means of warfare cannot be assessed in isolation from its expected method of use. The legal framework of the review is the international law applicable to the State, including international humanitarian law (IHL). In particular, this consists of the treaty and customary prohibitions and restrictions on specific weapons, as well as the general IHL rules applicable to all weapons, means and methods of warfare. General rules include the rules aimed at protecting civilians from the indiscriminate effects of weapons and combatants from unnecessary suffering. The assessment of a weapon in light of the relevant rules will require an examination of all relevant empirical information pertaining to the weapon, such as its technical description and actual performance, and its effects on health and the environment. This is the rationale for the involvement of experts of various disciplines in the review process.

Significant procedural issues that will merit consideration in establishing a review mechanism include determining which national authority is to be made responsible for the review, who should participate in the review process, the stages of the procurement process at which reviews should occur, and the procedures relating to decision-making and record-keeping. The Guide highlights the importance of ensuring that whatever the form of the mechanism, it is capable of taking an impartial and multidisciplinary approach to legal reviews of new weapons, and that States exchange information about their review procedures.

> *Therefore, those who are not thoroughly aware of the disadvantages in the use of arms cannot be thoroughly aware of the advantages in the use of arms.*
> –Sun Tzu, *The Art of War*, circa 500 BC

> *If the new and frightful weapons of destruction which are now at the disposal of the nations seem destined to abridge the duration of future wars, it appears likely, on the other hand, that future battles will only become more and more murderous.*
> – Henry Dunant, *Memory of Solferino*, 1862

> *[The International Military] Commission having by common agreement fixed the technical limits at which the necessities of war ought to yield to the requirements of humanity . . .*
> – St. Petersburg Declaration, 1868

Introduction

The right of combatants to choose their means and methods of warfare[1] is not unlimited.[2] This is a basic tenet of *international humanitarian law* (IHL), also known as the *law of armed conflict* or the *law of war*.

---

[1] The terms 'means and methods of warfare' designate the tools of war and the ways in which they are used. The Protocol Additional to the Geneva Conventions of 12 August 1949, and relating to the Protection of Victims of International Armed Conflicts (Protocol I), 8 June 1977 [hereinafter Additional Protocol I] refers alternately to 'methods or means of warfare' (Art. 35(1) and (3), Art. 51(5)(a), Art. 55(1)), 'methods and means of warfare' (titles of Part III and of Section I of Part III), 'means and methods of attack' (Art. 57(2)(a)(ii)), and 'weapon, means or method of warfare' (Art. 36).

[2] This principle is stipulated in e.g. Article 22 of the 1907 Hague Regulations Respecting the Laws and Customs of War on Land, and Article 35(1) of Additional Protocol I.

IHL consists of the body of rules that apply during armed conflict with the aim of protecting persons who do not, or no longer, participate in the hostilities (e.g. civilians and wounded, sick or captured combatants) and regulating the conduct of hostilities (i.e. the means and methods of warfare). IHL sets limits on armed violence in wartime in order to prevent, or at least reduce, suffering. It is based on norms as ancient as war itself, rooted in the traditions of all societies. The rules of IHL have been developed and codified over the last 150 years in international treaties, notably the 1949 Geneva Conventions and their Additional Protocols of 1977, complemented by a number of other treaties dealing with specific matters such as cultural property, child soldiers, international criminal justice, and use of certain weapons. Many of the rules of IHL are also considered part of *customary international law* based on widespread, representative and virtually uniform practice of States accepted as legal obligation and therefore mandatory for all parties to an armed conflict.

The combatants' right to choose their means and methods of warfare is limited by a number of basic IHL rules regarding the conduct of hostilities, many of which are found in Additional Protocol I of 1977 on the protection of victims of international armed conflicts.[3] Other treaties prohibit or restrict the use of specific weapons such as biological and chemical weapons, incendiary weapons, blinding laser weapons and landmines, among others. In addition, many of the basic rules and specific prohibitions and restrictions on means and methods of warfare may be found in customary international law.[4]

Reviewing the legality of new weapons, means and methods of warfare is not a novel concept. The first international instrument to refer to the legal assessment of emerging military technologies was the St Petersburg Declaration, adopted in 1868 by an International Military Commission. The Declaration addresses the development of future weapons in these terms:

> The Contracting or Acceding Parties reserve to themselves to come hereafter to an understanding whenever a precise proposition shall be drawn up in view of future improvements which science may effect in the armament of troops, in order to maintain the principles which they have established, and to conciliate the necessities of war with the laws of humanity.[5]

The only other reference in international treaties to the need to carry out legal reviews of new weapons, means and methods of warfare is found in Article 36 of Additional Protocol I of 1977:

> In the study, development, acquisition or adoption of a new weapon, means or method of warfare, a High Contracting Party is under an

---

[3] Additional Protocol I includes provisions imposing limits on the use of weapons, means and methods of warfare and protecting civilians from the effects of hostilities. See in particular Part III, Section I, and Part IV, Section I, Chapters I to IV.

[4] For a list of the general and specific treaty and customary IHL rules applicable to weapons, means and methods of warfare, see section 1.2 of this Guide, below.

[5] Declaration Renouncing the Use, in Time of War, of Explosive Projectiles Under 400 Grammes Weight, St Petersburg, 29 November/11 December 1868. The full text of the St Petersburg Declaration is reproduced in Annex II of this Guide.

obligation to determine whether its employment would, in some or all circumstances, be prohibited by this Protocol or by any other rule of international law applicable to the High Contracting Party.

The aim of Article 36 is to prevent the use of weapons that would violate international law in all circumstances and to impose restrictions on the use of weapons that would violate international law in some circumstances, by determining their lawfulness before they are developed, acquired or otherwise incorporated into a State's arsenal.

The requirement that the legality of all new weapons, means and methods of warfare be systematically assessed is arguably one that applies to *all* States, regardless of whether or not they are party to Additional Protocol I. It flows logically from the truism that States are prohibited from using illegal weapons, means and methods of warfare or from using weapons, means and methods of warfare in an illegal manner. The faithful and responsible application of its international law obligations would require a State to ensure that the new weapons, means and methods of warfare it develops or acquires will not violate these obligations.[6] Carrying out legal reviews of new weapons is of particular importance today in light of the rapid development of new weapons technologies.

Article 36 is complemented by Article 82 of Additional Protocol I, which requires that legal advisers be available at all times to advise military commanders on IHL and 'on the appropriate instruction to be given to the armed forces on this subject.' Both provisions establish a framework for ensuring that armed forces will be capable of conducting hostilities in strict accordance with IHL, through legal reviews of planned means and methods of warfare.

Article 36 does not specify how a determination of the legality of weapons, means and methods of warfare is to be carried out. A plain reading of Article 36 indicates that a State must assess the new weapon, means or method of warfare in light of the provisions of Additional Protocol I and of any other applicable rule of international law. According to the ICRC's Commentary on the Additional Protocols, Article 36 'implies the obligation to establish internal procedures for the purpose of elucidating the issue of legality, and the other Contracting Parties can ask to be informed on this point.'[7] But there is little by way of State practice to indicate what kind of 'internal procedures' should be established, as only a limited

---

[6]See, for example, the practice of Sweden and the United States, which established formal weapons review mechanisms as early as 1974, three years before the adoption of Additional Protocol I.

[7]Y. Sandoz, C. Swinarski, B. Zimmerman (eds.), Commentary on the Additional Protocols of 8 June 1977 to the Geneva Conventions of 12 August 1949, ICRC, Geneva, 1987 [hereinafter Commentary on the Additional Protocols], at paragraphs 1470 and 1482. States Parties would be required to share the procedures they adopt with other States Parties on the basis of Article 84 of Additional Protocol I: see below, note 96 and corresponding text.

number of States are known to have put in place mechanisms or procedures to conduct legal reviews of weapons.[8]

The importance of the legal review of weapons has been highlighted in a number of international fora. In 1999, the 27th International Conference of the Red Cross and Red Crescent encouraged States 'to establish mechanisms and procedures to determine whether the use of weapons, whether held in their inventories or being procured or developed, would conform to the obligations binding on them under international humanitarian law.' It also encouraged States 'to promote,

---

[8]States that are known to have in place national mechanisms to review the legality of weapons and that have made the instruments setting up these mechanisms available to the ICRC are: **Australia:** *Legal review of new weapons*, Australian Department of Defence Instruction (General) OPS 44-1, 2 June 2005 [hereinafter Australian Instruction]; **Belgium**: Défense, Etat-Major de la Défense, Ordre Général - J/836 (18 July 2002), establishing *La Commission d'Evaluation Juridique des nouvelles armes, des nouveaux moyens et des nouvelles méthodes de guerre* (Committee for the Legal Review of New Weapons, New Means and New Methods of Warfare) [hereinafter Belgian General Order]; **the Netherlands**: *Beschikking van de Minister van Defensie* (Directive of the Minister of Defence) nr. 458.614/A, 5 May 1978, establishing the *Adviescommissie Internationaal Recht en Conventioneel Wapengebruik* (Committee for International Law and the Use of Conventional Weapons) [hereinafter the Netherlands Directive]; **Norway**: *Direktiv om folkerettslig vurdering av vapen, krigforingsmetoder og krigforingsvirkemidler*, (Directive on the Legal Review on Weapons, Methods and Means of Warfare), Ministry of Defence, 18 June 2003 [hereinafter Norwegian Directive]; **Sweden**: *Förordning om folkrättslig granskning av vapenproject* (Ordinance on international law review of arms projects), Swedish Code of Statutes, SFS 1994:536 [hereinafter Swedish Monitoring Ordinance]; **the United States:** *Review of Legality of Weapons under International Law*, US Department of Defense Instruction 5500.15, 16 October 1974; *Weapons Review*, US Department of Air Force Instruction 51-402, 13 May 1994 [hereinafter US Air Force Instruction]; *Legal Services: Review of Legality of Weapons under International Law*, US Department of Army Regulation 27-53, 1 January 1979 [hereinafter US Army Regulation]; *Implementation and Operation of the Defense Acquisition System and the Joint Capabilities Integration and Development System*, US Department of Navy, Secretary of the Navy Instruction 5000.2C, 19 November 2004 [hereinafter US Navy Instruction]; *Policy for Non-Lethal Weapons*, US Department of Defense Directive 3000.3, 9 July 1996 [hereinafter Non-lethal Weapons Directive]; *The Defense Acquisition System*, US Department of Defense Directive 5000.1, 12 May 2003 [hereinafter US Acquisition Directive]. France and the United Kingdom have indicated to the ICRC that they carry out reviews pursuant to Ministry of Defence instructions, but these have not been made available. The United Kingdom's procedures are mentioned in the UK Ministry of Defence, *The Manual of the Law of Armed Conflict*, Oxford University Press, 2004, at p. 119, paragraph 6.20.1 [hereinafter referred to as 'UK Military Manual']. In Germany, the Federal Agency for Defence Procurement (BWB), upon instruction of the Defence Technology Department at the Federal Ministry of Defence, commissioned a 'Manual regarding a test of compliance with international law at the initial point of procurement – International arms control obligations and international humanitarian law' which was published in 2000: Rudolf Gridl, *Kriterienkatalog zur Überprüfung von Beschaffungsvorhaben im Geschäftsbereich des BWB/BMVg mit völkerrechtlichen Vereinbarungen: Internationale Rüstungskontrolle und humanitäres Völkerrecht*, Ebenhausen im Isartal: Stiftung Wissenschaft und Politik, 2000. For an overview of Article 36 and existing review mechanisms, see: Lt. Col. Justin McClelland, 'The review of weapons in accordance with Article 36 of Additional Protocol I', *International Review of the Red Cross*, Vol. 85, No. 850 (June 2003), pp. 397–415; I. Daoust, R. Coupland and R. Ishoey, 'New wars, new weapons? The obligation of States to assess the legality of means and methods of warfare', *International Review of the Red Cross*,, Vol. 84, No. 846 (June 2002) at pp. 359–361; Danish Red Cross, *Reviewing the Legality of New Weapons*, December 2000.

wherever possible, exchange of information and transparency in relation to these mechanisms, procedures and evaluations.'[9]

At the Second Review Conference of the Convention on Certain Conventional Weapons (CCW) in 2001, the States Parties urged 'States which do not already do so, to conduct reviews such as that provided for in Article 36 of Protocol I additional to the 1949 Geneva Conventions, to determine whether any new weapon, means or methods of warfare would be prohibited by international humanitarian law or other rules of international law applicable to them'.[10]

In December 2003, the 28th International Conference of the Red Cross and Red Crescent reaffirmed by consensus the goal of ensuring 'the legality of new weapons under international law,' this 'in light of the rapid developments of weapons technology and in order to protect civilians from the indiscriminate effects of weapons and combatants from unnecessary suffering and prohibited weapons.'[11] The Conference stated that all new weapons, means and methods of warfare 'should be subject to rigorous and multidisciplinary review', and in particular that such review 'should involve a multidisciplinary approach, including military, legal, environmental and health-related considerations.'[12] The Conference also encouraged States 'to review with particular scrutiny all new weapons, means and methods of warfare that cause health effects with which medical personnel are unfamiliar.'[13] Finally, the Conference invited States that have review procedures in place to cooperate with the ICRC with a view to facilitating the voluntary exchange of experience on review procedures.[14]

In this Guide, the terms 'weapons, means and methods of warfare' designate the means of warfare and the manner in which they are used. In order to lighten the text, the Guide will use the term 'weapons' as shorthand, but the terms 'means of warfare', 'methods of warfare', 'means and methods of warfare', and 'weapons, means and methods of warfare' will also be used as the context requires.[15]

Structure

This Guide is divided into two parts: the first deals with the substantive aspects of an Article 36 review, i.e. relating to its material scope of application, and the second deals with functional considerations, i.e. those of form and procedure.

---

[9]Section 21, Final Goal 1.5 of the Plan of Action for the years 2000–2003 adopted by the 27th International Conference of the Red Cross and Red Crescent, Geneva, 31 October to 6 November 1999. The Conference further stated that 'States and the ICRC may engage consultations to promote these mechanisms (...)'.

[10]Final Declaration of the Second Review Conference of the States Parties to the Convention on Certain Conventional Weapons, Geneva, 11–21 December 2001, CCW/CONF.II/2, at p. 11. Available at <http://disarmament.un.org:8080/ccw/ccwmeetings.html>.

[11]Final Goal 2.5 of the Agenda for Humanitarian Action adopted by the 28th International Conference of the Red Cross and Red Crescent, Geneva, 2–6 December 2003 [hereinafter Agenda for Humanitarian Action]. The full text of Final Goal 2.5 is reproduced in Annex I to this Guide. At the International Conference, two States – Canada and Denmark – made specific pledges to review their procedures concerning the development or acquisition of new weapons, means and methods of warfare.

[12]*Id.*, paragraph 2.5.1.

[13]*Id.*, paragraph 2.5.2.

[14]*Id.*, paragraph 2.5.3.

[15]See note 1 above and Section 1.1 below.

The material scope of application is dealt with before the functional considerations because determining the latter requires an understanding of the former. For example, it is difficult to determine the expertise that will be needed to conduct the review in advance of understanding what the review is required to do.

Part 1 on the review mechanism's material scope of application addresses three questions:

- What types of weapons must be subjected to a legal review? (**section 1.1**)
- What rules must the legal review apply to these weapons? (**section 1.2**)
- What kind of factors and empirical data should the legal review consider? (**section 1.3**)

Part 2 addresses the functional considerations of the review mechanism, in particular:

- The establishment of the review mechanism (**section 2.1**): by what type of constituent instrument and under whose authority?
- The structure and composition of the review mechanism (**section 2.2**): who is responsible for carrying out the review? what departments / sectors are represented? what kind of expertise should be considered in the review?
- The procedure for conducting a review (**section 2.3**): at what stage should the review of new weapons take place? how and by whom is the review procedure triggered? how is information about the weapon under review gathered?
- Decision-making (**section 2.4**): how are decisions reached? are decisions binding on the government or treated as recommendations? can decisions attach conditions to the approval of new weapons? is the review's decision final or can it be appealed?
- Record-keeping (**section 2.5**): should records be kept of the reviews that have been carried out and the decisions reached? who can have access to such records and under what conditions?

1 Material scope of application of the review mechanism

1.1 Types of weapons to be subjected to legal review

Article 36 of Additional Protocol I refers to 'weapons, means or methods of warfare'. According to the ICRC's Commentary on the Additional Protocols:

> the words 'methods and means' *include weapons in the widest sense, as well as the way in which they are used.* The use that is made of a weapon can be unlawful in itself, or it can be unlawful only under certain conditions. For example, poison is unlawful in itself, as would be any weapon which would, by its very nature, be so imprecise that it would inevitably cause indiscriminate damage. (...) However, a weapon that can be used with precision can also be abusively used against the civilian population. In this case, it is not the weapon which is prohibited, but the method or the way in which it is used.[16]

---

[16]Commentary on the Additional Protocols, paragraph 1402, emphasis added.

The material scope of the Article 36 legal review is therefore very broad. It would cover:

- weapons of all types - be they anti-personnel or anti-materiel, 'lethal', 'non-lethal' or 'less lethal' – and weapons systems;[17]
- the ways in which these weapons are to be used pursuant to military doctrine, tactics, rules of engagement, operating procedures and countermeasures;[18]
- all weapons to be acquired, be they procured further to research and development on the basis of military specifications, or purchased 'off the shelf';[19]
- a weapon which the State is intending to acquire for the first time, without necessarily being 'new' in a technical sense;[20]
- an existing weapon that is modified in a way that alters its function, or a weapon that has already passed a legal review but that is subsequently modified;[21]
- an existing weapon where a State has joined a new international treaty which may affect the legality of the weapon.[22]

---

[17]Subsection 3(a) of the Australian Instruction defines the term 'weapon' for the purposes of the Instruction, as 'an offensive or defensive instrument of combat used to destroy, injure, defeat or threaten. It includes weapon systems, munitions, sub-munitions, ammunition targeting devices, and other damaging or injuring mechanisms.' Subsection 1(a) of the Belgian General Order defines the term 'weapon' for the purposes of the General Order as 'any type of weapon, weapon system, projectile, munition, powder or explosive, designed to put out of combat persons and/or materiel' (unofficial translation from the French). Subsection 1.4 of the Norwegian Directive defines the word 'weapons', for the purposes of the Directive, as 'any means of warfare, weapons systems/project, substance, etc. which is particularly suited for use in combat, including ammunition and similar functional parts of a weapon.'. In the US, review of all 'weapons or weapons systems' is required: see the US Army Regulation, subsection 2(a); US Navy Instruction, p. 23, subsection 2.6; US Acquisition Directive, p. 8, subsection E.1.1.15. The US DOD Law of War Working Group has proposed standard definitions, pursuant to which the term 'weapons' refers to 'all arms, munitions, materiel, instruments, mechanisms, or devices that have an intended effect of injuring, damaging, destroying or disabling personnel or property', and the term 'weapon system' refers to 'the weapon itself and those components required for its operation, including new, advanced or emerging technologies which may lead to development of weapons or weapon systems and which have significant legal and policy implications. Weapons systems are limited to those components or technologies having direct injury or damaging effect on people or property (including all munitions and technologies such as projectiles, small arms, mines, explosives, and all other devices and technologies that are physically destructive or injury producing).' See W. Hays Parks, Office of The Judge Advocate General of the Army, 'Weapons Review Programme of the United States', presented at the Expert Meeting on Legal Reviews of Weapons and the SIrUS Project, Jongny sur Vevey, Switzerland, 29–31 January 2001. (Both this presentation and the report of the meeting are on file with the ICRC.)

[18]See, for example, Norwegian Directive, subsections 1.4 and 2.4.

[19]See also subsection 2.3.1 below.

[20]Commentary on the Additional Protocols, paragraph 1472.

[21]See, for example, Australian Instruction, Section 2 and subsection 3(b) and footnote 3 thereof; Belgian General Order, subsection 5(i) and (j); Norwegian Directive, subsection 2.3 *in fine*; US Air Force Instruction, subsections 1.1.1, 1.1.2, 1.1.3; and US Army Regulation, subsection 6(a)(3).

[22]See, for example, Norwegian Directive, subsections 2.2 ('To the extent necessary, legal review shall also be done with regard to existing weapons, methods and means of warfare, in particular when Norway commits to new international legal obligations.') and 2.6 ('In addition, relevant rules of International Law that may be expected to enter into force for Norway in the near future, shall also be taken into consideration. Furthermore, particular emphasis shall be put on views on International Law put forward by Norway internationally.'). See also the US Air Force Instruction, subsection 1.1.3.

When in doubt as to whether the device or system proposed for study, development or acquisition is a 'weapon', legal advice should be sought from the weapons review authority.

A weapon or means of warfare cannot be assessed in isolation from the method of warfare by which it is to be used. It follows that the legality of a weapon does not depend solely on its design or intended purpose, but also on the manner in which it is expected to be used on the battlefield. In addition, a weapon used in one manner may 'pass' the Article 36 'test', but may fail it when used in another manner. This is why Article 36 requires a State 'to determine whether its employment would, *in some or all circumstances*, be prohibited' by international law (emphasis added).

As noted in the ICRC's Commentary on the Additional Protocols, a State need only determine 'whether the employment of a weapon *for its normal or expected use* would be prohibited under some or all circumstances. A State is not required to foresee or analyse all possible misuses of a weapon, for almost any weapon can be misused in a way that would be prohibited.'[23]

1.2 Legal framework: rules to be applied to new weapons, means and methods of warfare

In determining the legality of a new weapon, the reviewing authority must apply existing international law rules which bind the State – be they treaty-based or customary. Article 36 of Additional Protocol I refers in particular to the Protocol and to 'any other rule of international law applicable' to the State. The relevant rules include *general* rules of IHL applying to *all weapons, means and methods of warfare*, and *particular* rules of IHL and international law prohibiting the use of *specific weapons and means of warfare* or restricting the methods by which they can be used.

The first step is to determine whether employment of the particular weapon or means of warfare under review is prohibited or restricted by a treaty which binds the reviewing State or by customary international law (subsection **1.2.1** below). If there is no such specific prohibition, the next step is to determine whether employment of the *weapon or means of warfare* under review and the normal or expected *methods* by which it is to be used would comply with the general rules applicable to all weapons, means and methods of warfare found in Additional Protocol I and other treaties that bind the reviewing State or in customary international law (subsection **1.2.2** below). In the absence of relevant treaty or customary rules, the reviewing authority should consider the proposed weapon in light of the principles of humanity and the dictates of public conscience (subsection **1.2.2.3** below).

Of those States that have established formal mechanisms to review the legality of new weapons, some have empowered the reviewing authority to take into consideration not only the law as it stands at the time of the review, but also likely

---

[23]Commentary on the Additional Protocols, paragraph 1469, emphasis added.

future developments of the law.[24] This approach is meant to avoid the costly consequences of approving and procuring a weapon the use of which is likely to be restricted or prohibited in the near future.

The sections below list the relevant treaties and customary rules without specifying in which situations these apply – i.e. whether they apply in international or non-international armed conflicts, or in all situations. This is to be determined by reference to the relevant treaty or customary rule, bearing in mind that most of the rules apply to all types of armed conflict. Besides, as stated in the *Tadic* decision of the Appeals Chamber of the International Criminal Tribunal for the former Yugoslavia in relation to prohibited means and methods of warfare, 'what is inhumane, and consequently proscribed, in international wars, cannot but be inhumane and inadmissible in civil strife'.[25]

1.2.1   Prohibitions or restrictions on specific weapons

1.2.1.1   Prohibitions or restrictions on specific weapons under international treaty law

In conducting reviews, a State must consider the international instruments to which it is a party that prohibit the use of specific weapons and means of warfare, or that impose limitations on the way in which specific weapons may be used. These instruments include (in chronological order):[26]

- Declaration Renouncing the Use, in Time of War, of Explosive Projectiles Under 400 Grammes Weight, St-Petersburg, 29 November/11 December 1868 (hereafter the 1868 St-Petersburg Declaration).
- Declaration (2) concerning Asphyxiating Gases. The Hague, 29 July 1899.
- Declaration (3) concerning the Prohibition of Using Bullets which Expand or Flatten Easily in the Human Body, The Hague, 29 July 1899.
- Convention (IV) respecting the Laws and Customs of War on Land and its annex: Regulations concerning the Laws and Customs of War on Land, The Hague, 18 October 1907, Article 23 (a), pursuant to which it is forbidden to employ poison or poisoned weapons.
- Convention (VIII) relative to the Laying of Automatic Submarine Contact Mines. The Hague, 18 October 1907.
- Protocol for the Prohibition of the Use of Asphyxiating, Poisonous or Other Gases, and of Bacteriological Methods of Warfare, Geneva, 17 June 1925.

---

[24]See, for example, UK Military Manual, p. 119, paragraph 6.20.1, which states: 'The review process takes into account not only the law as it stands at the time of the review but also attempts to take account of likely future developments in the law of armed conflict.' See also Norwegian Directive, at paragraph 2.6, which states that 'relevant rules of International Law that may be expected to enter into force for Norway in the near future shall also be taken into consideration.' The same provision adds that 'particular emphasis shall be put on views on International Law put forward by Norway internationally.'
[25]ICTY, *Prosecutor v. Tadic*, Decision on the Defence Motion for Interlocutory Appeal on Jurisdiction (Appeals Chamber), 2 October 1995, Case no. IT-94-1, para. 119 and 127.
[26]Reference is made only to the instruments and not to the specific prohibitions or restrictions contained therein, except in the case of the Rome Statute of the International Criminal Court.

- Convention on the Prohibition of the Development, Production and Stock-piling of Bacteriological (Biological) and Toxin Weapons and on their Destruction. Opened for Signature at London, Moscow and Washington, 10 April 1972.
- Convention on the Prohibition of Military or Any Hostile Use of Environmental Modification Techniques, 10 December 1976 (ENMOD Convention).
- Convention on Prohibitions or Restrictions on the Use of Certain Conventional Weapons Which May be Deemed to be Excessively Injurious or to Have Indiscriminate Effects (CCW), Geneva, 10 October 1980, and Amendment to Article 1, 21 December 2001. The Convention has five Protocols:
  - Protocol on Non-Detectable Fragments (Protocol I), Geneva, 10 October 1980;
  - Protocol on Prohibitions or Restrictions on the Use of Mines, Booby-Traps and Other Devices (Protocol II). Geneva, 10 October 1980; or Protocol on Prohibitions or Restrictions on the Use of Mines, Booby-Traps and Other Devices as amended on 3 May 1996 (Protocol II to the 1980 Convention as amended on 3 May 1996);
  - Protocol on Prohibitions or Restrictions on the Use of Incendiary Weapons (Protocol III), Geneva, 10 October 1980;
  - Protocol on Blinding Laser Weapons (Protocol IV to the 1980 Convention), 13 October 1995;
  - Protocol on Explosive Remnants of War (Protocol V), 28 November 2003.[27]
- Convention on the Prohibition of the Development, Production, Stockpiling and Use of Chemical Weapons and on their Destruction, Paris, 13 January 1993.
- Convention on the Prohibition of the Use, Stockpiling, Production and Transfer of Anti-Personnel Mines and on their Destruction, 18 September 1997.
- Rome Statute of the International Criminal Court, 17 July 1998, Article 8(2)(b), paragraphs (xvii) to (xx), which include in the definition of war crimes for the purpose of the Statute the following acts committed in international armed conflict:[28]

  '(xvii) Employing poison or poisoned weapons';

  '(xviii) Employing asphyxiating, poisonous or other gases, and all analogous liquids, materials or devices';

  '(xix) Employing bullets which expand or flatten easily in the human body, such as bullets with a hard envelope which does not entirely cover the core or is pierced with incisions';

---

[27]The Protocol on Explosive Remnants of War does not prohibit or restrict the use of weapons, but stipulates the responsibilities for dealing with the post-hostilities effects of weapons that are considered legal *per se*. However, Article 9 of the Protocol encourages each State Party to take 'generic preventive measures aimed at minimising the occurrence of explosive remnants of war, including, but not limited to, those referred to in Part 3 of the Technical Annex.'

[28]These are not new rules of IHL, but instead criminalize prohibitions that exist pursuant to other treaties and to customary international law.

'(xx) Employing weapons, projectiles and material and methods of warfare which are of a nature to cause superfluous injury or unnecessary suffering or which are inherently indiscriminate in violation of the international law of armed conflict, provided that such weapons, projectiles and material and methods of warfare are the subject of a comprehensive prohibition and are included in an annex to this Statute, by an amendment in accordance with the relevant provisions set forth in articles 121 and 123.'[29]

1.2.1.2   Prohibitions or restrictions on specific weapons under customary international law

In conducting reviews, a State must also consider the prohibitions or restrictions on the use of specific weapons, means and methods of warfare pursuant to customary international law. According to the ICRC study on *Customary International Humanitarian Law*,[30] these prohibitions or restrictions would include the following:

- The use of poison or poisoned weapons is prohibited.[31]
- The use of biological weapons is prohibited.[32]
- The use of chemical weapons is prohibited.[33]
- The use of riot-control agents as a method of warfare is prohibited.[34]
- The use of herbicides as a method of warfare is prohibited under certain conditions.[35]
- The use of bullets which expand or flatten easily in the human body is prohibited.[36]
- The anti-personnel use of bullets which explode within the human body is prohibited.[37]
- The use of weapons, the primary effect of which is to injure by fragments which are not detectable by x-ray in the human body is prohibited.[38]
- The use of booby-traps which are in any way attached to or associated with objects or persons entitled to special protection under international humanitarian law or with objects that are likely to attract civilians is prohibited.[39]

---

[29]At the time of writing, there is no such annex to the Statute.

[30]J.-M. Henckaerts and L. Doswald-Beck (eds.), *Customary International Humanitarian Law*, Cambridge: Cambridge University Press, 2005.

[31]*Id.*, Vol. I, Rule 72, at 251.

[32]*Id.*, Rule 73, at 256.

[33]*Id.*, Rule 74, at 259.

[34]*Id.*, Rule 75, at 263.

[35]*Id.*, Rule 76, at 265. The rule sets out the conditions under which the use of herbicides as a method of warfare is prohibited as follows: 'if they: a) are of a nature to be prohibited chemical weapons; b) are of a nature to be prohibited biological weapons; c) are aimed at vegetation that is not a military objective; d) would cause incidental loss of civilian life, injury to civilians, damage to civilian objects, or a combination thereof, which may be expected to be excessive in relation to the concrete and direct military advantage anticipated; or e) would cause widespread, long-term and severe damage to the natural environment.'

[36]*Id.*, Rule 77, at 268.

[37]*Id.*, Rule 78, at 272.

[38]*Id.*, Rule 79, at 275.

[39]*Id.*, Rule 80, at 278.

- When landmines are used, particular care must be taken to minimise their indiscriminate effects. At the end of active hostilities, a party to the conflict which has used landmines must remove or otherwise render them harmless to civilians, or facilitate their removal.[40]

- If incendiary weapons are used, particular care must be taken to avoid, and in any event to minimise, incidental loss of civilian life, injury to civilians and damage to civilian objects. The anti-personnel use of incendiary weapons is prohibited, unless it is not feasible to use a less harmful weapon to render a person *hors de combat*.[41]

- The use of laser weapons that are specifically designed, as their sole combat function or as one of their combat functions, to cause permanent blindness to unenhanced vision is prohibited.[42]

1.2.2   General prohibitions or restrictions on weapons, means and methods of warfare

If no specific prohibition or restriction is found to apply, the weapon or means of warfare under review and the normal or expected methods by which it is to be used must be assessed in light of the general prohibitions or restrictions provided by treaties and by customary international law applying to all weapons, means and methods of warfare.

A number of the rules listed below are primarily context-dependent, in that their application is typically determined at field level by military commanders on a case-by-case basis taking into consideration the conflict environment in which they are operating at the time and the weapons, means and methods of warfare at their disposal. But these rules are also relevant to the assessment of the legality of a new weapon before it has been used on the battlefield, to the extent that the characteristics, expected use and foreseeable effects of the weapon allow the reviewing authority to determine whether or not the weapon will be capable of being used lawfully in certain foreseeable situations and under certain conditions. For example, if the weapon's destructive radius is very wide, it may be difficult to use it against one or several military targets located in a concentration of civilians without violating the prohibition on the use of indiscriminate means and methods of warfare[43] and/or the rule of proportionality.[44] In this regard, when approving such a weapon, the reviewing authority should attach conditions or comments to the approval, to be integrated into the rules of engagement or operating procedures associated with the weapon.

---

[40]*Id.*, Rules 81–83, at 280, 283, and 285 respectively. Rule 82 specifies that a party to the conflict using landmines must record their placement as far as possible.

[41]*Id.*, Rules 84 and 85, at 287 and 289, respectively.

[42]*Id.*, Rule 86, at 292.

[43]See Additional Protocol I, Article 51(4)(b) and (c), referred to under subsection 1.2.2.1 below, and the rule of customary international law prohibiting indiscriminate attacks, under subsection 1.2.2.2 below.

[44]See Article 51(5)(b) of Additional Protocol I, referred to under subsection 1.2.2.1 below, and rule of proportionality under customary international law, under subsection 1.2.2.2 below.

1.2.2.1   *General prohibitions or restrictions on weapons, means and methods of warfare under international treaty law*

A number of treaty-based general prohibitions or restrictions on weapons, means and methods of warfare must be considered. In particular, States party to Additional Protocol I must consider the rules under that treaty, as required by Article 36. These include:[45]

- Prohibition to employ weapons, projectiles and material and methods of warfare of a nature to cause superfluous injury or unnecessary suffering (Art. 35(2)).
- Prohibition to employ methods or means of warfare which are intended, or may be expected to cause widespread, long-term and severe damage to the natural environment (Articles 35(3) and 55).
- Prohibition to employ a method or means of warfare which cannot be directed at a specific military objective and consequently, that is of a nature to strike military objectives and civilians or civilian objects without distinction (Art. 51(4)(b)).
- Prohibition to employ a method or means of warfare the effects of which cannot be limited as required by Additional Protocol I and consequently, that is of a nature to strike military objectives and civilians or civilian objects without distinction (Art. 51(4)(c)).
- Prohibition of attacks by bombardment by any methods or means which treats as a single military objective a number of clearly separated and distinct military objectives located in a city, town, village or other area containing a similar concentration of civilians or civilian objects (Art. 51(5)(a)).
- Prohibition of attacks which may be expected to cause incidental loss of civilian life, injury to civilians, damage to civilian objects, or a combination thereof, which would be excessive in relation to the concrete and direct military advantage anticipated (proportionality rule) (Art. 51(5)(b)).

1.2.2.2   *General prohibitions or restrictions on weapons, means and methods of warfare under customary international law*

General prohibitions or restrictions on the use of weapons, means and methods of warfare pursuant to customary international law must also be considered. These would include:

- Prohibition to use means and methods of warfare which are of a nature to cause superfluous injury or unnecessary suffering.[46]
- Prohibition to use weapons which are by nature indiscriminate.[47] This includes means of warfare which cannot be directed at a specific military objective, and means of warfare the effects of which cannot be limited as required by IHL.[48]

---

[45]Selected provisions of Additional Protocol I are reproduced in Annex III to this Guide.
[46]Henckaerts and Doswald-Beck (eds.), note 30 above, Rule 70, at 237.
[47]*Id.*, Rule 71, at 244. See also Rule 11, at 37.
[48]*Id.*, Rule 12, at 40.

- Prohibition of attacks by bombardment by any method or means which treats as a single military objective a number of clearly separated and distinct military objectives located in a city, town, village or other area containing a similar concentration of civilians or civilian objects.[49]
- Prohibition to use methods or means of warfare that are intended, or may be expected, to cause widespread, long-term and severe damage to the natural environment. Destruction of the natural environment may not be used as a weapon.[50]
- Prohibition to launch an attack which may be expected to cause incidental loss of civilian life, injury to civilians, damage to civilian objects, or a combination thereof, which would be excessive in relation to the concrete and direct military advantage anticipated (proportionality rule).[51]

1.2.2.3   Prohibitions or restrictions based on the principles of humanity and the dictates of public conscience (the 'Martens clause')

Consideration should be given to whether the weapon accords with the principles of humanity and the dictates of public conscience, as stipulated in Article 1(2) of Additional Protocol I, in the preamble to the 1907 Hague Convention (IV), and in the preamble to the 1899 Hague Convention (II). This refers to the so-called 'Martens clause', which Article 1(2) of Additional Protocol I formulates as follows:

> In cases not covered by this Protocol or by other international agreements, civilians and combatants remain under the protection and authority of the principles of international law derived from established custom, from the principles of humanity and from dictates of public conscience.

The International Court of Justice in the case of the Legality of the Threat or Use of Nuclear Weapons affirmed the importance of the Martens clause 'whose continuing existence and applicability is not to be doubted'[52] and stated that it 'had proved to be an effective means of addressing rapid evolution of military technology.'[53] The Court also found that the Martens clause represents customary international law.[54]

A weapon which is not covered by the existing rules of the international humanitarian law would be considered contrary to the Martens clause if it is determined *per se* to contravene the principles of humanity or the dictates of public conscience.

---

[49]*Id.*, Rule 13, at 43.

[50]*Id.*, Rule 45, at 151. The summary of the rule notes that: 'It appears that the United States is a "persistent objector" to the first part of this rule. In addition, France, the United Kingdom and the United States are persistent objectors with regard to the application of the first part of this rule to the use of nuclear weapons.' See also Rule 44.

[51]*Id.*, Rule 14, at 46.

[52]*Legality of the Threat or Use of Nuclear Weapons,* Advisory Opinion, 8 July 1996, paragraph 87.

[53]*Id.*, paragraph 78.

[54]*Id.*, paragraph 84.

1.3   Empirical data to be considered by the review

In assessing the legality of a particular weapon, the reviewing authority must examine not only the weapon's design and characteristics (the 'means' of warfare) but also how it is to be used (the 'method' of warfare), bearing in mind that the weapon's effects will result from a combination of its design *and* the manner in which it is to be used.

In order to be capable of assessing whether the weapon under review is subject to specific prohibitions or restrictions (listed in subsection 1.2.1 above) or whether it contravenes one or more of the general rules of IHL applicable to weapons, means and methods of warfare (listed in subsection 1.2.2 above), the reviewing authority will have to take into consideration a wide range of military, technical, health and environmental factors. This is the rationale for the involvement of experts from various disciplines in the review process.[55]

For each category of factors described below, the relevant general rule of IHL is referred to, where appropriate.

1.3.1   Technical description of the weapon

An assessment will logically begin by considering the weapon's technical description and characteristics, including:

- a full technical description of the weapon;[56]
- the use for which the weapon is designed or intended, including the types of targets (e.g. personnel or materiel; specific target or area, etc.);[57]
- its means of destruction, damage or injury.

1.3.2   Technical performance of the weapon

The technical performance of the weapon under review is of particular relevance in determining whether its use may cause *indiscriminate effects*. The relevant factors would include:

- the accuracy and reliability of the targeting mechanism (including e.g. failure rates, sensitivity of unexploded ordnance, etc.);
- the area covered by the weapon;

---

[55]The importance of ensuring a multidisciplinary approach to the legal review of weapons is emphasised in Action 2.5.2 of Agenda for Humanitarian Action adopted by the 28th International Conference of the Red Cross and Red Crescent and was noted by the Expert Meeting on Legal Reviews of Weapons and the SIrUS Project referred to in note 17 above. See also section 2.2 below.

[56]In addition to the design, material composition and fusing system of the weapon, the technical description would include 'range, speed, shape, materials, fragments, accuracy, desired effect, and nature of system or subsystem employed for firing, launching, releasing or dispensing': see the US Department of Air Force Instruction 51–402, Weapons Review, 13 May 1994 (implementing US Department of Air Force Policy Directive 51-4, Compliance with the Law of Armed Conflict, 26 April 1993 and US Department of Defence Directive 5100.77, DoD Law of War Program, 9 December 1998), at subsection 1.2.1.

[57]This is referred to by some as the weapon's 'mission' or 'military purpose'.

- whether the weapons' foreseeable effects are capable of being limited to the target or of being controlled in time or space (including the degree to which a weapon will present a risk to the civilian population after its military purpose is served).

1.3.3 Health-related considerations

Directly related to the weapon's mechanism of injury (damage mechanism) is the question of what types of injuries the new weapon will be capable of inflicting. The factors to be considered in this regard could include:[58]

- the size of the wound expected when the weapon is used for its intended purpose (as determined by wound ballistics);
- the likely mortality rate among the victims when the weapon is used for its intended purpose;
- whether the weapon would cause anatomical injury or anatomical disability or disfigurement which are specific to the design of the weapon.

If a new weapon injures by means other than explosive or projectile force, or otherwise causes health effects that are qualitatively or quantitatively different from those of existing lawful weapons and means of warfare, additional factors to be considered could include:[59]

- whether all relevant scientific evidence pertaining to the foreseeable effects on humans has been gathered;
- how the mechanism of injury is expected to impact on the health of victims;
- when used in the context of armed conflict, what is the expected field mortality and whether the later mortality (in hospital) is expected to be high;
- whether there is any predictable or expected long term or permanent alteration to the victims' psychology or physiology;
- whether the effects would be recognised by health professionals, be manageable under field conditions and be treatable in a reasonably equipped medical facility.

These and other health-related considerations are important to assist the reviewing authority in determining whether the weapon in question can be expected to cause *superfluous injury or unnecessary suffering*. Assessing the legality of a

---

[58]See, for example, US Air Force Instruction, subsection 1.2.1, which requires that the reviewer be provided with information *inter alia* on the 'nature of the expected injury to persons (including medical data, as available)'.

[59]The 28th International Conference of the Red Cross and Red Crescent encouraged States 'to review with particular scrutiny all new weapons, means and methods of warfare that cause health effects with which medical personnel are unfamiliar': paragraph 2.5.2 of Agenda for Humanitarian Action. In addition, the Expert Meeting on Legal Reviews of Weapons and the SIrUS Project noted that 'we are familiar with the effects of weapons which injure by explosives, projectile force or burns and weapons causing these effects need to be reviewed accordingly' and that 'there is a need for particularly rigorous legal reviews of weapons which injure by means and cause effects with which we are not familiar' (report of the meeting at p. 8, note 17 above).

weapon in light of this rule involves weighing the relevant health factors together against the intended military purpose or expected military advantage of the new weapon.[60]

1.3.4   Environment-related considerations

In determining the effects of the weapon under review on the natural environment, and in particular whether they are expected to cause excessive incidental damage to the natural environment or widespread, long-term and severe damage to the natural environment,[61] the relevant questions to be considered would include:

- have adequate scientific studies on the effects on the natural environment been conducted and examined?
- what type and extent of damage are expected to be directly or indirectly caused to the natural environment?
- for how long is the damage expected to last; is it practically/economically possible to reverse the damage, i.e. to restore the environment to its original state; and what would be the time needed to do so?
- what is the direct or indirect impact of the environmental damage on the civilian population?
- is the weapon specifically designed to destroy or damage the natural environment,[62] or to cause environmental modification?[63]

2   Functional aspects of the review mechanism

In setting up a weapons review mechanism, a number of decisions need to be made relating to the manner in which it is to be established, its structure and composition, the procedure for conducting a review, decision-making and record-keeping.

---

[60]According to the ICRC Study on Customary International Humanitarian Law, 'The prohibition of means of warfare which are of a nature to cause superfluous injury or unnecessary suffering refers to the effect of a weapon on combatants. Although there is general agreement on the existence of the rule, views differ on how it can actually be determined that a weapon causes superfluous injury or unnecessary suffering. States generally agree that suffering that has no military purpose violates this rule. Many States point out that the rule requires that a balance be struck between military necessity, on the one hand, and the expected injury or suffering inflicted on a person, on the other hand, and that excessive injury or suffering, i.e., that which is out of proportion to the military advantage sought, therefore violates the rule. Some States also refer to the availability of alternative means as an element that has to go into the assessment of whether a weapon causes unnecessary suffering or superfluous injury.' Henckaerts and Doswald-Beck (eds.), note 30 above, under Rule 70, at 240 (footnotes omitted).

[61]See Articles 35(3) and 55 of Additional Protocol I, referred to above under subsection 1.2.2.1, and rules of customary international law under subsection 1.2.2.2. Of relevance to the consideration of environmental factors is Rule 44 of ICRC Study on Customary International Humanitarian Law, which states *inter alia*: 'Lack of scientific certainty as to the effects on the environment of certain military operations does not absolve a party to the conflict from taking' all feasible precautions 'to avoid, and in any event to minimise, incidental damage to the environment'. See Henckaerts and Doswald-Beck (eds.), note 30 above.

[62]See the customary international law rule referenced in note 50 above.

[63]See ENMOD Convention, listed under subsection 1.2.1.1 above.

The following questions are indicative of the elements to be considered. Reference to State practice is limited to published procedures only.

2.1   How should the review mechanism be established?

2.1.1   By legislation, regulation, administrative order, instruction or guidelines?

Article 36 of Additional Protocol I does not specify in what manner and under what authority reviews of the legality of new weapons are to be constituted. It is the responsibility of each State to adopt legislative, administrative, regulatory and/or other appropriate measures to effectively implement this obligation. At a minimum, Article 36 requires that each State Party set up a formal procedure and, in accordance with Article 84 of Additional Protocol I, other States parties to the Protocol may ask to be informed about this procedure.[64] The establishment of a formal procedure implies that there be a standing mechanism ready to carry out reviews of new weapons whenever these are being studied, developed, acquired or adopted.

Of the six States that have made available their weapons review procedures, one has established its review mechanism pursuant to a government ordinance[65] and five have done so pursuant to instructions, directives or orders of their Ministry of Defence.[66]

2.1.2   Under which authority should the review mechanism be established?

The review mechanism can be established by, and made accountable to, the government department responsible for the study, development, acquisition or adoption of new weapons, typically the Ministry of Defence or its equivalent. This has the advantage that the Ministry of Defence is also the same authority that issues weapon handling instructions. Most States that have established review mechanisms have done so under the authority of their Ministry of Defence.

Alternatively, the review mechanism could be established by the government itself and implemented by an inter-departmental entity, which is the option preferred by one State.[67] It is also conceivable that another relevant department be entrusted with the establishment of the review mechanism, such as, for example, the authority responsible for government procurement.

---

[64]See note 7 above and note 96 below.

[65]See Swedish Monitoring Ordinance.

[66]The Ministries of Defence of the Netherlands and Norway and the Department of Defence of the United States have adopted 'Directives' to establish their legal review mechanisms. The US Directive has been implemented through separate instructions by each of the three military departments (Army, Navy and Air Force). The Ministry of Defence of Belgium has adopted a 'General Order' to establish its legal review mechanism. The Department of Defence of Australia has adopted a general 'Defence Instruction' to establish its legal review mechanism. See note 8 above for complete references.

[67]In Sweden, the *Delegation for international law monitoring of arms projects* is established by the Government, which also appoints its members. See Section 8 of the Swedish Monitoring Ordinance.

Whatever the establishing authority, care should be taken to ensure that the reviewing body is capable of carrying out its work in an impartial manner, based on the law and on relevant expertise.[68]

2.2   *Structure and composition of the review mechanism*

2.2.1   Who should be responsible for carrying out the review?

The responsibility for carrying out the legal review may be entrusted to a special body or committee made up of permanent representatives of relevant sectors and departments. This is the option taken by four of the States that have made known their review mechanisms.[69] Two of these have adopted a 'mixed' system, whereby a single official – the head of defence – is advised by a standing committee that carries out the review.[70]

In the two other States, the review is the responsibility of a single official (the Director-General of the Defence Force Legal Service in one State, and the Judge-Advocate General of the military department responsible for acquiring a given weapon in the other State). In carrying out the review, the official consults the concerned sectors and relevant experts.[71]

The material scope of the review requires that it consider a wide range of expertise and viewpoints. The review of weapons by a committee may have the advantage of ensuring that the relevant sectors and fields of expertise are involved in the assessment.[72]

Whether the reviewing authority is an individual or a committee, it must have the appropriate qualifications, and in particular a thorough knowledge and understanding of IHL. In this regard, it would be appropriate for the legal advisers appointed to the armed forces to take part in the review, or to head the committee responsible for the review.

2.2.2   What departments or sectors should be involved in the review? What kinds of experts should participate in the review?

Whether it is conducted by a committee or by an individual, the review should draw on the views of the relevant sectors and departments, and a wide range of expertise. As seen under section 1 of this Guide, a multidisciplinary approach, including the

---

[68]See subsection 2.2.2 below.

[69]Belgium, the Netherlands, Norway and Sweden: see note 8 above.

[70]Belgium has a committee that advises the Head of Defence, who is responsible for 'taking action required by international law' based on the committee's advice: see the Belgian General Order, at section 2(b). Norway has a committee that advises the Chief of Defence, who in turn is responsible for advising and reporting to the Defence Military Organisation: see Norwegian Directive, at section 2.1.

[71]See Australian Instruction, section 6, and the US Department of Defence Instruction 5500.15, sub-section IV.A. In the United States, when the Office of the Judge Advocate General of one military department conducts a legal review of a new weapon, it generally coordinates the legal review with the other military departments and services, as well as the office of the General Counsel, Department of Defence, to ensure consistency in interpretation.

[72]See Lt. Col. McClelland, 'The review of weapons in accordance with Article 36 of Additional Protocol I', note 8 above, at p. 403.

relevant legal, military, health, arms technology and environmental experts, is essential in order to assess fully the information relating to the new weapon and make a determination on its legality.[73] In this regard, in addition to the relevant sectors of the Ministry of Defence and the Armed Forces, the review may need to draw on experts from the departments of foreign affairs (in particular international law experts), health, and the environment, and possibly on expert advice from outside of the administration.

In three of the States that have made available their review mechanisms, the permanent membership is taken from the relevant sectors of the Ministry of Defence or equivalent. In addition to legal officers responsible for advising the Ministry (e.g. from the Judge-Advocate General's office), permanent members include a military doctor from the medical services of the armed forces,[74] and representatives of the departments responsible for operative planning, logistics and military engineering.[75] These mechanisms also provide the possibility for ad-hoc participation by experts drawn from other Ministries or external experts.[76]

Another State has included as permanent members of its review body officials outside of the Ministry of Defence – in particular researchers in weapons technology, members of the Surgeon-General's office and an international law expert of the Ministry of Foreign Affairs.[77]

Of the two States that vest the authority to review weapons in a single official, one requires defence agencies responsible for health, capability development, and science and technology (among other fields) to provide the official with 'technical guidance, ballistics information, analysis and assessments of weapons effects, and appropriate... experts', while in the other State, the reviewing authority may consult with medical officers and other relevant experts.[78]

2.3   Review process

2.3.1   At what stage should the review of the new weapon take place?

The temporal application of Article 36 is very broad. It requires an assessment of the legality of new weapons at the stages of their '*study, development, acquisition or adoption*'. This covers all stages of the weapons procurement process, in particular the initial stages of the research phase (i.e. conception, study), the development phase

---

[73]See note 55 above and corresponding text.

[74]See, for example, Belgian General Order, subsection 4(a)(1).

[75]For example, the Norwegian Committee, which includes in the Committee representatives of the Section for Operative Planning of the Department of Operational and Emergency Response Planning, the Joint Operative Headquarters, the Defence Staff College, the Defence Logistical Organisation and the Defence Research Institute: see Norwegian Directive, subsection 4.2.

[76]See, for example, Belgian General Order, subsection 4(c) and Norwegian Directive, subsection 4.3.

[77]Sweden: see Danish Red Cross, note 8 above, at p. 28 and website of 'Government Offices of Sweden' at www.sweden.gov.se.

[78]See the Australian Instruction, section 6, and for the US, see, for example, US Army Regulation, subsection 5(d) ('Upon request of [the Judge Advocate General], [the Surgeon General] provides the medical consultation needed to complete the legal review of weapons or weapon systems').

(i.e. development and testing of prototypes) and the acquisition phase (including 'off-the-shelf' procurement).[79]

In practical terms this means that:

- For a State producing weapons itself, be it for its own use or for export, reviews should take place at the stage of the conception/design of the weapon, and thereafter at the stages of its technological development (development of prototypes and testing), and in any case before entering into the production contract.[80]
- For a State purchasing weapons, either from another State or from the commercial market including through 'off the shelf' procurement, the review should take place at the stage of the study of the weapon proposed for purchase, and in any case before entering into the purchasing agreement. It should be emphasized that the purchasing State is under an obligation to conduct its own review of the weapon it is considering to acquire, and cannot simply rely on the vendor or manufacturer's position as to the legality of the weapon, nor on another State's evaluation.[81] For this purpose, all relevant information and data about the weapon should be obtained from the vendor prior to purchasing the weapon.
- For a State adopting a technical modification or a field modification to an existing weapon,[82] a review of the proposed modification should also take place at the earliest stage.

At each stage of the review, the reviewing authority should take into consideration how the weapon is proposed or expected to be used, i.e. the methods of warfare associated with the weapon.

---

[79]See, for example, Australian Instruction, section 7 ('For Major Capital Investment Projects, [the Chief of Capability Development Group] is responsible for requesting legal reviews as these projects progress through the major project approval process.'); Belgian General Order, subsection 5(a) ('When the Armed Forces study, develop, or wish to acquire or adopt a new weapon, a new means or a new method of warfare, this weapon, means or method must be submitted to the Committee for a legal review at the earliest possible stage and in any case before the acquisition or adoption' (unofficial translation)); Norwegian Directive, subsection 2.3 ('The reviews shall be made as early as possible, normally already in the concept/study phase, when operational needs are identified, the military objectives are defined, the technical, resources and financial conditions are settled.'); UK Military Manual at p. 119, paragraph 6.20.1 ('In the UK the weapons review process is conducted in a progressive manner as concepts for new means and methods of warfare are developed and as the conceptual process moves towards procurement.'); US Air Force Instruction 51-402, at subsections 1.1.1 ('The Judge Advocate General (TJAG) will ensure all weapons being developed, bought, built or otherwise acquired, and those modified by the Air Force are reviewed for legality under international law prior to use in a conflict') and 1.1.2 ('at the earliest possible stage of the acquisition process, including the research and development stage').

[80]See, for example, Belgian General Order, subsection 5(a) ('...at the earliest possible stage and in any case before the acquisition or adoption'); US Department of Defence Directive 5500.15 at subsection IV. A.1. ('The legal review will take place prior to the award of an initial contract for production.')

[81]See Commentary on the Additional Protocols, paragraph 1473. See also the UK Military Manual at p. 119, paragraph 6.20.1 ('This obligation [Article 36 of Additional Protocol I] is imposed on all states party, not only those that produce weapons').

[82]See, for example, the US Air Force Instruction, at subsection 1.1.1: the Judge Advocate General 'will ensure all weapons being developed, bought, built, or otherwise acquired, and those *modified* by the Air Force are reviewed for legality under international law prior to use in a conflict.' (emphasis added). See also Australian Instruction, section 10 ('Any proposal to make field modifications to weapons shall be vetted in accordance with this instruction'). See also note 21 above.

In addition to being required by Article 36, the rationale for conducting legal reviews at the earliest possible stage is to avoid costly advances in the procurement process (which can take several years) for a weapon which may end up being unusable because illegal. The same rationale underlies the need for conducting reviews at different stages of the procurement process, bearing in mind that the technical characteristics of the weapon and its expected uses can change in the course of the weapon's development. In this connection, a new review should be carried out when new evidence comes to light on the operational performance or effects of the weapon both during and after the procurement process.[83]

2.3.2 How and by whom is the legal review mechanism triggered?

Each of the authorities responsible for the study, development, acquisition, modification or adoption of a weapon should be required to submit the matter to the reviewing authority for a legal review at the stages identified above. This can be done through, for example, a notification[84] or a request for an advisory opinion[85] or for a legal review.[86]

In addition, the reviewing authority could itself be empowered to undertake assessments of its own initiative.[87]

2.3.3 How does the review mechanism obtain information on the weapon in question, and from what sources?

At each stage of any given case, the authorities responsible for studying, developing, acquiring or adopting the new weapon should make available to the reviewing authority all relevant information on the weapon, in particular the information described in section 1.3 above.

The reviewing authority should be empowered to request and obtain any additional information and to order any tests or experiments needed to carry out and complete the review, from the relevant government departments or external actors as appropriate.[88]

---

[83]See, for example, Belgian General Order, subsection 5(i) ('If new relevant information is made known after the file has been processed by the Committee, the weapon, means or method of warfare shall be re-submitted to the Committee for legal review pursuant to the above-mentioned procedure' (unofficial translation)) and Norwegian Directive, subsection 2.3 *in fine* ('Should circumstances at a later stage change significantly, the international legal aspects shall be re-assessed').

[84]See, for example, Swedish Monitoring Ordinance, section 9.

[85]See, for example, Norwegian Directive, subsection 4.6.

[86]See, for example, Australian Instruction, sections 7 and 8, and Belgian General Order, subsection 5(b).

[87]As in the case of Norwegian Directive, subsection 4.3. The Swedish reviewing body also has a right of initiative: see Danish Red Cross, note 8 above, at p. 28 and I. Daoust *et al.*, *id.*, at p. 355.

[88]See for example US Army Regulation, subsections 5(b)(3) and (5), which require the Materiel Developer, when requested by the Judge Advocate General, to provide 'specific additional information pertaining to each weapon or weapon system', and to conduct 'experiments, including wound ballistics studies, on weapons or weapons systems subject to review...'. See also Australian Instruction, sections 6–8, and Belgian General Order, subsection 5(e).

2.4   Decision-making

2.4.1   How does the review mechanism reach decisions?

This question is relevant to cases where the reviewing authority is a committee. Ideally, decisions should be reached by consensus, but another decision-making procedure should be provided in cases where consensus is not possible, either through a voting system, majority and minority reports, or by vesting in the chair of the committee final decision-making authority.

2.4.2   Should the reviewing authority's decision be binding or should it be treated only as a recommendation?

As the reviewing authority is making a determination on the conformity of the new weapon with the State's international legal obligations, it is difficult to justify the proposition that acquisition of a new weapon can proceed without a favourable determination by the reviewing authority. For example, if the reviewing authority finds that the new weapon is prohibited by IHL applicable to the concerned State, the development or acquisition of the weapon should be halted on this basis as a matter of law.[89]

2.4.3   May the reviewing authority attach conditions to its approval of a new weapon?

The reviewing authority is required by the terms of Article 36 to determine whether the employment of the weapon under consideration would 'in some or all circumstances' be legal.[90] Therefore it may find that the use of the new weapon is prohibited in certain situations. In such a case the authority could either approve the weapon on condition that restrictions be placed on its operational use, in which case such restrictions should be incorporated into the rules of engagement or standard operating procedures relevant to the weapon, or it could request modifications to the weapon which must be met before approval can be granted.[91]

2.4.4   Should the reviewing authority's decision be final or should it be subject to appeal or review?

Of the States that have made known their review mechanisms, two expressly provide for the possibility of appeal or review of its decisions.[92] If an appeal mechanism is provided, care should be taken to ensure that the appellate or

---

[89]In the United States, a weapon cannot be acquired unless it has been subjected to a legal review: see, for example, US Navy Instruction, section 2.6 ('No weapon or weapon system may be acquired or fielded without a legal review'). See also Australian Instruction, sections 5 and 11.

[90]See section 1.1 above.

[91]For example, section 7 of the Swedish Review Ordinance states: 'If the arms project does not meet the requirement of international humanitarian law, the Delegation shall urge the party that has submitted the matter to the Delegation to undertake construction changes, consider alternative arms projects or issue limitations on the operative use of weapons.'

[92]See the US Department of Defence Directive 5500.15, at subsection IV.C, pursuant to which an opinion of the Judge Advocate General will be reviewed by the General Counsel of the Department of Defence when requested by the Secretary of Defence, the Secretary of a Military Department, the Director of Defence Research and Engineering, the Assistant Secretary of Defence (Installations and Logistics) or any Judge Advocate General; see also Swedish Monitoring Ordinance, section 10, which provides that a decision may be appealed 'to the Government'.

reviewing body is also qualified in IHL and conducts its review on the basis of legal considerations, taking into account the relevant multidisciplinary elements.

2.5 Record-keeping

2.5.1 Should records be kept of the decisions of the review mechanism?

The reviewing authority's work will be more effective over time if it maintains an archive of all its opinions and decisions on the weapons it has reviewed. By enabling the reviewing authority to refer to its previous decisions, the archive also facilitates consistency in decision-making. It is also particularly useful where the weapon under review is a modified version of a weapon previously reviewed.

Of the States that have made known their review mechanisms, two require the reviewing authority to maintain permanent files of the legal reviews.[93] At least one other has an obligation to maintain permanent files under a general obligation of the administration to archive decisions.[94]

2.5.2 To whom and under what conditions should these records be accessible?

It is up to each State to decide whether to allow access to the review records, in whole or in part, and to whom. The State's decision will be influenced by whether in a given case the weapon itself is considered confidential.

Amongst others, the following factors could be taken into account when deciding on whether to disclose reviews, and to whom:

• the value of transparency among different government departments, and towards external experts and the public;
• the value of sharing experience with other States;
• the obligation for all States to ensure respect for IHL in all circumstances, in particular in cases where it is determined that the use of the weapon under review would contravene IHL.

In at least four of the States that have made known their review mechanisms, decisions of the reviewing authority are known to be subject to legislation governing public access to information, which applies equally to other governmental bodies.[95]

---

[93] See the Australian Instruction, section 13, which requires the Director-General Australian Defence Force Legal Service to 'maintain a Weapons Review Register [that] will include a copy of all legal reviews and be the formal record of all weapons that have been reviewed', and the US Department of Defence Instruction 5500.15, subsection IV.A.2, which requires each Judge Advocate General to 'maintain permanent files or opinions issued by him'. See, in this regard, paragraph 1.1.3 of the US Air Force Instruction, paragraph 5(e)(2) of the US Army Regulation, and paragraph 2.6 of US Navy Instruction.
[94] See Belgium, Law on Archives, 24 June 1955.
[95] In the United States, the majority of review reports are unclassified and accessible to the public pursuant to the Freedom of Information Act: see H. Parks, note 17 above. In Sweden, the reports of the Delegation are subject to the Freedom of the Press Act: see Danish Red Cross, note 8 above, at p. 28 and I. Daoust *et al.*, *id.* at p. 355. See also Belgium, Law of 11 April 1994 regarding publicity of the Administration, and Australia, Freedom of Information Act 1982.

Pursuant to such legislation, access to information is subject to exemptions which include the non-disclosure of sensitive information affecting national security.

While there is no obligation on the reviewing State to make the substantive findings of its review public nor to share them with other States, it would be required to share its review procedures with other States Parties to Additional Protocol I, in accordance with Article 84 of the Protocol.[96] In this regard, both the 27th and the 28th International Conference of the Red Cross and the Red Crescent, which includes all of the States Parties to the Geneva Conventions, have encouraged States to exchange information on their review mechanisms and procedures, and have called upon the ICRC to facilitate such exchanges.[97]

CONTACTS

The International Committee of the Red Cross (ICRC) provides advice, support and documentation to governments on national implementation of international humanitarian law. It can be contacted through the nearest delegation or at the address given below.

International Committee of the Red Cross
19, Avenue de la Paix
1202 Geneva, Switzerland
Tel.: +41 22 734 6001 (Switchboard)
+41 22 730 2667 (Arms Unit)
+41 22 730 2321 (Advisory Service)
Email:weapons.gva@icrc.org
http://www.icrc.org

ANNEX I

28th International Conference of the Red Cross and Red Crescent Geneva, 2-6 December 2003: Agenda for Humanitarian Action, Final Goal 2.5

Final Goal 2.5 – Ensure the legality of new weapons under international law

In light of the rapid development of weapons technology and in order to protect civilians from the indiscriminate effects of weapons and combatants from unnecessary suffering and prohibited weapons, all new weapons, means and methods of warfare should be subject to rigorous and multidisciplinary review.

---

[96]See Commentary on the Additional Protocols, paragraph 1470 and footnote 12 thereof. Article 84 reads: 'The High Contracting Parties shall communicate to one another, as soon as possible, through the depositary and, as appropriate, through the Protecting Powers, their official translations of this Protocol, as well as the laws and regulations which they may adopt to ensure its application.'
[97]See the Agenda for Humanitarian Action, paragraph 2.5.3.

Actions proposed

2.5.1 In accordance with 1977 Additional Protocol I (Article 36), States Parties are urged to establish review procedures to determine the legality of new weapons, means and methods of warfare. Other States should consider establishing such review procedures. Reviews should involve a multidisciplinary approach, including military, legal, environmental and health-related considerations.

2.5.2 States are encouraged to review with particular scrutiny all new weapons, means and methods of warfare that cause health effects with which medical personnel are unfamiliar.

2.5.3 The ICRC will facilitate the voluntary exchange of experience on review procedures. States that have review procedures in place are invited to cooperate with the ICRC in this regard. The ICRC will organize, in cooperation with government experts, a training workshop for States that do not yet have review procedures.

ANNEX II

Declaration Renouncing the Use, in Time of War, of Explosive Projectiles Under 400 Grammes Weight, Saint Petersburg, 29 November/11 December 1868

On the proposition of the Imperial Cabinet of Russia, an International Military Commission having assembled at St. Petersburg in order to examine the expediency of forbidding the use of certain projectiles in time of war between civilized nations, and that Commission having by common agreement fixed the technical limits at which the necessities of war ought to yield to the requirements of humanity, the Undersigned are authorized by the orders of their Governments to declare as follows:

Considering:

- That the progress of civilization should have the effect of alleviating as much as possible the calamities of war;
- That the only legitimate object which States should endeavour to accomplish during war is to weaken the military forces of the enemy;
- That for this purpose it is sufficient to disable the greatest possible number of men;
- That this object would be exceeded by the employment of arms which uselessly aggravate the sufferings of disabled men, or render their death inevitable;
- That the employment of such arms would, therefore, be contrary to the laws of humanity;

The Contracting Parties engage mutually to renounce, in case of war among themselves, the employment by their military or naval troops of any projectile of a weight below 400 grammes, which is either explosive or charged with fulminating or inflammable substances.

They will invite all the States which have not taken part in the deliberations of the International Military Commission assembled at St. Petersburg by sending Delegates thereto, to accede to the present engagement.

This engagement is compulsory only upon the Contracting or Acceding Parties thereto in case of war between two or more of themselves; it is not applicable to non-Contracting Parties, or Parties who shall not have acceded to it.

It will also cease to be compulsory from the moment when, in a war between Contracting or Acceding Parties, a non-Contracting Party or a non-Acceding Party shall join one of the belligerents.

The Contracting or Acceding Parties reserve to themselves to come hereafter to an understanding whenever a precise proposition shall be drawn up in view of future improvements which science may effect in the armament of troops, in order to maintain the principles which they have established, and to conciliate the necessities of war with the laws of humanity.

ANNEX III

Selected provisions of Additional Protocol I (Protocol additional I to the Geneva Conventions of 1949 and relating to the protection of victims of international armed conflict (Protocol I), 8 June 1977)

Article 1, paragraph 2 [the 'Martens clause']

2. In cases not covered by this Protocol or by other international agreements, civilians and combatants remain under the protection and authority of the principles of international law derived from established custom, from the principles of humanity, and from the dictates of public conscience.

Article 35 – Basic rules

1. In any armed conflict, the right of the Parties to the conflict to choose methods or means of warfare is not unlimited.
2. It is prohibited to employ weapons, projectiles and material and methods of warfare of a nature to cause superfluous injury or unnecessary suffering.
3. It is prohibited to employ methods or means of warfare which are intended, or may be expected, to cause widespread, long-term and severe damage to the natural environment.

Article 36 – New weapons

In the study, development, acquisition or adoption of a new weapon, means or method of warfare, a High Contracting Party is under an obligation to determine whether its employment would, in some or all circumstances, be prohibited by this Protocol or by any other rule of international law applicable to the High Contracting Party.

Article 48 – Basic rule

In order to ensure respect for and protection of the civilian population and civilian objects, the Parties to the conflict shall at all times distinguish between the civilian population and combatants and between civilian objects and military objectives and accordingly shall direct their operations only against military objectives.

Article 51 – Protection of the civilian population

1. The civilian population and individual civilians shall enjoy general protection against dangers arising from military operations. To give effect to this protection, the following rules, which are additional to other applicable rules of international law, shall be observed in all circumstances.
2. The civilian population as such, as well as individual civilians, shall not be the object of attack. Acts or threats of violence the primary purpose of which is to spread terror among the civilian population are prohibited.

(...)

4. Indiscriminate attacks are prohibited. Indiscriminate attacks are:
    1. those which are not directed at a specific military objective;
    2. those which employ a method or means of combat which cannot be directed at a specific military objective; or
    3. those which employ a method or means of combat the effects of which cannot be limited as required by this Protocol;
    and consequently, in each such case, are of a nature to strike military objectives and civilians or civilian objects without distinction.
5. Among others, the following types of attacks are to be considered as indiscriminate:
    (a) an attack by bombardment by any methods or means which treats as a single military objective a number of clearly separated and distinct military objectives located in a city, town, village or other area containing a similar concentration of civilians or civilian objects; and
    (b) an attack which may be expected to cause incidental loss of civilian life, injury to civilians, damage to civilian objects, or a combination thereof, which would be excessive in relation to the concrete and direct military advantage anticipated.

(...)

Article 55 – Protection of the natural environment

1. Care shall be taken in warfare to protect the natural environment against widespread, long-term and severe damage. This protection includes a prohibition of the use of methods or means of warfare which are intended or may be expected to cause such damage to the natural environment and thereby to prejudice the health or survival of the population.

(...)

Index

Automatic Braking System (ABS) 6, 44, 412
Automatic Identification System (AIS) transponders 241
automatic operation 4–5, 44, 67
automation bias 254
autonomous (sub)system design 139
    CADMID cycle 140–1
    design validation and verification (V&V) 141
        reasons for V&V 141–4
    integration and test 147–9
    validation of an autonomous system 146
    verification of an autonomous system 146–7
    weapon systems, general principles for 139–40
autonomous capabilities 341
    Article 36 review questions for system of systems 344–5
    Article 36 reviews for incremental upgrades 364
        architecture 364–7
        behaviour generator 374–5
        knowledge database 375–6
        sensory processing 367–70
        value judgement 372–4
        world modelling 370–2
    configuration and control during the OODA process 350
        command and control (C2) 354–6
        configurations during OODA 350–4
    example capability with upgrades for autonomous operation 345
        functionality in 347–9
        incremental approach to autonomy 349–50
        top-down approach 345–7
    introducing autonomy into systems 342–3
    limits to autonomy 376–7
    technical evidence required for top-down approach 356

knowledge database 363–4
    sensory processing 356
    value judgement, world modelling and behaviour generator 362–3
autonomous cars 43
Autonomous Control Levels (ACLs) 49, 73
autonomous functioning 68
Autonomous Levels for Unmanned Systems (ALFUS) 51, 76–8
autonomous operation 5, 7–8, 44
    example capability with upgrades for 345–50
autonomous strike capability 347
autonomous system 3–5, 44, 69
    in a complex environment 393–5
    functional elements, requirements for 409
        behaviour generator requirements 413
        knowledge database requirements 410–11
        legal criteria 414
        operator interface requirements 410
        overarching requirements 409–10
        sensor and sensory-processing requirements 411
        test, validation and verification (V&V) 414–15
        value judgement requirements 412–13
        world model requirements 411–12
    general principles for 383–4
        Asimov's laws of robotics 391–2
        ethical principles in design 384
        five 'EPSRC' rules 388–90
        principles from assertions 390–1
        United Nations Declaration of Human Rights 385–7
        value-based design 384–5
    insurance and litigation issues for 395–6
    misuse of 423–4